W9-AXS-902

Inspire Science

Science Handbook

Grades 4-5

mheducation.com/prek-12

Send all inquiries to:
McGraw-Hill Education
8787 Orion Place
Columbus, OH 43240

ISBN: 978-0-07-679238-2
MHID: 0-07-679238-2

Printed in the United States of America.

1 2 3 4 5 6 7 8 9 DOW 21 20 19 18 17 16

Table of Contents

OVERVIEW

The Inspire Science Handbook is a book all about science. The Handbook has information on many science topics. You can use this book to look up something that you want to know. For example, you can use it to learn about different living things and their ecosystems. You can also use this book to learn how to act like a scientist, such as being able to understand how to read different weather instruments.

(t-b) Jodi Jacobson/E+/Getty Images; releon8211/iStock/Getty Images; nelic/E+/Getty Images; zokara/E+/Getty Images

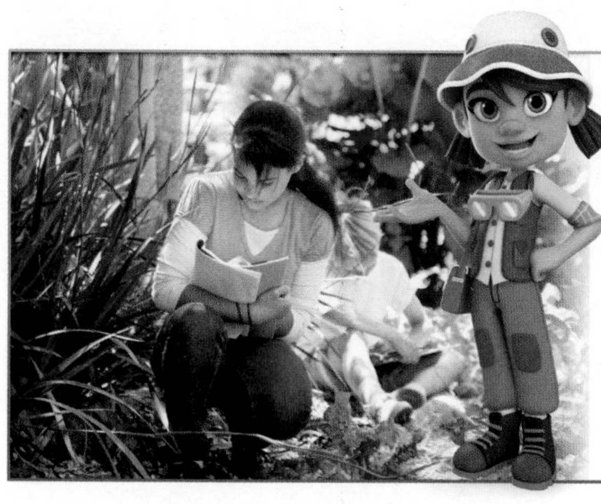

(t) Comstock/Stockbyte/Getty Images; (c)NASA/Aubrey Gemignani; (b) Alistair Berg/Digital Vision/Getty Images

How to Use This Book

Looking at the Front Pages

At the front of the book you will find the Table of Contents. This section of the book will help you quickly locate information that you are looking for. For example, if you are looking for information about how energy is created, you would look in the Physical Science section of the Table of Contents.

Sections

Throughout the book, you will see many different titles and headings. The main focus of each section is called out in large text. Each main focus then has smaller sections with more detail about the topic. Each section also uses a different color: Life Science is green, Earth and Space Science is orange, and Physical Science is purple.

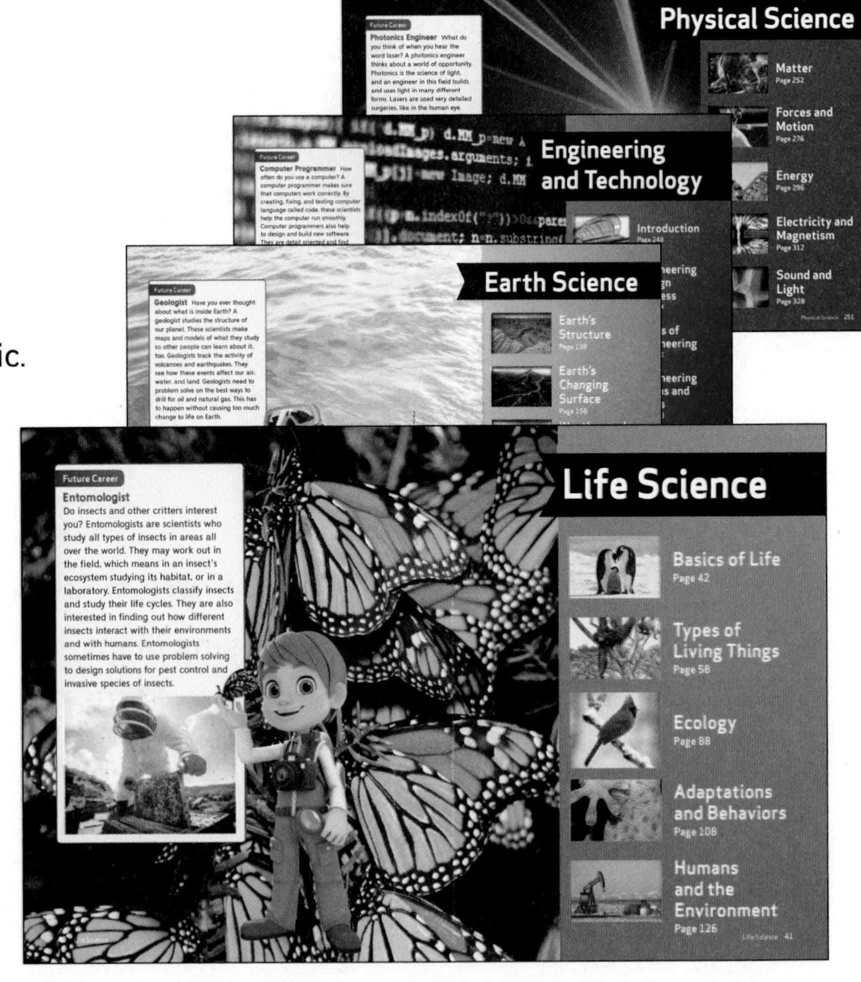

Page Features

What Could I Be? In this box, you will find careers that relate to the content on the page. Flip to the Careers section in the back of the book to learn more about careers that interest.

What Could I Be? Bacteriologist

Interested in becoming an expert on single-celled living things? A bacteriologist studies effects of these tiny organisms on plants and animals.

The **Make Connections** box will tell you about other sections of the book that relate to what you are currently reading. Turn to the page number listed to learn more about that topic.

Make Connections

Jump to the *Human Body* section to learn about your organ systems.

The **Fact Checker** box will help you correct any misconceptions you might have about science. By knowing the correct science, you can build on your learning to understand more complex content.

Fact Checker

Some kinds of bacteria make us sick, but others are helpful. They help your body break down the food you eat.

Did You Know? In this box, you will learn additional science facts that will expand your learning. You might also find ideas to research or plan your own investigations.

Did You Know?

The human body contains many levers. Your head and neck act as a first-class lever. Your foot acts as a second-class lever. Your forearm acts as a third-class lever.

The **Word Study** box will break words down to their different parts, making it easier for you to understand their meaning.

Word Study

Bio comes from the Greek word for *life* as in *biology*.

In the **Skill Builder** box, you will learn how to read diagrams, charts, graphs, tables, and other graphics that can help you understand the concepts you need as you learn science.

Skill Builder Read a Diagram

Notice how the part of the Moon we can see from Earth changes. The lit area visible to us grows larger, and then smaller.

Career Kids

Throughout the book you will see the Career Kids. They will help you learn about different jobs that grown-ups have that relate to specific science topics. For example, these kids can help you understand what it would be like to be a park ranger or entomologist. Make sure to check out the Career Kids and learn about their jobs in STEM!

Poppy is a nine-year-old girl. She has been to five National Parks and hopes to go to visit all 58 of them! Her personal hero is naturalist John Muir (the man who helped start the National Parks Service and the Sierra Club). Like Muir, she wants to help keep the national parks open for animals to roam free and for people to learn to appreciate nature. "Then, they will fight to protect it!" She hopes to be a park ranger one day.

Malik is a nine-year-old boy. He has always been interested in lasers, especially the types he sees at concerts and sporting events. He read that lasers are used in tools that can heal hearts and eyes, transmit information around the world, and even cut different materials. He wants to be a photonics engineer and work with lasers one day.

Hannah is a nine-year-old girl. She likes to take things apart, put them back together, and repair things that are broken. She spends a lot of time on her father's job site and watches him cut, shape, and weld things together for all types of industries. She wants to be like her dad and become a welder, too.

Antonio is a ten-year-old boy. He became fascinated with robots ever since he watched *Lost in Space* reruns with his dad. Recently, he went to a F.I.R.S.T. (For the Inspiration and Recognition of System and Technology) robotics competition where student-built robots competed in games! He wants to be on a F.I.R.S.T. team and become a robotics engineer one day.

Owen is an eight-year-old boy. He loves exploring his backyard for insects. He can never get enough of them! His love of bugs all started when he caught his first firefly. Now, he catches insects with his camera and sketchbook. He has learned how important insects are and how necessary they are to our lives. He wants to become an entomologist when he grows up.

Grace is a ten-year-old girl. She loves computers. She likes to do research, play games, and watch videos on them. She is interested in learning how to write computer code and program computers. She recently read a biography about Grace Hopper, a computer scientist and inventor of COBOL (one of the first high-level programming languages) and wants to be just like her! She hopes to be a computer programmer when she grows up and learn how to program computers to do all kinds of things.

Maya is a ten-year-old girl. She spends most of her time hiking and exploring outdoors with her sister, Marisol. On each hike, she collects rocks and pebbles. She has a large collection. She is curious about how rocks form, the layers of the Earth, and fossils, too. She wants to be a geologist one day.

Marisol is a six-year-old girl. She comes from a long line of firefighters and paramedics. She loves spending time at her father's fire station learning how they protect the neighborhood and save lives. She is always ready with a bandage if her sister Maya scrapes her knee on a hike. She wants to follow in her family footsteps and be a paramedic/EMT.

Hiro is an eight-year-old boy and swims like a fish. His favorite things to do are swimming and snorkeling. If he could live in the water, he would! Recently, he was on a trip to Florida and saw all kinds of fish while snorkeling, and went to an oceanarium to learn about the ocean and its animals. He wants to become an ocean engineer to help protect oceans.

Looking at the Back Pages

The Inspire Science Handbook has several resources in the back of the book that will help you learn more about science, technology, engineering, and mathematics. You will find useful information about STEM careers and ways to practice math and reading skills in science, too!

Career Connections

Have you ever wondered what you might do as a career? Careers in STEM are very exciting and are also in high demand. These types of careers are important to solve problems such as keeping our world safe and clean while making advancements in technology. Check out the Careers section in the back of the book to learn about some of the awesome STEM careers that are out there in Life, Earth, and Physical science as well as Engineering.

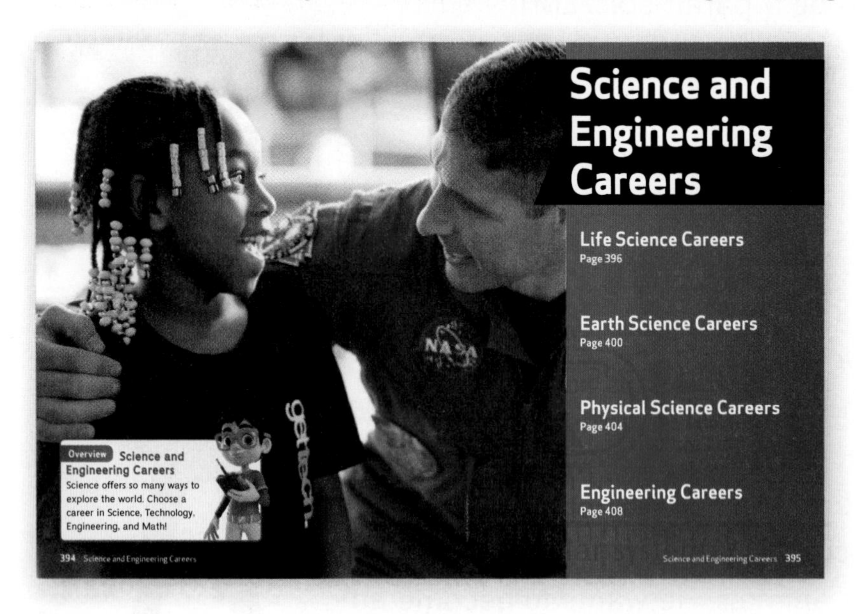

Science Guide

There are many skills that are needed in order to be a scientist or engineer. They even use math and reading skills just like you do in school! The Science Guide pages will give tips on how to use the tools and strategies used by scientists and engineers.

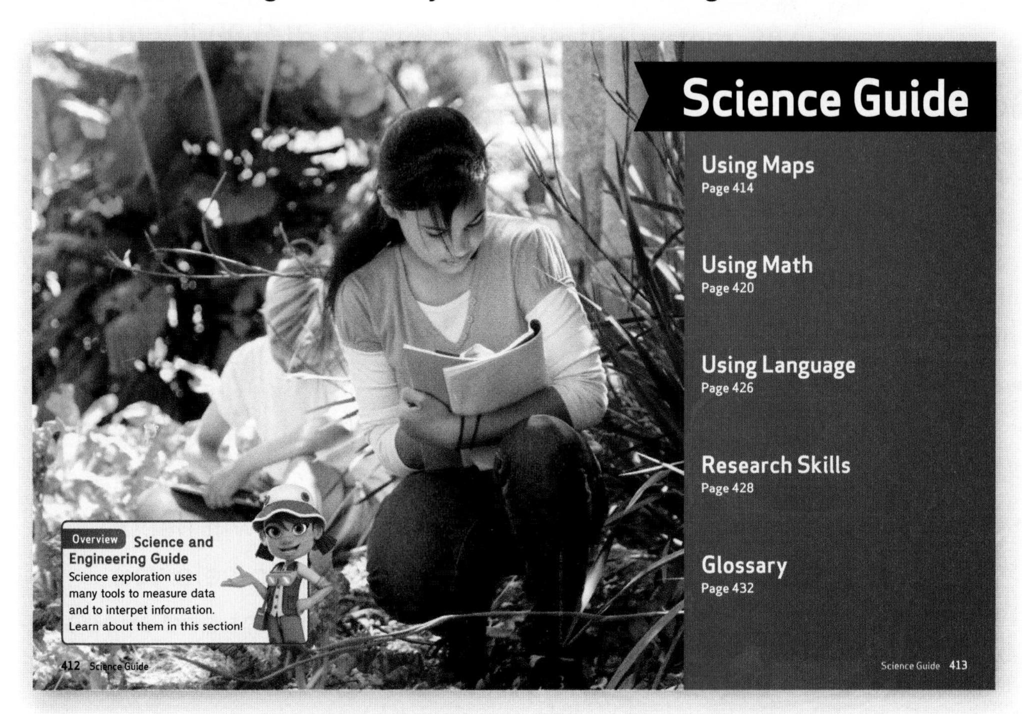

Science Guide

Overview Science and Engineering Guide
Science exploration uses many tools to measure data and to interpret information. Learn about them in this section!

412 Science Guide

Science Guide 413

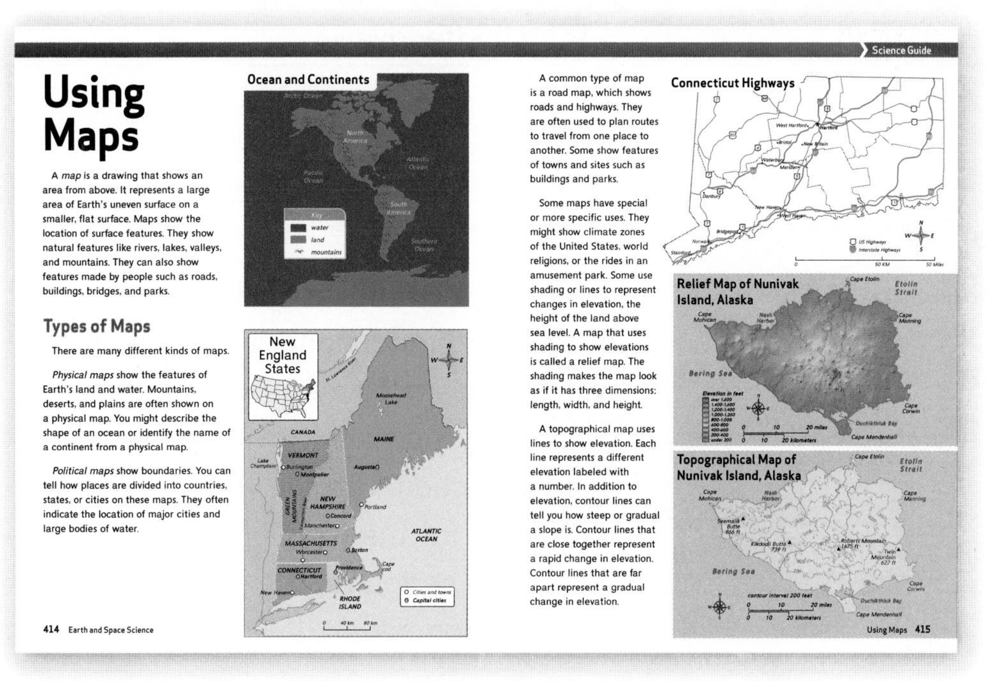

Science Guide

Using Maps

A *map* is a drawing that shows an area from above. It represents a large area of Earth's uneven surface on a smaller, flat surface. Maps show the location of surface features. They show natural features like rivers, lakes, valleys, and mountains. They can also show features made by people such as roads, buildings, bridges, and parks.

Types of Maps

There are many different kinds of maps.

Physical maps show the features of Earth's land and water. Mountains, deserts, and plains are often shown on a physical map. You might describe the shape of an ocean or identify the name of a continent from a physical map.

Political maps show boundaries. You can tell how places are divided into countries, states, or cities on these maps. They often indicate the location of major cities and large bodies of water.

A common type of map is a road map, which shows roads and highways. They are often used to plan routes to travel from one place to another. Some show features of towns and sites such as buildings and parks.

Some maps have special or more specific uses. They might show climate zones of the United States, world religions, or the rides in an amusement park. Some use shading or lines to represent changes in elevation, the height of the land above sea level. A map that uses shading to show elevations is called a relief map. The shading makes the map look as if it has three dimensions: length, width, and height.

A topographical map uses lines to show elevation. Each line represents a different elevation labeled with a number. In addition to elevation, contour lines can tell you how steep or gradual a slope is. Contour lines that are close together represent a rapid change in elevation. Contour lines that are far apart represent a gradual change in elevation.

414 Earth and Space Science

Using Maps 415

Maps Maps are helpful in knowing your location and the location of other places on Earth. There are several different kinds of maps that can even tell you the climate and landforms found in a region.

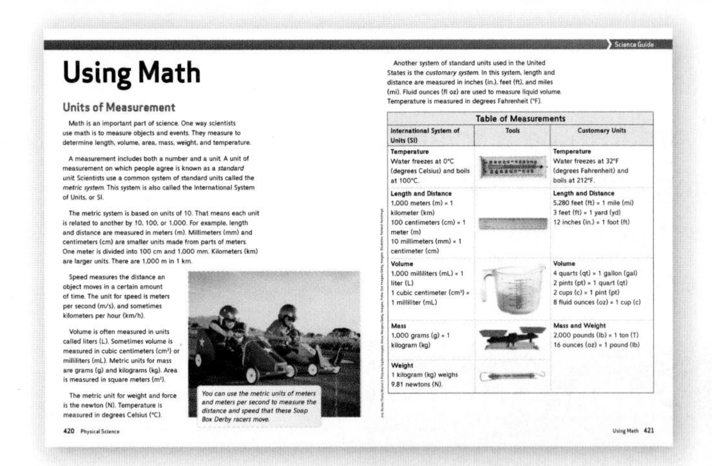

Using Math

Units of Measurement

Math is an important part of science. One way scientists use math is to measure objects and events. They measure to determine length, volume, area, mass, weight, and temperature.

A measurement includes both a number and a unit. A unit of measurement on which people agree is known as a *standard unit*. Scientists use a common system of standard units called the *metric system*. This system is also called the International System of Units, or SI.

The metric system is based on units of 10. That means each unit is related to another by 10, 100, or 1,000. For example, length and distance are measured in meters (m). Millimeters (mm) and centimeters (cm) are smaller units made from parts of meters. One meter is divided into 100 cm and 1,000 mm. Kilometers (km) are larger units. There are 1,000 m in 1 km.

Speed measures the distance an object moves in a certain amount of time. The unit for speed is meters per second (m/s), and sometimes kilometers per hour (km/h).

Volume is often measured in units called liters (L). Sometimes volume is measured in cubic centimeters (cm^3) or milliliters (mL). Metric units for mass are grams (g) and kilograms (kg). Area is measured in square meters (m^2).

The metric unit for weight and force is the newton (N). Temperature is measured in degrees Celsius (°C).

You can use the metric units of meters and meters per second to measure the distance and speed that these Soap Box Derby racers move.

Another system of standard units used in the United States is the *customary system*. In this system, length and distance are measured in inches (in.), feet (ft), and miles (mi). Fluid ounces (fl oz) are used to measure liquid volume. Temperature is measured in degrees Fahrenheit (°F).

Table of Measurements

International System of Units (SI)	Tools	Customary Units
Temperature Water freezes at 0°C (degrees Celsius) and boils at 100°C.		**Temperature** Water freezes at 32°F (degrees Fahrenheit) and boils at 212°F.
Length and Distance 1,000 meters (m) = 1 kilometer (km) 100 centimeters (cm) = 1 meter (m) 10 millimeters (mm) = 1 centimeter (cm)		**Length and Distance** 5,280 feet (ft) = 1 mile (mi) 3 feet (ft) = 1 yard (yd) 12 inches (in.) = 1 foot (ft)
Volume 1,000 milliliters (mL) = 1 liter (L) 1 cubic centimeter (cm^3) = 1 milliliter (mL)		**Volume** 4 quarts (qt) = 1 gallon (gal) 2 pints (pt) = 1 quart (qt) 2 cups (c) = 1 pint (pt) 8 fluid ounces (oz) = 1 cup (c)
Mass 1,000 grams (g) = 1 kilogram (kg)		**Mass and Weight** 2,000 pounds (lb) = 1 ton (T) 16 ounces (oz) = 1 pound (lb)
Weight 1 kilogram (kg) weighs 9.81 newtons (N).		

Math in Science Science and math are very closely related. Math skills are needed for making measurements and analyzing data in experiments and in everyday observations. This section will help you understand the math systems that apply to science and engineering practices.

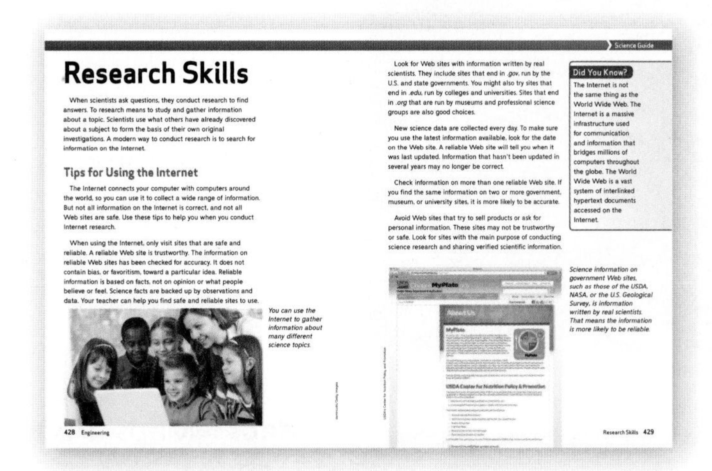

Research Skills

When scientists ask questions, they conduct research to find answers. To research means to study and gather information about a topic. Scientists use what others have already discovered about a subject to form the basis of their own original investigations. A modern way to conduct research is to search for information on the Internet.

Tips for Using the Internet

The Internet connects your computer with computers around the world, so you can use it to collect a wide range of information. But not all information on the Internet is correct, and not all Web sites are safe. Use these tips to help you when you conduct Internet research.

When using the Internet, only visit sites that are safe and reliable. A reliable Web site is trustworthy. The information on reliable Web sites has been checked for accuracy. It does not contain bias, or favoritism, toward a particular idea. Reliable information is based on facts, not on opinion or what people believe or feel. Science facts are backed up by observations and data. Your teacher can help you find safe and reliable sites to use.

You can use the Internet to gather information about many different science topics.

Look for Web sites with information written by real scientists. They include sites that end in *.gov*, run by the U.S. and state governments. You might also try sites that end in *.edu*, run by colleges and universities. Sites that end in *.org* that are run by museums and professional science groups are also good choices.

New science data are collected every day. To make sure you use the latest information available, look for the date on the Web site. A reliable Web site will tell you when it was last updated. Information that hasn't been updated in several years may no longer be correct.

Check information on more than one reliable Web site. If you find the same information on two or more government, museum, or university sites, it is more likely to be accurate.

Avoid Web sites that try to sell products or ask for personal information. These sites may not be trustworthy or safe. Look for sites with the main purpose of conducting science research and sharing verified scientific information.

Did You Know?
The Internet is not the same thing as the World Wide Web. The Internet is a massive infrastructure used for communication and information that bridges millions of computers throughout the globe. The World Wide Web is a vast system of interlinked hypertext documents accessed on the Internet.

Science information on government Web sites, such as those of the USDA, NASA, or the U.S. Geological Survey, is information written by real scientists. That means the information is more likely to be reliable.

Research Skills Being able to read, write, and record information is essential to being a scientist or an engineer. Sometimes you will need to research information to help you understand science content and support your observations. This section will provide ways that you can perform research effectively.

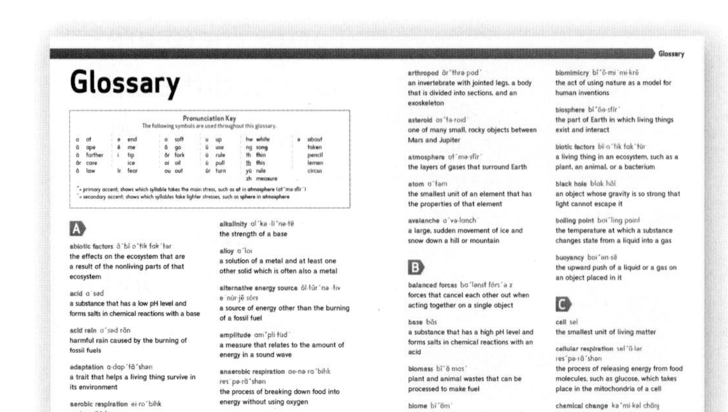

Glossary

Pronunciation Key
The following symbols are used throughout this glossary.

A

abiotic factors the effects on the ecosystem that are a result of the nonliving parts of that ecosystem

acid a substance that has a low pH level and forms salts in chemical reactions with a base

acid rain harmful rain caused by the burning of fossil fuels

adaptation a trait that helps a living thing survive in its environment

aerobic respiration the process of using oxygen to break down food into energy

alkalinity the strength of a base

alloy a solution of a metal and at least one other solid which is often also a metal

alternative energy source a source of energy other than the burning of a fossil fuel

amplitude a measure that relates to the amount of energy in a sound wave

anaerobic respiration the process of breaking down food into energy without using oxygen

aquifer an underground layer of rock or soil filled with water

arthropod an invertebrate with jointed legs, a body that is divided into sections, and an exoskeleton

asteroid one of many small, rocky objects between Mars and Jupiter

atmosphere the layers of gases that surround Earth

atom the smallest unit of an element that has the properties of that element

avalanche a large, sudden movement of ice and snow down a hill or mountain

B

balanced forces forces that cancel each other out when acting together on a single object

base a substance that has a high pH level and forms salts in chemical reactions with an acid

biomass plant and animal wastes that can be processed to make fuel

biome one of Earth's large ecosystems, each with its own kind of climate, soil, plants, and animals

biomimicry the act of using nature as a model for human inventions

biosphere the part of Earth in which living things exist and interact

biotic factors a living thing in an ecosystem, such as a plant, an animal, or a bacterium

black hole an object whose gravity is so strong that light cannot escape it

boiling point the temperature at which a substance changes state from a liquid into a gas

buoyancy the upward push of a liquid or a gas on an object placed in it

C

cell the smallest unit of living matter

cellular respiration the process of releasing energy from food molecules, such as glucose, which takes place in the mitochondria of a cell

chemical change a change of matter that occurs when atoms link together in a new way, creating a new substance different from the original substances

Glossary Want to know the meaning of the word mitochondria? The glossary will be your resource for definitions of vocabulary words that will be presented in each section. Use the glossary like a dictionary for the highlighted words you encounter as you read the information in the Handbook.

Inspire Science

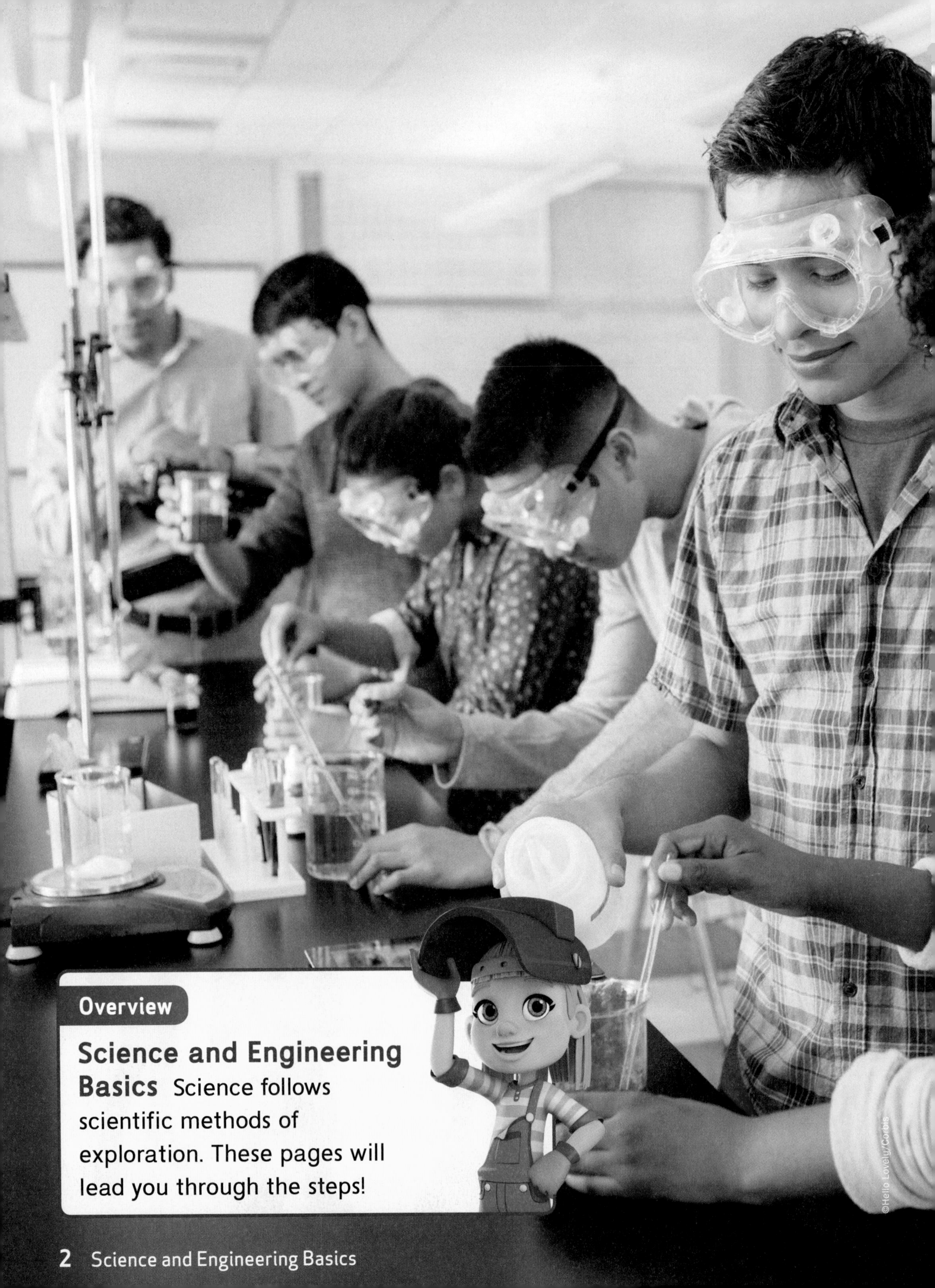

Overview

Science and Engineering Basics Science follows scientific methods of exploration. These pages will lead you through the steps!

Science and Engineering Basics

What are Science and Engineering?

The fields of science and engineering involve collection of data and logical thinking. Science and engineering are often used together.

The microscope was invented using science and engineering. People used science to understand lenses. They used what they learned to produce a tool that used lenses. They used engineering to create the tool.

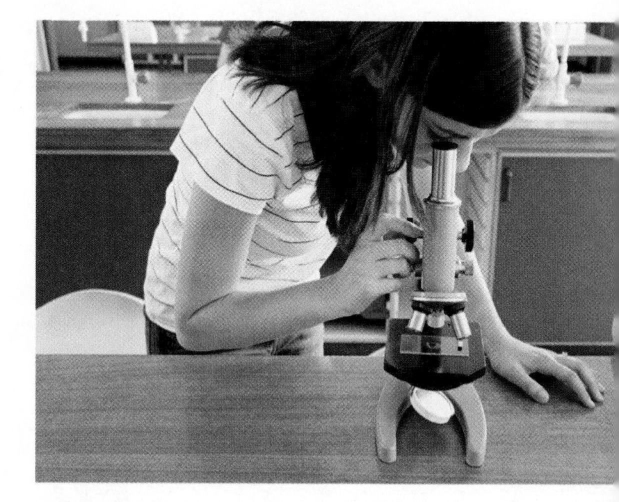

The microscope was invented using science and engineering.

Science and engineering are a part of your daily life. Every time you ask a question or construct an explanation, you are using a skill that scientists and engineers use as well. Learning to evaluate data and being able to think logically are also important to scientists and engineers. This section will help you become familiar with the methods that scientists and engineers use.

Science

The word *science* may bring to mind microscopes and people in lab coats. These are all parts of science, but what is science, really? Science has been used for thousands of years, even before microscopes. The first person who wondered what plants were made of and then tried to figure it out was using science. *Science* is trying to answer questions about the world by using evidence.

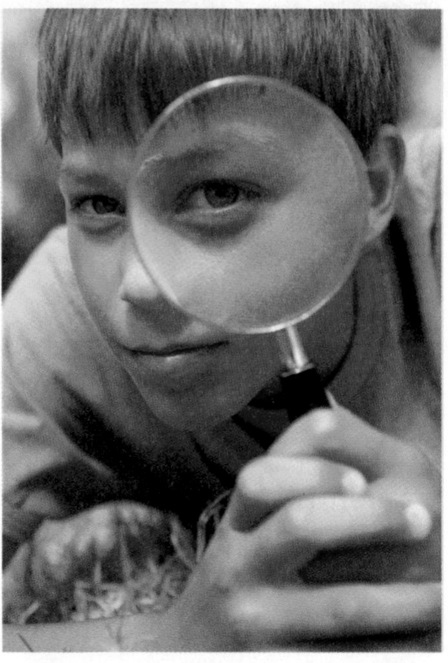

A lens can make something look larger.

People use observations to learn how lenses and mirrors change the way they saw something. They applied what they learned to build a tool that would help them. Microscopes were invented to help people see tiny things. They couldn't see these things with just their eyes. Using microscopes, they were able to answer more questions about the world.

However, answering science questions does not create new tools. That creation is done through engineering.

Engineering

Engineering is using science and math to solve a problem or meet a need.

For example, engineers can test ways different lenses work together when making a microscope. Sometimes they make an image larger. Sometimes they make it smaller. An engineer would design and test several different combinations of lenses to find the best way to get the largest image. Discoveries in science lead to advancements in engineering. Answering questions can help solve problems, and solving problems can help answer questions.

Science and Engineering Practices

Scientists and engineers use certain practices. *Practices* are sets of skills and knowledge that are used to do a job.

Many of these practices are things that you do during everyday life, even if you do not realize it. Most of the practices are used in both science and engineering. A few are used just for science or just for engineering.

A cricket makes a chirping noise by rubbing its wings together.

Asking Questions and Defining Problems

Asking Questions

Asking questions is a science practice that drives every scientific investigation. Asking questions is not as simple as it may seem. A poorly worded question does not help a scientist gather much information. A good question can lead to important information and connections, or even more questions.

Questions in science can come from observations. Making observations involves using your senses to gather information and take note of what occurs. For example, if you hear a chirping noise, you might ask: "What is that sound?" You can use observations to ask more specific questions. You follow the noise and find that it is coming from a cricket. Then you may ask: "Why is the cricket making that sound? How is it making that sound?" These are clearer questions coming from observation.

Defining Problems

Defining problems is an engineering practice that must happen before any solution can be found. Engineers study how people do things and try to make the experience better. If people don not have a way to do something yet, engineers explore ways to make it happen.

Engineers identify problems for people and society and then design solutions to those problems. The solution could be a process, a system, or an object, such as a tool. Space suits worn by astronauts are technological solutions designed by engineers. Astronauts needed a way to explore outer space safely, so engineers worked to find a solution.

Defining a problem involves thinking about a lot of questions, types of information, and procedures. You also need to think about any requirements for the solution. Defining a problem also includes thinking about *constraints*, or possible limitations to the problem being able to be solved. A well-defined problem contains all the requirements for a solution.

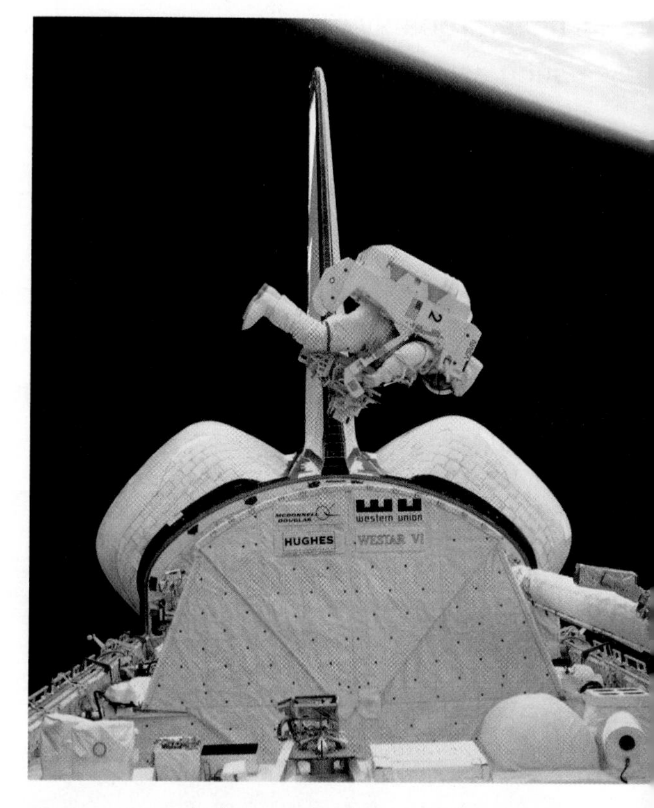

A space suit is a solution to the problem of humans being able to survive in space.

NASA/JSC

Developing and Using Models

Developing a Model

If you are have ever built a model spaceship, car, or castle with construction toys, then you have used a model. What about scientific models? In your science classes, you have probably seen or even made models of information, such as of the solar system. A *model* is a representation of a part of an idea, event, process, structure, or object.

Models allow scientists to answer questions and engineers to solve problems because they help people understand how things are made and how they work. Models come in many forms. They can be diagrams, math problems, computer simulations, maps, or physical reproductions.

This model of the solar system shows how the planets orbit around the Sun.

Anthony Bradshaw/Getty Images

Describing the Very Large or the Very Small

Models can describe processes that cannot be observed with our senses. Sometimes processes are too large or too small to see.

When you get a cut on your finger, you know that a scab forms, but you cannot see the tiny processes going on that help it to form. A digital model can help you visualize what is happening. Many processes in the universe, the solar system, or even on Earth are too big to directly observe. A model helps you see them on a smaller scale.

Generating Data

Models may represent data from an investigation. They can be used to generate data that are useful for answering questions or finding solutions. *Data* are the information used to answer a question.

Data are often numbers, but they do not have to be. Models can be used to generate data to test ideas. Suppose your class is walking to a field trip location. A computer mapping tool would allow you to put in your destination. Then it would provide you with data, which would include the various choices for routes and the amount of time each will take.

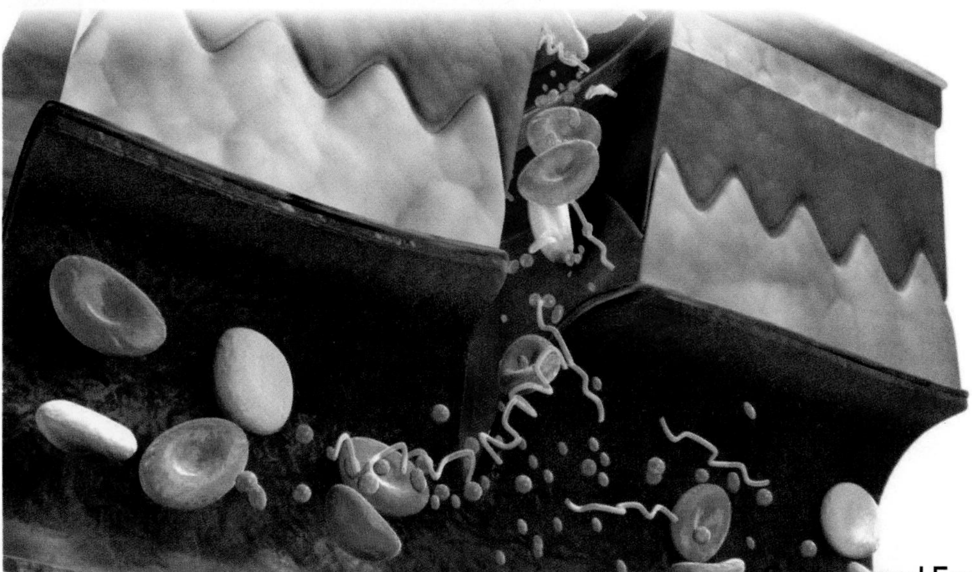

This image of blood clotting comes from a digital model showing what is too small for your eyes to see.

MedicalRF.com

Planning and Carrying Out Investigations

Planning and carrying out investigations is an important science practice. In fact, you can not do science without it!

Forming a Hypothesis

Scientists and engineers rely on carefully planned investigations to do their work. First a scientist asks a question. Next, he or she forms a hypothesis based on experience and research. A *hypothesis* is a predictive statement that can be tested.

An investigation must be done to test the hypothesis. During the investigation, the scientist or engineer will collect data. The data can be used to determine if the hypothesis is supported.

The investigation process can progress based on the new information. If the hypothesis was not supported, the scientist can review their question to see if something else needs to be tested. If the hypothesis was supported, scientists can work with engineers to then design possible solutions to the original question. Scientists and engineers often use the steps shown to the right when carrying out investigations.

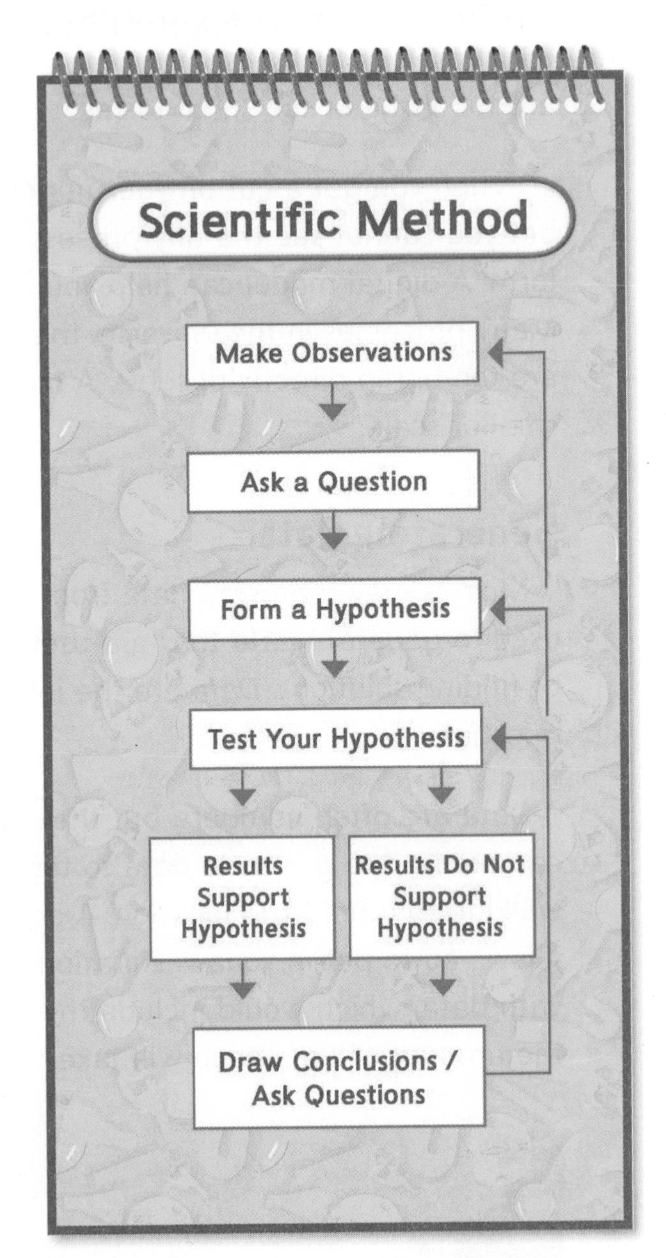

Components of an Investigation

When planning an investigation, the first step is identifying the variables. Variables are components of an investigation. A good investigation includes three types of variables. Independent variables are the things that you change to test your hypothesis. Dependent variables are the things that change as a result of the changes to the independent variables. Control variables are the things you keep the same in each test. The dependent variable is measured or observed. The independent variable is tested or changed to see how it affects the dependent variable.

For example, you could do an experiment to determine what amount of water helps a plant grow the best. You want to make sure the only difference between the plants is the amount of water they receive. You control all the other variables. The plants should get the same amount of sunlight, have the same air temperature, and be in the same type of soil. These are the control variables. The independent variable is the amount of water. Each plant should receive different measured amounts. The dependent variable is how much each plant grows in a certain amount of time.

These plants grew differently because the amount of water they received was different.

Analyzing and Interpreting Data

Analyzing and interpreting data is very important to scientists and engineers. The data will help answer a question or determine if a solution solves a problem. Scientists and engineers use data to understand relationships between variables and events. Being able to understand these relationships requires knowledge of how to interpret the data correctly.

Using Graphs

Graphs are often useful for analyzing, interpreting, and communicating data. They give a visual representation of how the data is related. Graphs work best when the data collected uses numbers or measurements. You can learn more about using graphs in the Doing Science and Engineering section.

How Variables Relate

One thing that data can show is the relationship between the variables in a scientific investigation. Any change in the dependent variable is a resultwill change because the independent variable has been changed. The data that you gather should show this relationship. You can compare data from investigations with other variables to look for patterns.

You can use graph paper to make a graph.

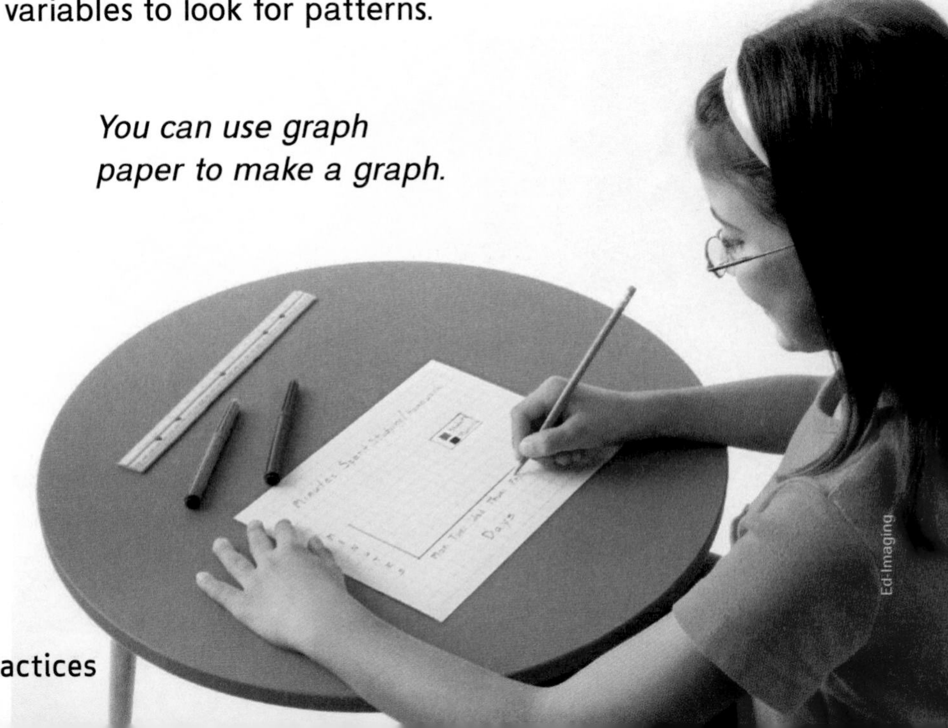

Evaluating Data

Data that is accurate is close to the true value. Data that is precise is very close to other measured values, even if these are not close to the true value. Measured data should be both accurate and precise to be reliable. Using the correct tools and reading them correctly results in reliable data. For example, liquids can be accurately measured using a graduated cylinder. When you fill a beaker or graduated cylinder, you need to read the level of the liquid correctly. If you were measuring time, how could you make your measurements more accurate? A greater number of trials can improve accuracy.

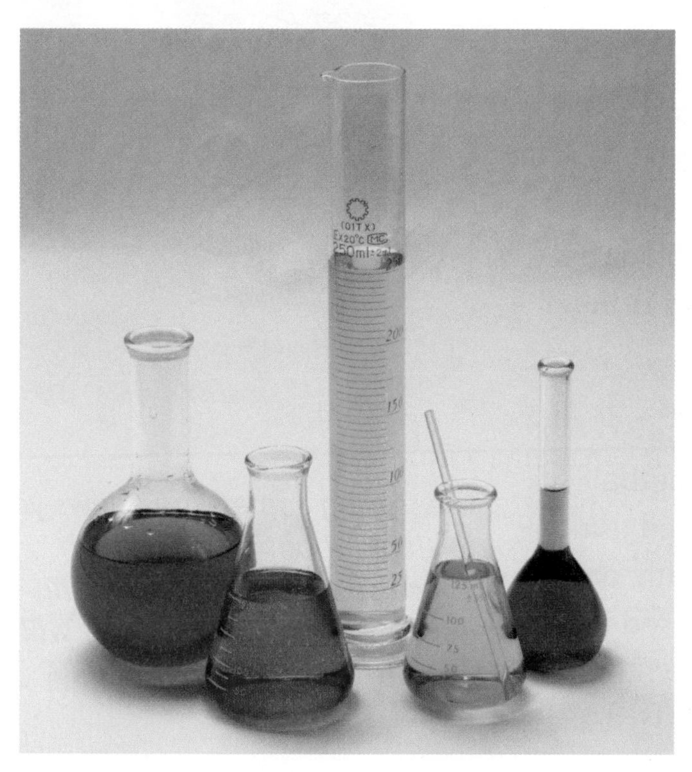

Make sure to measure liquids correctly.

A stopwatch can measure time accurately.

Using Math and Computational Thinking

Math is used in science and engineering to represent variables and their relationships. It is also used to make predictions about events and results. Engineers apply mathematical equations to design systems. Scientists use computers and programs developed by engineers. Both fields work together to advance technology and achieve results. Using math and computational thinking is a key practice for both science and engineering.

Using Math

Math is the study of amount, structure, and change. It uses numbers to describe amounts and relationships of objects, processes, and events. The ideas and equations used in math are developed to explain the world around us and everything that happens. It could be said that mathematics is the "language" of science.

Everything in the world around you can be represented using math. Shopping malls, football stadiums, and shoe boxes can be represented using math by their measurements of length, width, and height. These measurements can then be used to calculate their areas and volumes.

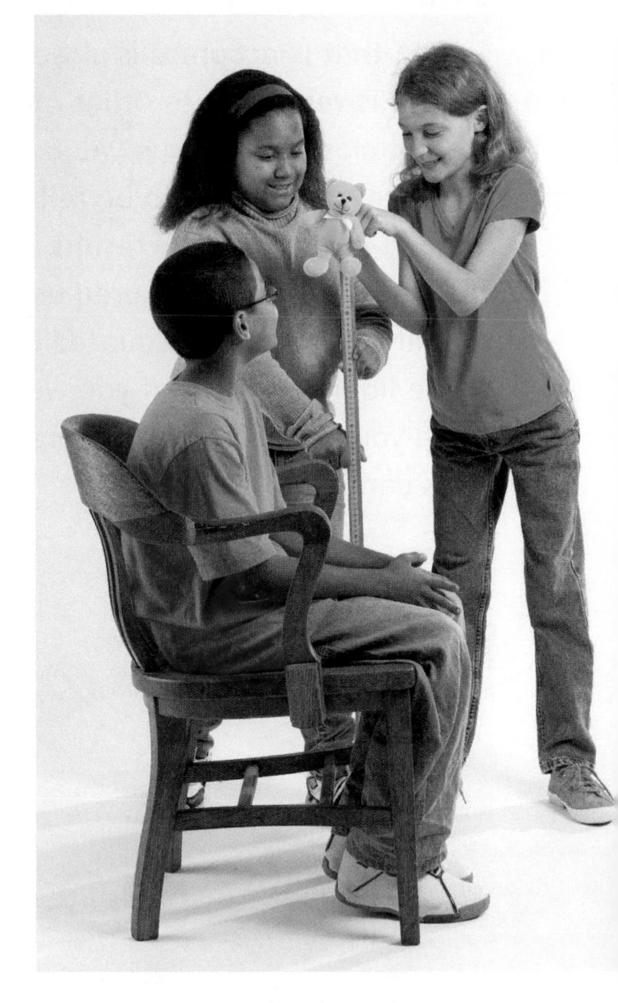

Measuring something allows you to think about it using math. These students are measuring how far the stuffed animal will fall when they drop it.

Computational Thinking

Scientific investigations and engineering designs often rely on computer calculations and simulations. These computer applications need to be developed, tested, and implemented, all of which require computational thinking. The term *computational* refers to problem-solving performed using computer-science reasoning.

Computational thinking involves the ability to recognize the relationship between events and mathematics. This sort of thinking can allow you to collect useful data and do an analysis. Computer programs make it easier to analyze large amounts of data.

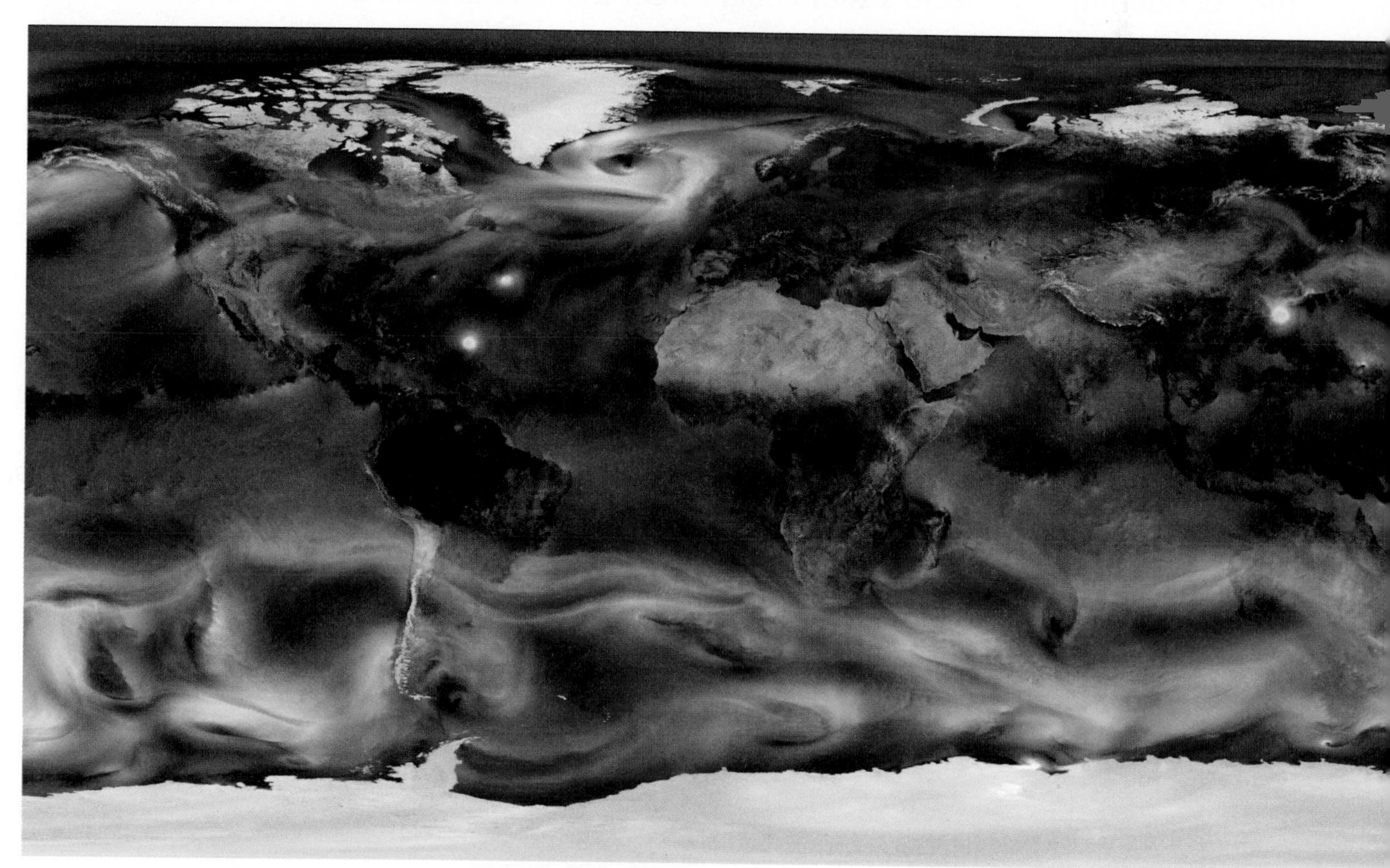

William Putman/NASA Goddard Space Flight Center

This computer program simulates global winds. It allows scientists to make predictions about the affects those winds will have.

Constructing Explanations and Designing Solutions

Constructing Explanations

Once a scientist has asked a question, conducted an investigation, and analyzed the data, his or her goal is to come up with an explanation. The explanation must be based on evidence produced from the investigation. Constructing explanations is a science practice that relates directly to the practice of asking questions.

To explain how something works, scientists must identify how variables relate to each other. Understanding these relationships and considering other factors that could affect the results enable scientists to make valid explanations.

Explanations can also be supported by existing scientific information. The data scientists collect from their investigations provide evidence to support these existing science ideas.

You can use the law of gravity to help explain why different objects fall at different speeds.

Jill Braaten/McGraw-Hill Education

Designing Solutions

Recall that after conducting research and analyzing the data, an engineer's ultimate goal is to design a solution to a problem. The solution must be based on evidence from research. Its reliability is based on testing and revising solutions while following design specifications and constraints, or limits. Designing solutions is an engineering practice that relates directly to the practice of defining problems.

Designing a solution usually involves meeting certain requirements. There may also be limits on the project, such as a certain cost or time. A successful solution will meet those requirements within the limits of the project.

An electric car is the result of a design solution. It is a car that does not use as much gas, but still can be made in a cost effective manner.

Engaging in Argument from Evidence

Have you ever gotten into an argument with a sibling or a friend? In everyday life, arguments are often negative events because they are linked with angry or sad emotions. In science and engineering, though, arguments are necessary for advancements to be made. In science, an *argument* is a statement based on logic or evidence that supports or opposes something.

Argumentation ensures that all aspects of an explanation or design solution are considered. Engaging in argument from evidence is a science and engineering practice that serves to identify the best explanation or solution.

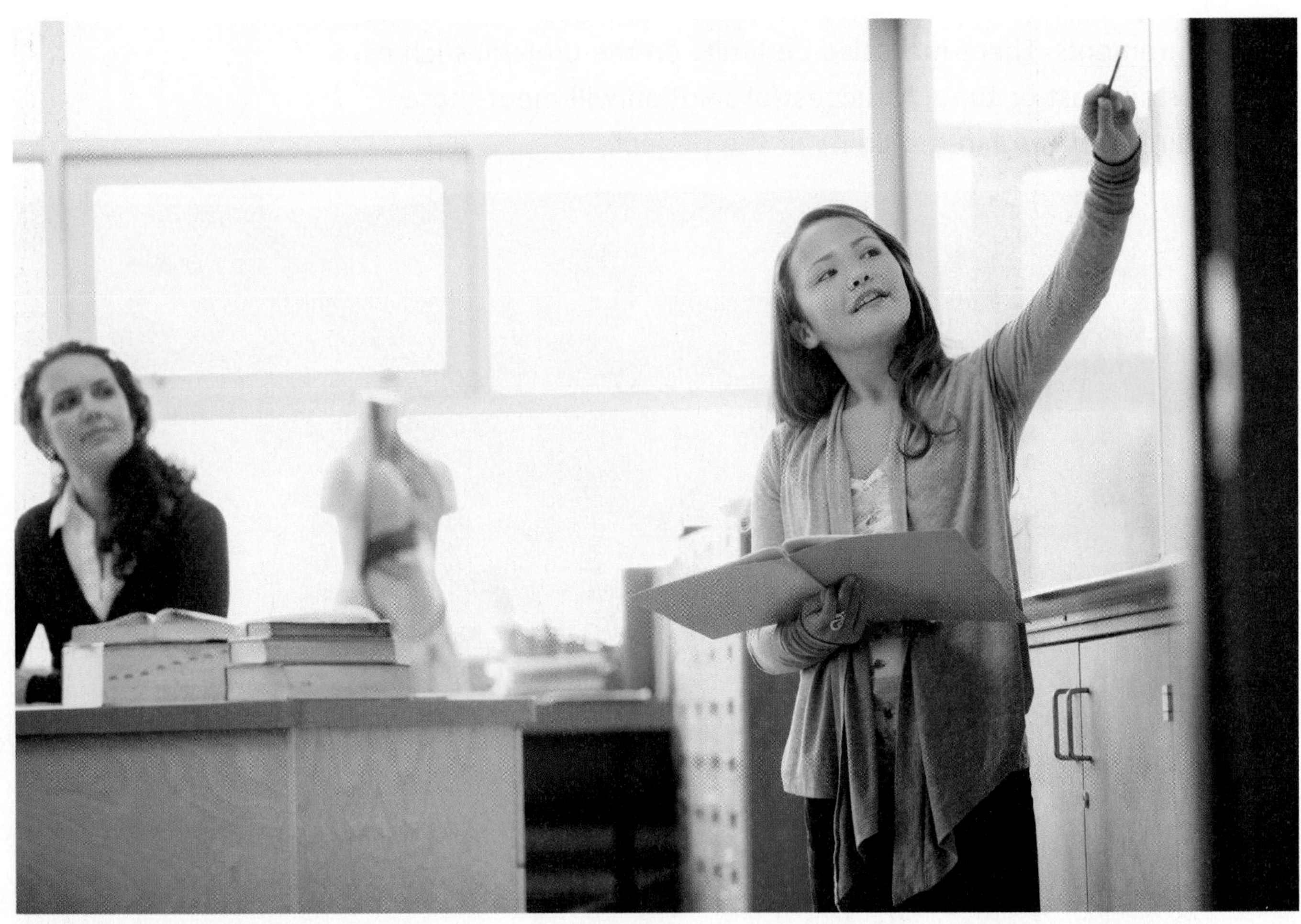

You should support your argument with evidence from your investigations.

The Right Way to Argue

People often have opposing views on how best to conduct an investigation, interpret results, or communicate information. These views should have reliable data and scientific principles as evidence in order to be valid. When you plan and conduct an investigation, it is important to evaluate procedures and results, keeping in mind that you need to defend, or give logical reasons for, your choices and results.

The science practices include arguing explanations, defending data, and recommending designs. For example, engineers need to consider constraints and impacts on society when recommending a design. Often, solutions lead to other problems that are not easy to predict. It is important that you listen to, compare, and evaluate competing ideas and methods.

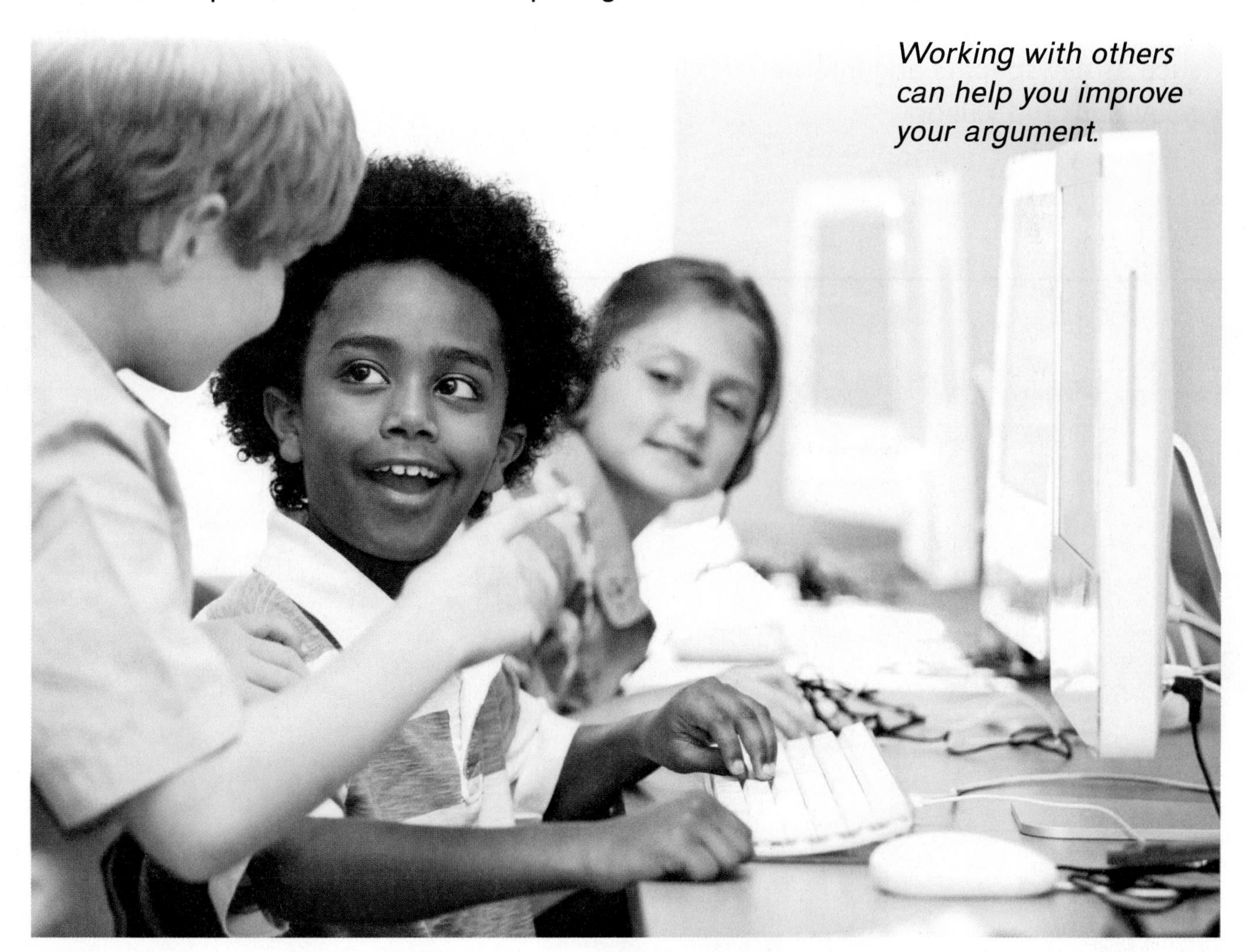

Working with others can help you improve your argument.

Obtaining, Evaluating, and Communicating Information

Communication is important in every aspect of our lives. In science and engineering, it allows for new explanations and design solutions to be shared with society and the science and engineering communities. These communications can be written or verbal. They can be online, in books, on the news, or in public or private discussion. Learning how to understand and evaluate communicated information is valuable if you are interested in pursuing a career in science.

Obtaining Information

Scientific information is valuable not only for scientists and engineers, but also for the general public. Integrating information from various media sources helps to give you a better understanding. Types of *media* include written texts, such as newspapers and Internet articles, as well as radio, television, and videos.

You can use Internet or book resources to gather information.

Evaluating Information

You must think critically about the information you read in order to evaluate it. When obtaining scientific information, you should critique the source to determine if it is credible or biased. Information is considered *credible* if it can be trusted to be true. *Biased* information reflects a strong interest of the writer or sponsor. You need to evaluate the information you gather and decide whether more information is needed.

Communicating Information

Communicating the information you have obtained and evaluated will help to educate others. This communication will include sharing the findings of your own investigations as well.

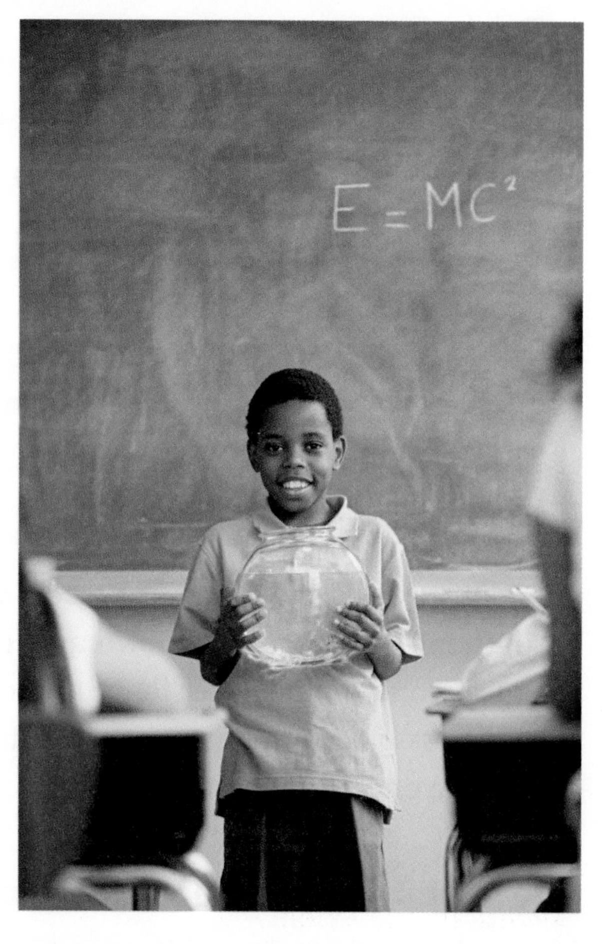

Share information that you have obtained with other people.

Science and Engineering Processes

The Scientific Process

Scientists all follow a similar process when they are investigating a question. This process is known as the *scientific process.* It allows scientists to maintain consistency and reliability in their findings. The process guides them in their investigations and does not always have to happen in order.

The Steps of the Scientific Process

Make Observations One of the most important skills a scientist must have is being able to make careful observations. This step of the process is where scientists find reasons to explore new ideas.

Ask a Question The scientific process is driven by a question that the scientist is trying to answer. This question comes from a scientist's observations of something they want to know more about, such as, "What type of biomass contains the most energy?"

Form a Hypothesis Scientists need to form a hypothesis when investigating a question. A hypothesis is a statement that can betested to answer a question. Hypotheses are educated predictions about the results of an investigation.

Test the Hypothesis This part of the scientific process has several steps. First, scientists have to select a strategy for how they will test their hypothesis. They might plan an experiment or collect information through a survey, depending on the topic. Then, scientists need to plan their procedure. This procedure needs to be clear enough for other scientists to follow in case the experiment needs to be recreated. Lastly, scientists carry out their plan to test their hypothesis. They record data and make observations throughout their plan.

Using Results to Draw Conclusions After scientists finish collecting data by testing their hypothesis, they need to analyze the data to decide if their hypothesis was or was not supported by the results. If their hypothesis is supported by the results, they can move on to draw conclusions about how it might answer their question. If their hypothesis is not supported by the results, they need to go back and come up with a new question or hypothesis to try and get the information they need.

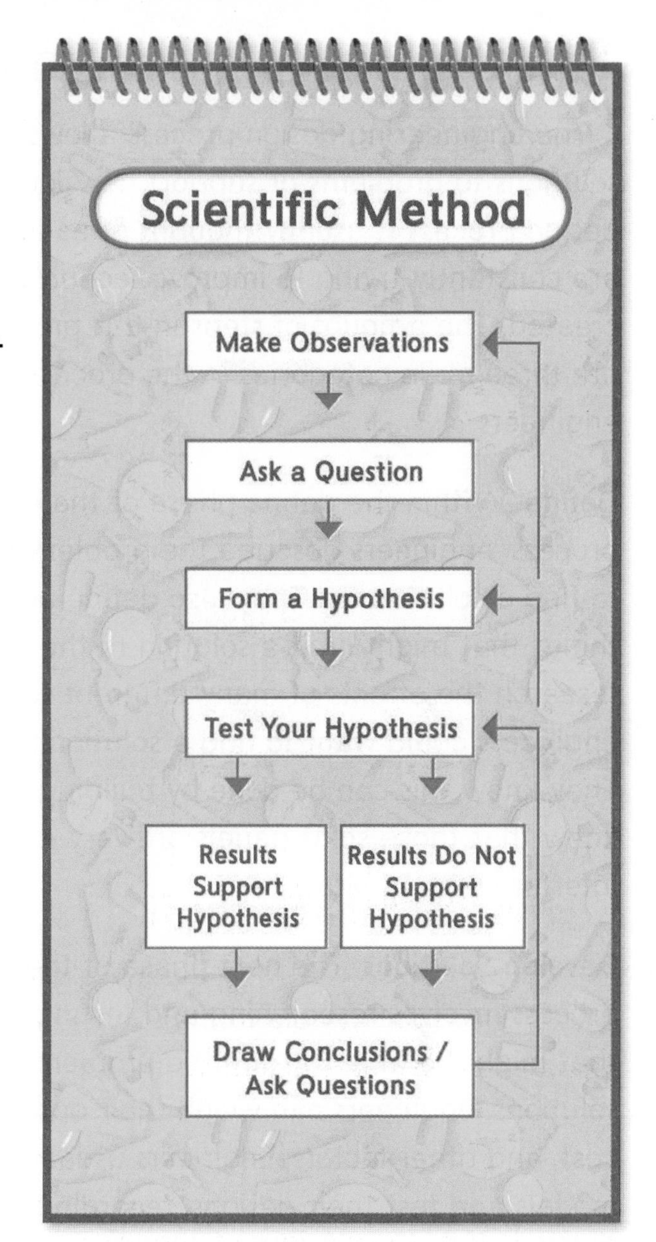

The Engineering Design Process

The engineering design process allows researchers to find solutions to problems or support new ideas. The engineering design process is usually thought of as a loop, since engineers are constantly trying to improve technologies once they are created. The amount of steps in this process can vary, but there are three main categories in the process that guides the actions of engineers.

Define Within the define phase of the engineering design process, engineers describe the problem to which they are trying to find a solution for. They also define possible constraints, or things that might limit a solution to the problem. They might research the effects of many different solutions. For example, engineers could want to find a solution for using less fossil fuel. They know this can be done by building solar panels, but also know that those solar panels are very expensive and might interfere with habitats.

Develop Solutions The next phase of the engineering design process involves researching and testing multiple solutions that might be able to solve the problem. By exploring multiple solutions, engineers can weigh their options in terms of materials, cost, and other factors. Engineers usually work as a team to build models and test their designs, recording data as they go so that they can make informed decisions.

Optimize Solutions The engineering design process is one that is continuous and focuses on improving solutions to make them better with each iteration. This phase involves improving successful solutions based on tests. Even if a solution fails, engineers and other scientists can use this information to improve the design! Often times, even simple solutions need to be revised several times before it is reliable enough to use again and again.

The Engineering Design Process

Engineering Design Loop

Identify a problem.

Define the project limits.

Research and brainstorm.

Make a model.

Test your idea.

Evaluate and present.

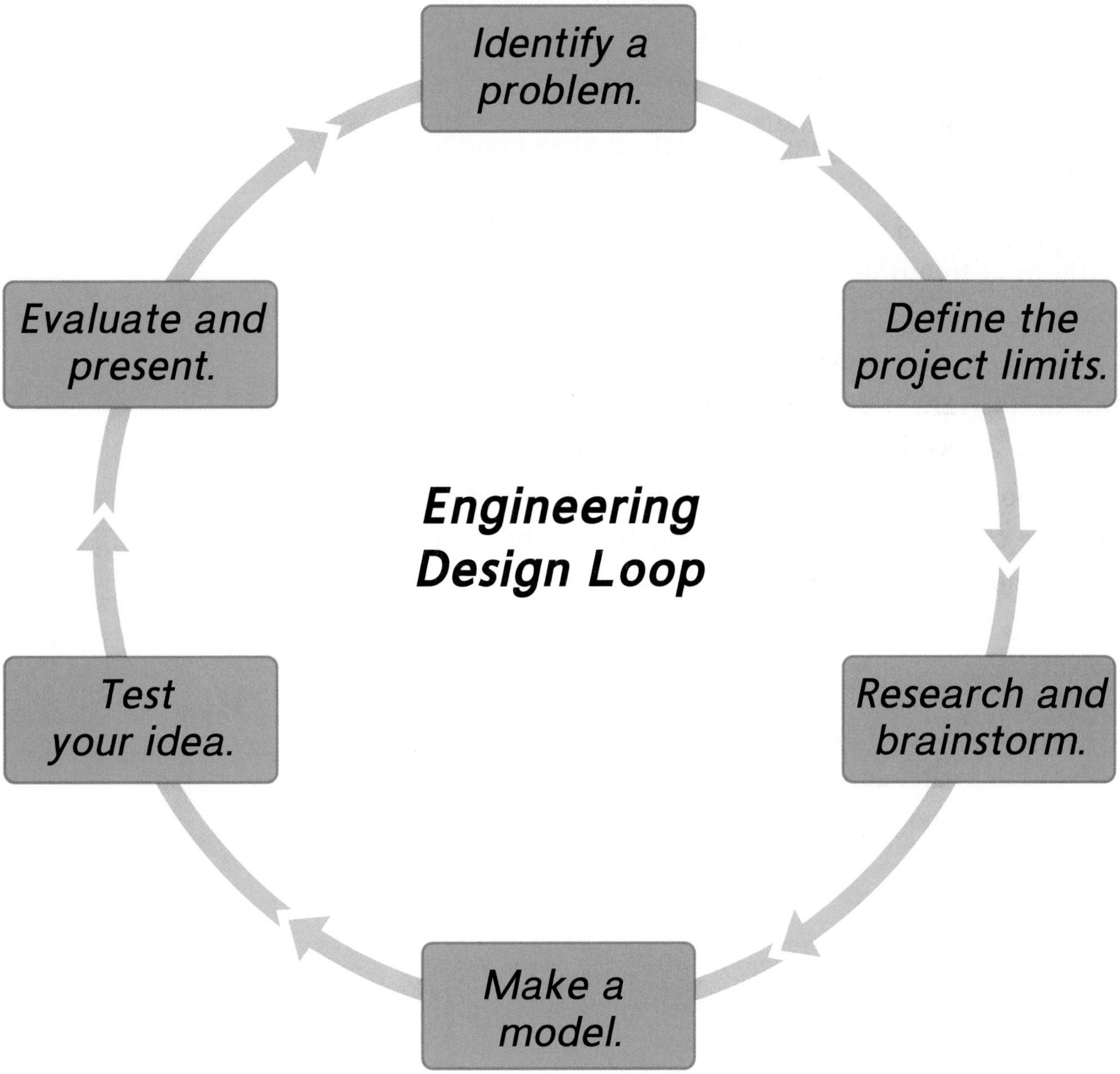

Doing Science and Engineering

In order to put science and engineering into practice, you must have several skills that include being able to organize data, work safely, and use tools. Using these skills, you can gain information needed to solve a problem or positively influence society.

Organizing Data

It is important to organize any data that you find through science and engineering investigations so that it can be communicated to others. Organizing data can also help you find patterns or trends that may tell you something new. There are many ways that data can be organized so it is more useful.

Make a Chart

Charts are useful for recording information during an experiment and for communicating information. In a chart, only the column or row has meaning, but not both. In this chart, one column lists living things, while the second lists nonliving things. There is no relationship across the rows of objects.

Living	Nonliving
tree	rock
chipmunk	puddle
bird	cloud

Organizing data from a survey can be done with a chart. The question and corresponding answers can be displayed in the columns and rows.

Make a Table or Web

Tables Tables can help you organize and record data during an investigation. The columns and rows have headings that tell you what kind of information goes in each part.

The table below shows the properties of certain minerals. Follow each row and column to learn about each mineral's characteristics.

Mineral Identification Table

	Hardness	Luster	Streak	Color	Other
pyrite	6–6.5	metallic	greenish-black	brassy yellow	called "fool's gold"
quartz	7	nonmetallic	none	colorless, white, rose, smoky, purple, brown	
mica	2–2.5	nonmetallic	none	dark brown, black, or silver-white	flakes when peeled
feldspar	6	nonmetallic	none	colorless, beige, pink	
calcite	3	nonmetallic	white	colorless, white	bubbles when acid is placed on it

Idea Webs An idea web shows how ideas or concepts are connected to each other. Idea maps help you organize information about a topic. The idea map shown here connects different ideas about rocks.

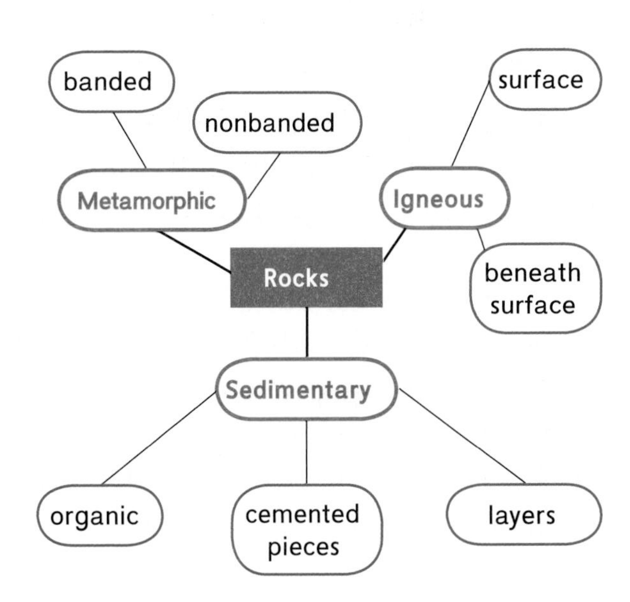

Organizing Data

Make a Graph

Graphs are very helpful in organizing and analyzing data. This next level of organization can make it easy to notice patterns, trends, and relationships within the data that was collected.

Bar Graphs A bar graph uses bars to show the quantity of a type of data. For example, the warmest and coldest months for a city can be graphed. The average temperature for each month can be found online or in the newspaper. The data can then be organized in a chart and then used to make a bar graph.

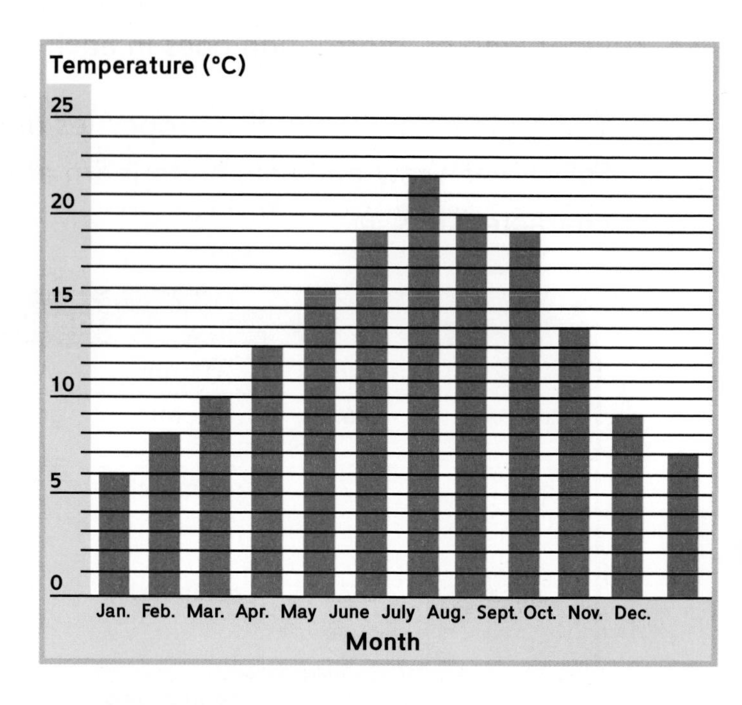

Month	Temperature (°C)
January	6
February	8
March	10
April	13
May	16
June	19
July	22
August	20
September	19
October	14
November	9
December	7

1. Look at the bar for a month on the graph. Put your finger at the top of the bar. Move your finger straight to the left to find the average temperature for that month.
2. Find the highest bar on the graph. This represents the month with the highest average temperature.
3. Find the lowest bar on the graph. This represents the month with the lowest average temperature.
4. Do you see any patterns in the average monthly temperature for this town?

Pictographs A pictograph uses symbols or pictures to show quantities of information. It is different from a bar graph because it shows data in increments. For example, estimated water usage each day is shown in the pictograph to the right.

Water Used Daily (liters)	
drinking	10
showering	100
bathing	120
brushing teeth	40
washing dishes	80
washing hands	30
washing clothes	160
flushing toilet	50

Line Graphs A line graph can show how information changes over time. Information can be recorded over the course of an hour, a day, a year, or a decade. Temperature is often recorded over time for weather forecasts and can then easily be represented as a line graph.

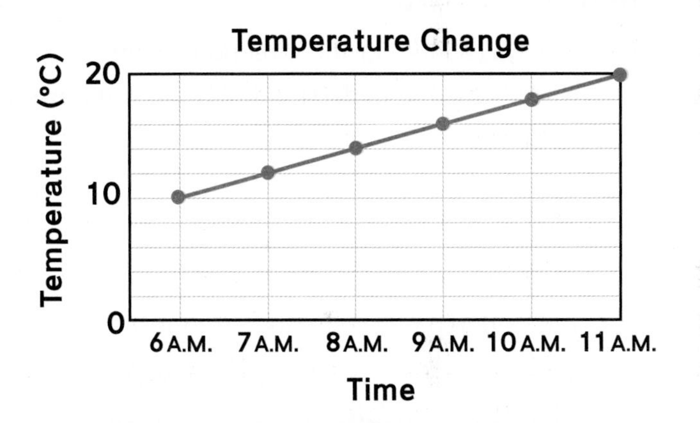

Time	Temperature (°C)
6 A.M.	10
7 A.M.	12
8 A.M.	14
9 A.M.	16
10 A.M.	18
11A.M.	20

Working Safely

Think about the times in your life when there were rules that you needed to follow. We know that rules are important to keep us safe and to keep situations fair. It is also very important for scientists and engineers follow certain safety rules while they do their jobs. Some safety rules might be different based on what kind of role they have, but they are just as necessary to make sure that no one becomes ill or injured. Safety rules are important whether they are applied in a classroom or laboratory setting or out in the field.

In the Classroom

Before an Activity

- Always read all of the directions before starting an investigation. Make sure you understand them. If you have questions about the directions, ask your teacher to help you understand.
- When you see a warning in the directions, such as BE CAREFUL:, be extra careful to follow all of the safety rules.
- Listen to your teacher for special safety directions as well. If you do not understand something, ask for help.
- Wash your hands with soap and water before (and after!) an activity.
- Always pay attention to your surroundings when you gather materials for the activity.

During an Activity

- Wear a safety apron if you work with anything messy or with something that might spill.
- Wear safety goggles when your teacher tells you to wear them. Wear them when working with anything that can fly into your eyes or when working with liquids that might splash.
- Keep your hair and clothes away from materials. Tie your hair back if it is long and roll up long sleeves that might get in the way.
- Do not eat or drink anything during the experiment. Sometimes, a material can be safe to touch but is dangerous to eat.

(t) Tetra Images/SuperStock; (b) McGraw-Hill Education/Eclipse Studios

After an Activity

- Dispose of materials the way your teacher tells you to.
- Put equipment and extra materials back following your teacher's instructions.
- Clean up your work area and wash your hands if needed.

Special Safety Rules

Sometimes, science and engineering investigations require special equipment. Sometimes you need to study something outside the classroom. Keep these safety rules in mind along with the ones that you saw on the previous pages.

Handling Materials

- If something spills, clean it up right away so that no one slips and falls and the material doesn't damage anything. If you are not sure how to clean something up correctly, ask your teacher.
- Tell your teacher if something breaks. If glass breaks, do not clean it up yourself.
- Keep your hands dry when using equipment that uses electricity.

Outside the Classroom

- Always travel outside the classroom with a trusted adult such as your teacher, parent, or guardian.
- Do not touch animals or plants without an adult's approval. The animal might bite, and the plant might be poisonous.
- Remember to treat living things, the environment, and one another with respect.

(t) Ken Karp/McGraw-Hill Education; (b) Dave & Les Jacobs/Getty Images

Measurements

Units of Measurement

Temperature The temperature on this thermometer reads 46 degrees Fahrenheit. That is the same as 8 degrees Celsius.	
Length This girl is 4 feet and 11 inches tall. That is the same 1 meter and 50 centimeters.	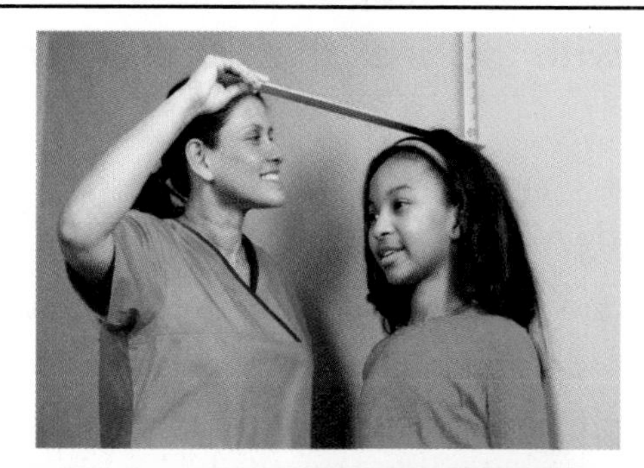
Mass You can measure the mass of this box of crayons in grams.	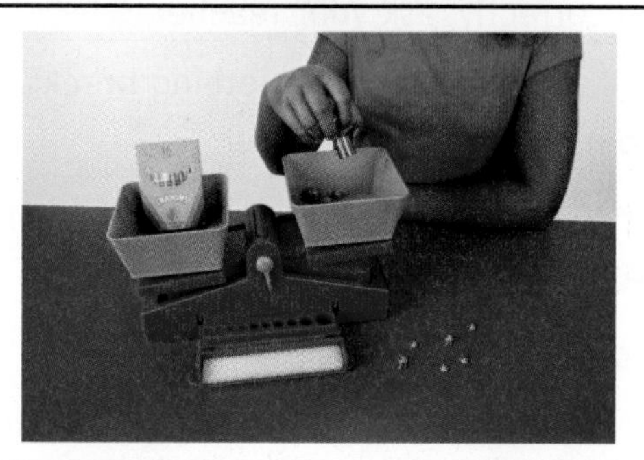
Volume of Fluids These plastic bottles hold about 2 liters of soda. That is a little more than 2 quarts. Some milk cartons hold about 1 quart of milk.	

(t) McGraw-Hill Education; (tc) McGraw-Hill Education/Eclipse Studios; (bc) McGraw-Hill Education/Matt Meadows; (b) McGraw-Hill Education

Weight/Force This little bowl of fruit weighs about 2.5 pounds. That means that the force of gravity is 11.8 newtons.	
Speed This woman can ride her bike 100 meters in 50 seconds. That means her speed is 2 meters per second.	

Table of Measures

SI International Units/Metric Units	Customary Units
Temperature Water freezes at 0 degrees Celsius (°C) and boils at 100°C.	**Temperature** Water freezes at 32 degrees Fahrenheit (°F) and boils at 212°F.
Length and Distance 10 millimeters (mm) = 1 centimeter (cm) 100 centimeters = 1 meter (m) 1,000 meters = 1 kilometer (km)	**Length and Distance** 12 inches (in.) = 1 foot (ft) 3 feet = 1 yard (yd) 5,280 feet = 1 mile (mi)
Volume 1 cubic centimeter (cm^3) = 1 milliliter (mL) 1,000 milliliters = 1 liter (L)	**Volume of Fluids** 8 fluid ounces (fl oz) = 1 cup (c) 2 cups = 1 pint (pt) 2 pints = 1 quart (qt) 4 quarts = 1 gallon (gal)
Mass 1,000 milligrams (mg) = 1 gram (g) 1,000 grams = 1 kilogram (kg)	
Area 1 square meter (m^2) = 1 m x 1 m 10,000 square meters (m^2) = 1 hectare	**Area** 1 square foot (ft^2) = 1 ft x 1 ft 43,560 square feet (ft^2) = 1 acre
Speed meters per second (m/s) kilometers per hour (km/h)	**Speed** miles per hour (mph)
Weight/Force 1 newton (N) = 1 kg x $1 m/s^2$	**Weight/Force** 16 ounces (oz) = 1 pound (lb) 2,000 pounds = 1 ton (T)

(t) lynx/iconotec.com/Glow Images; (b) Table Mesa Prod./Getty Images

Measurements

Measure Time

You measure time to find out how long something takes to happen. Stopwatches and clocks are tools you can use to measure time. Seconds, minutes, hours, days, and years are some units of time.

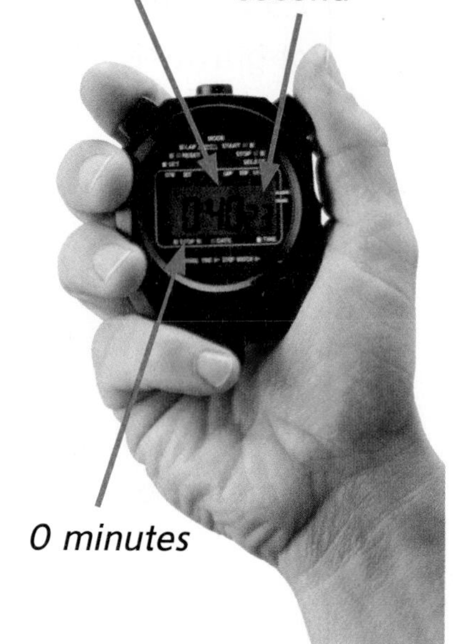

Push the button on the top right of the stopwatch to start timing. Push the button again to stop timing.

©Pixtal/SuperStock

Use a Stopwatch to Measure Time

Get a cup of water and an antacid tablet from your teacher. Tell your partner to place the tablet in the cup of water. Start the stopwatch when the tablet touches the water. Stop the stopwatch when the tablet completely dissolves. Record the time shown on the stopwatch.

Measure Length

You measure length to find out how long or how far away something is. Rulers, tape measures, and meter sticks are some tools you can use to measure length. You can measure length using units called meters. Smaller units are made from parts of meters. Larger units are made of many meters.

Look at the ruler above. Each number represents 1 centimeter (cm). There are 100 centimeters in 1 meter. In between each number are 10 lines. The distance between each line is equal to 1 millimeter (mm). There are 10 millimeters in 1 centimeter.

Measure Liquid Volume

Volume is the amount of space something takes up. Beakers, measuring cups, and graduated cylinders are tools you can use to measure liquid volume. These containers are marked in units called milliliters (mL).

Measurements

Measure Mass

Mass is the amount of matter an object has. You use a balance to measure mass. To find the mass of an object, you compare it with objects whose masses you know. Grams are units used to measure mass.

Measure Force/Weight

You measure force to find the strength of a push or pull. Force can be measured in units called newtons (N). A spring scale is a tool used to measure force.

Weight is a measure of the force of gravity pulling down on an object. To find an object's weight using a spring scale, hold the scale by the top and hang the object by the hook. One pound is equal to about 4.5 N.

Measure Temperature

Temperature is how hot or cold something is. You use a tool called a thermometer to measure temperature. In the United States, temperature is often measured in degrees Fahrenheit (°F). However, temperature is also measured in degrees Celsius (°C).

McGraw-Hill Education/Ken Karp

Tools of Science

Use a Microscope

A microscope is a tool that magnifies objects, or makes them look larger. A microscope can make an object look hundreds or thousands of times larger. Look at the photo to learn the different parts of a microscope.

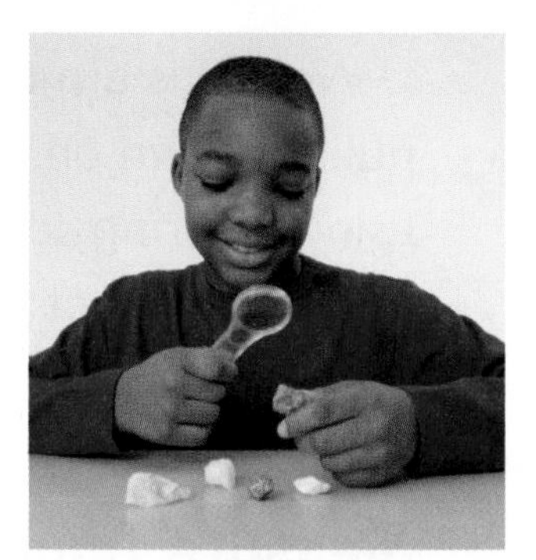

Use a Hand Lens

A hand lens is another tool that magnifies objects. It is not as powerful as a microscope. However, a hand lens still allows you to see details of an object that you cannot see with your eyes alone. As you move a hand lens away from an object, you can see more details. If you move a hand lens too far away, the object will look blurry.

Use a Calculator

Sometimes during an investigation, you have to add, subtract, multiply, or divide numbers. A calculator can help you carry out these operations.

Tools of Science

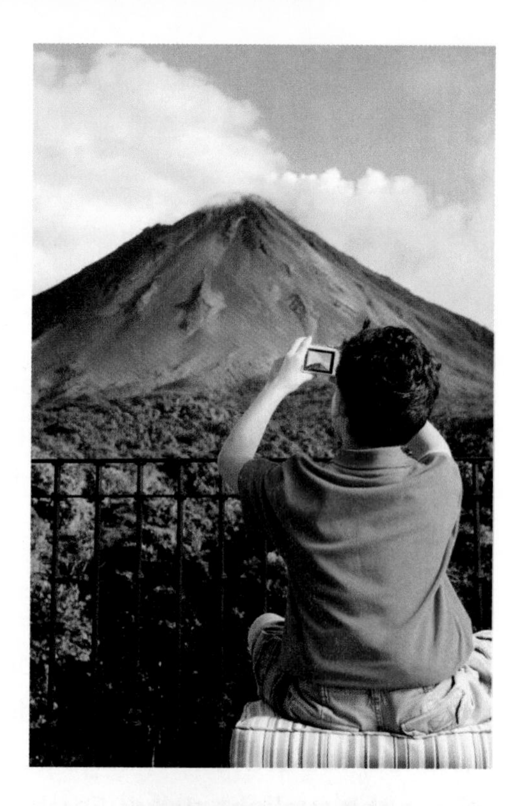

Use a Camera

During an investigation or nature study, it helps to observe and record changes that happen over time. Sometimes it can be difficult to see these changes if they happen quickly or very slowly. A camera can help you keep track of visible changes. Studying photos can help you understand what happens over the course of time.

Use a Computer

A computer has many uses. You can use a computer to get information from compact discs (CDs), digital video discs (DVDs), and jump drives. You can also use a computer to write reports and to show information.

The Internet connects our computer with computers around the world, so you can collect all kinds of information. When using the Internet, visit only Web sites that are safe and reliable. Your teacher can help you find safe and reliable sites to use. Whenever you are online, never give any information about yourself to others.

Crosscutting Concepts

Themes that appear throughout science and engineering are called crosscutting concepts. These themes "cut across" different disciplines, tying them together in various ways. There are seven major crosscutting concepts of science and engineering.

Patterns You see a pattern when something appears over and over again in particular conditions. Sometimes this allows you to predict what will happen under similar circumstances. For instance, the seasons change in a repeating pattern according to where the sunlight strikes Earth as it orbits the Sun. Knowing this helps scientists predict whether other planets have similar seasons based on their tilt and orbit. Graphs allow you to see whether patterns emerge.

Cause and effect The concept of cause and effect is an important one in science and engineering. A cause-and-effect relationship may be as simple as a bowling ball knocking over pins or as complex as evaluating how a volcanic eruption can affect weather, soil, organisms, and the atmosphere.

Scale, proportion, and quantity Since science studies what occurs at the cellular, molecular, or atomic level, knowledge of scale and proportion is essential. For instance, DNA is so small that a scale model of it is helpful in understanding its structure. Scale is also important when studying large systems like the solar system

Engineers build models to scale in order to get realistic results from their tests. Exact quantities are necessary so that scientific investigations can be repeated.

Systems and systems models Systems show how things interact with each other. The body's organ systems, ecosystems, the Sun-Earth-Moon system, and complex machines such as bicycles are examples of systems found throughout science. System models are used in both engineering and science. People can manipulate different components of a system to see how the rest of the components are affected. For instance, an engineer might make a model of a transportation system to examine traffic flow at different times of the day.

Energy and matter: Flows, cycles, and conservation Energy and matter are interwoven into nearly every aspect of science in some form. Energy flows and matter cycles, and they are both conserved as they do so. Scientists observe how the Sun's energy flows through food webs and how water cycles between Earth and the atmosphere.

Structure and function Learning how something works involves understanding its structures and how each functions. Structure and function can apply to the very large (galaxies) and the very small (atoms).

An engineer developing a prosthetic limb must understand the function of each structure in the natural limb in order to mimic its flexibility and strength.

Stability and change One of the aspects of investigating systems is whether they are stable or changing. A system can be said to have stability if most of its components are unchanging—and possibly even if they are changing, but in a predictable, cyclic way. The system changes when something happens to upset the stability. Stability and change are conditions that can apply to multiple systems in science. A system's stability might be of particular importance to engineers working at the citywide or countrywide level, as a change in the system could affect millions of people.

Entomologist

Do insects and other critters interest you? Entomologists are scientists who study all types of insects in areas all over the world. They may work out in the field, which means in an insect's ecosystem studying its habitat, or in a laboratory. Entomologists classify insects and study their life cycles. They are also interested in finding out how different insects interact with their environments and with humans. Entomologists sometimes have to use problem solving to design solutions for pest control and invasive species of insects.

Life Science

Basics of Life

Characteristics of Living Things

There are millions of different kinds of organisms on Earth. *Organism* is another word for a living thing. All organisms share certain characteristics. They are all made of cells. They all have basic needs. They all reproduce, grow, and develop. They all respond to their surroundings.

Living Things Are Made of Cells

All organisms are made of small building blocks called cells. Cells carry out all basic life processes, such as converting food into energy and getting rid of waste materials. An organism can be one cell or can be made of many cells.

Living Things Have Basic Needs

Every living thing has basic needs. They all need water, food, air, and shelter. Organisms cannot survive unless these needs are met.

Food provides energy for living things.

Water All organisms need water in order for their bodies to function. Water moves materials through an organism's body. It also helps remove wastes.

Food All organisms need a source of energy to live. Organisms get energy from food. Animals must eat to get energy. Plants make their own food.

Gases Most living things need oxygen, a gas found in air and in water. Oxygen is needed to release the energy stored in food. Plants also need another gas—carbon dioxide.

Shelter Shelters protect animals from harsh conditions and weather. Shelters also protect animals from predators.

Living Things Reproduce

All organisms must *reproduce*, or make more of their own kind. The survival of a species depends on its ability to produce offspring, the young of living organisms. Through reproduction, genetic material is transferred from parent to offspring. The genetic material gives offspring specific traits and controls how the offspring will look, grow, and develop.

Many multi-celled organisms require two parents to reproduce. Each parent contributes one special cell. Those two cells combine to form a single new cell. This cell grows into a new organism with characteristics from both parents.

Did You Know?

After the female emperor penguin lays a single egg, the male penguin keeps the egg warm between his feet until it hatches. These actions help the penguins successfully reproduce.

Living Things Grow and Develop

Growth is the process of becoming larger. Development is the process of change that occurs during an organism's life. Organisms have life cycles that involve changes in their size, shape, ability to move, and feeding behavior. As organisms grow and develop, they use energy to make new cells.

Living Things Respond to Their Surroundings

All organisms react to changes in their environment. The change that causes the reaction is called a **stimulus**. The reaction or change in behavior of an organism is a **response**. Sights, sounds, smells, tastes, and the things that animals sense through touch are stimuli. Gravity, water, and light are three stimuli that plants respond to.

Adaptation and Change

Every ecosystem presents challenges to the organisms that live there. A desert animal must be able to withstand extreme heat. A fish must be able to obtain oxygen from the water. A plant must be able to store water. An **adaptation** is a physical trait or a behavior that helps an organism survive in its environment.

The chameleon's color changes with the varying colors of its environment.

Physical Adaptations

Many plants and animals have body parts that are physical adaptations. Teeth, body color, claws, gills, and beaks are some physical adaptations that allow animals to survive in certain environments. For example, desert plants and animals have adaptations for living in a hot, dry environment. Cacti have thick, waxy stems that store water. A kangaroo rat's teeth can cut through the tough stems to get the water.

Physical adaptations also help animals hunt and obtain food. A cheetah's sharp teeth allow it to tear the flesh of its prey. The long, narrow bill of a hummingbird helps it to feed deep inside narrow flowers. The sharp talons of a hawk capture prey.

Many animals have *protective coloration*—skin patterns or fur coloring that help the animal blend into the environment. For example, a zebra's stripes help it hide from predators, while a cheetah's spots help it stalk its prey without being seen. Some animals can actually change color or texture, allowing them to blend into a changing environment.

In winter, the fur of the arctic fox is white. In summer, its coat changes to brown.

Behavioral Adaptations and Animal Movement

Some adaptations are behavioral. When fall changes to winter, animals such as bears *hibernate* to withstand the cold. Hibernating animals live off their body fat and use very little energy. Bats, snakes, and turtles are some other animals that hibernate during cold winter months.

Seasonal changes in temperature can also result in movement known as *migration*. When the weather changes, some animals migrate to places where they can find better sources of food. Others migrate to reproduce. Hummingbirds, gray whales, monarch butterflies, Canada geese, caribou, and salmon are examples of animals that migrate.

Many animals, such as elephants, sheep, and caribou, travel in herds for protection. An unlucky animal that is separated from a herd may become a predator's target. Other animals, such as wolves, hunt in packs. The wolves' ability to work together enables them to surround and capture much larger prey.

Snow geese migrate south during the winter months.

U.S. Fish & Wildlife Service/Dave Menkhe

Plant Behavior

Plants cannot move around as animals can, yet plants also react to changes in their environments. Plants respond to stimuli such as sunlight, water, or gravity. A plant responds to a stimulus by changing its pattern of growth. This response is known as a *tropism*.

Word Study

The word *tropism* comes from the Greek word *tropos*, which means "a turning."

Plant Tropisms

Phototropism Plants respond to light by growing toward the light source. In this photo, sunlight is the stimulus. In response, the plant produces a chemical that causes cells on different sides of its trunk to grow at different rates. The plant bend toward the light. This response is known as phototropism. The word part *photo* means "light."

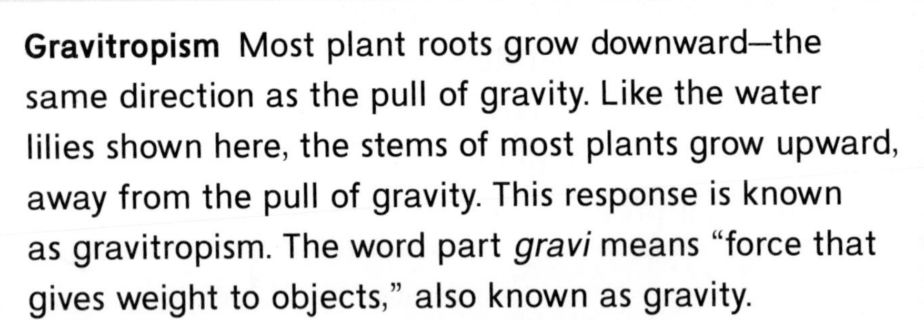

Gravitropism Most plant roots grow downward–the same direction as the pull of gravity. Like the water lilies shown here, the stems of most plants grow upward, away from the pull of gravity. This response is known as gravitropism. The word part *gravi* means "force that gives weight to objects," also known as gravity.

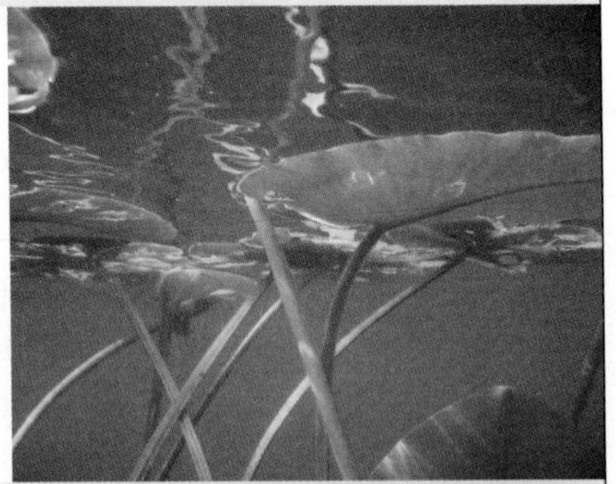

Hydrotropism Roots sense water in the soil and grow toward it. This response is known as hydrotropism. The word part *hydro* means "water." These roots are showing both hydrotropism by searching for water and gravitropism by searching for soil.

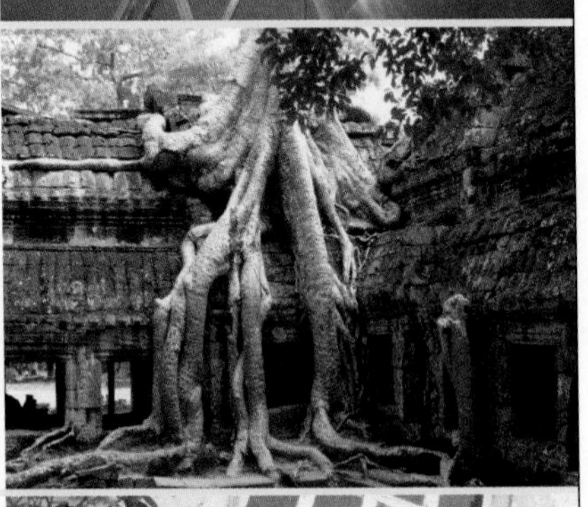

Thigmotropism Some plants respond to touch or contact with an object by curling around that object or clinging to it. This response is known as thigmotropism. The word part *thigmo* means "touch."

(t) hxdbzxy/iStock/Getty Images; (tc) marck140/iStock/Getty Images; (bc) Lissa Harrison; (b) ©JUPITERIMAGES/PHOTOS.COM/Alamy

Animal Senses

Animals use their senses of sight, hearing, touch, taste, and smell to detect changes in the environment. Some animals have highly developed senses. Cats have keen night vision that helps them hunt for prey. The position of an animal's eyes may keep it safe. A rabbit's eyes are positioned on the sides of the animal's head. This gives the rabbit better side vision, allowing it to spot predators from all around. A shark's eyes are very sensitive to light, allowing it to spot prey in murky water.

Other animals have a highly developed sense of hearing. They use their hearing and their ability to make sounds to hunt or communicate. Bats, for example, make sounds that bounce off their prey. The echo allows the bat to navigate and hunt in the dark. Finding food or other objects in this manner is called **echolocation**. Whales and dolphins also use echolocation to find food and communicate with each other in the water.

Animals use their senses of taste and smell to survive. A highly developed sense of smell allows animals to detect prey or to stay safe from predators. A skunk wards off predators using a powerful-smelling spray. Many plants and animals are poisonous, which keeps predators away.

The poison arrow frog's skin secretes a poisonous substance. Its colorful skin warns predators to stay away.

Design Pics/Corey Hochachka

Getting and Using Energy

Like all organisms, plants need energy to grow and survive. They get this energy from the Sun. The specialized parts of the plant, including the leaves, roots, and stems, work together to provide the materials that the plant needs to survive.

Photosynthesis

Plants use the energy in sunlight to make food from water and carbon dioxide in a process called **photosynthesis.** This process takes place in the leaves of the plant. Because most leaves are broad and flat, they are able to collect sunlight efficiently. Specialized structures in the leaves called **chloroplasts** absorb sunlight. The chloroplasts contain **chlorophyll,** which traps the energy of the Sun. Here, the food-making process begins.

In order to complete the process of photosynthesis, the leaves must get the remaining raw materials. Water is absorbed into the roots of the plant. It travels up the stem through specialized tissues called **xylem** that make up the veins of the leaf.

The leaves get the raw material carbon dioxide from the air. Air enters and moves out of the leaves through tiny openings on the underside of the leaves called *stomata*. A single opening is called a *stoma*. The stomata are surrounded by *guard cells*. When the leaf has plenty of water, the guard cells swell and pull the stomata open. This opening allows water and air to leave the plant. When the plant is low on water, the guard cells shrink. This shrinkage causes the stomata to close, which prevents water from escaping.

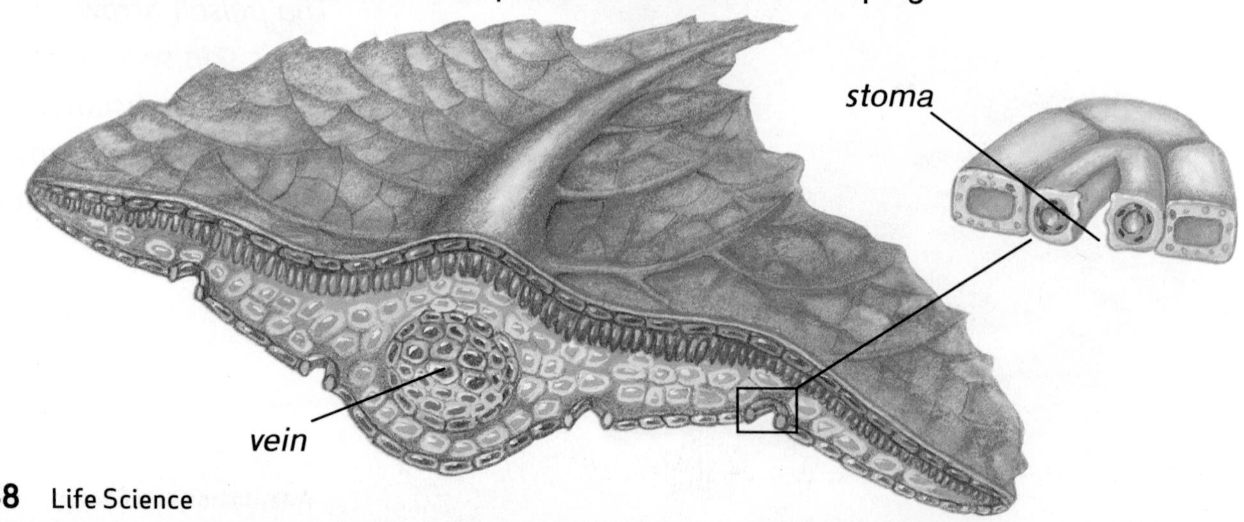

As the plant obtains its raw materials, the process of photosynthesis takes place. Carbon dioxide and water enter the chloroplasts. There they combine in the presence of the trapped light energy to produce sugar and oxygen. The sugars are transported to all parts of the plant through specialized tissues called *phloem*. Energy from the sunlight is now contained in the sugars. The sugars are what the plant uses for food. Oxygen leaves the plant through its stomata. The process of photosynthesis can be written like this:

carbon dioxide + water + energy → sugar + oxygen

Word Study

Photosynthesis comes from the Greek words, *photo,* meaning "light," and *synthesis,* meaning "put together."

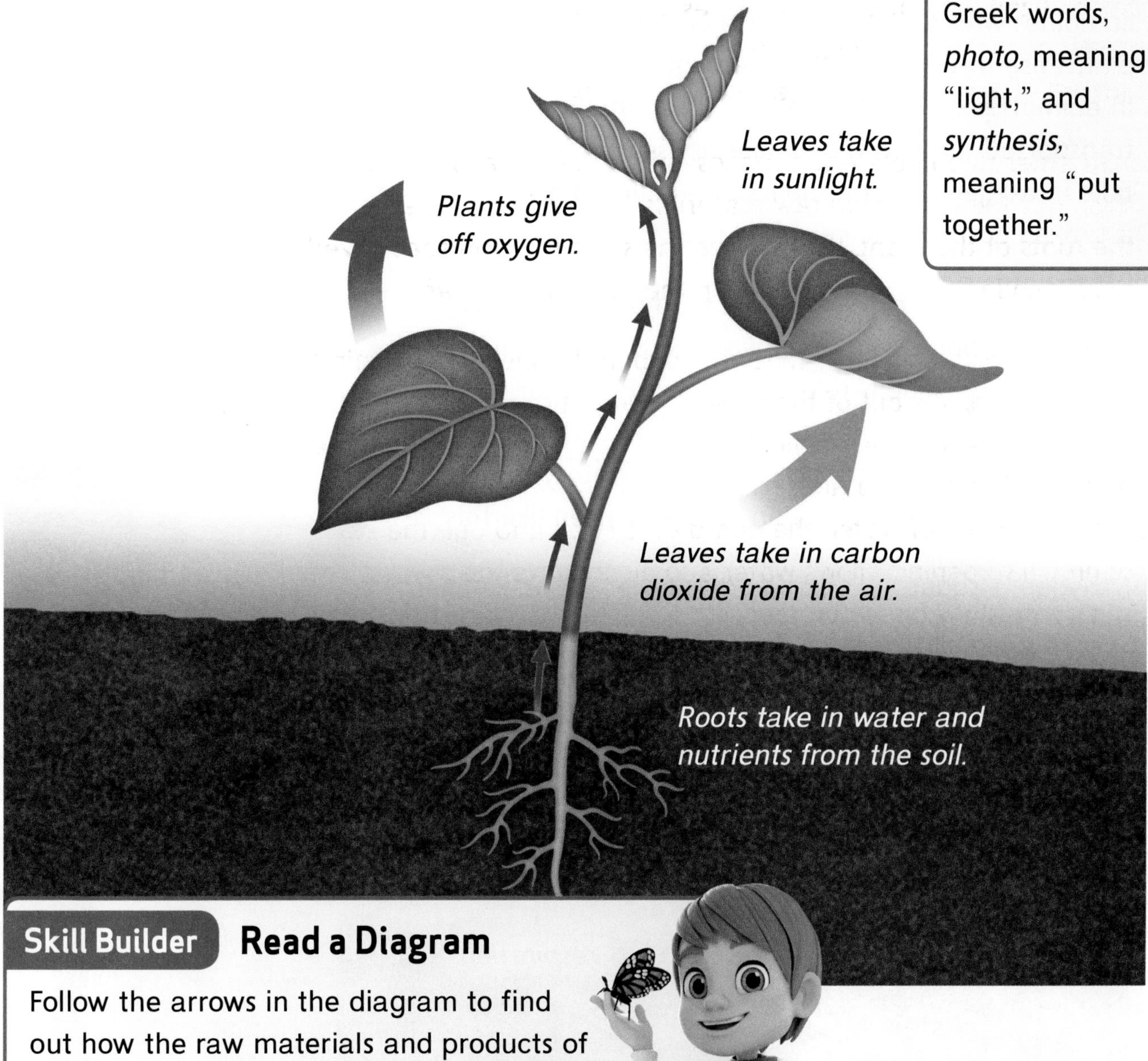

Skill Builder Read a Diagram

Follow the arrows in the diagram to find out how the raw materials and products of photosynthesis move in and out of the plant.

Cellular Respiration

The most important source of energy for Earth is the Sun. Plants store the Sun's energy in the sugars they make. Other organisms also use this energy when they eat plants or eat organisms that have eaten plants. The energy is released when cells use oxygen to break down these sugars in the process of **cellular respiration.**

Cellular respiration takes place in plant and animal cells. It also takes place in the cells of protists, fungi, and some bacteria. The process takes place in a specialized part inside the cell known as the *mitochondria*. Here, oxygen combines with sugars to release energy. Both animals and plants use this energy for movement, growth, and repair. In animals, oxygen is taken in through the lungs and transported through the circulatory system to cells. Plants take in oxygen through their stomata. In both cases, sugar is broken down inside cells and used for energy.

During cellular respiration, plant and animal cells produce carbon dioxide and water as waste products. The carbon dioxide is released back into the air when living things respire. Plants then use the carbon dioxide and water to carry out the process of photosynthesis. The process of cellular respiration can be written like this:

oxygen + sugar → carbon dioxide + water + energy

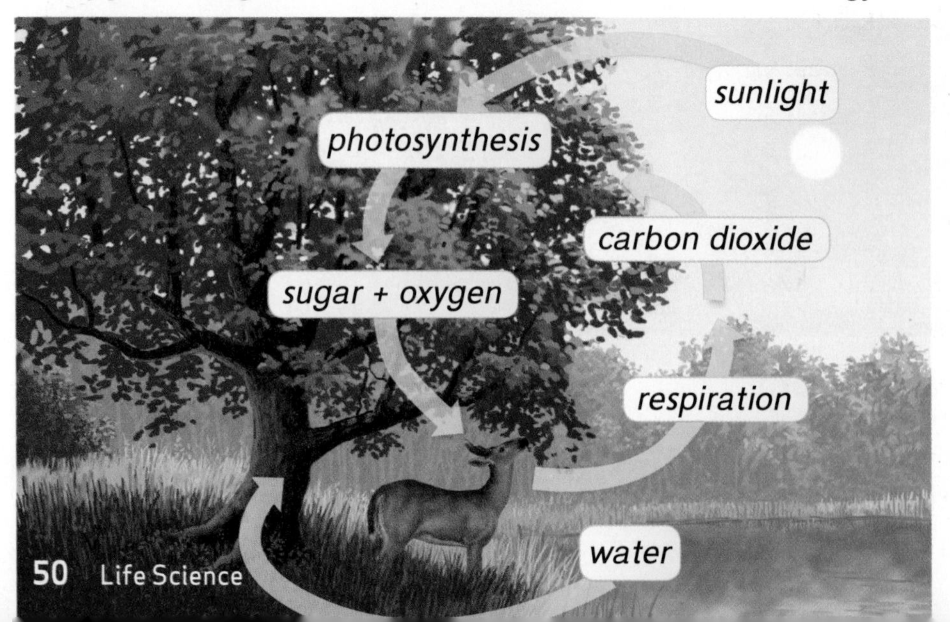

Make Connections

Jump to the Cells section to see a photograph of mitochondria.

Skill Builder

Read a Diagram

Follow the arrows to find out how photosynthesis and cellular respiration are related.

Energy from Food

Organisms get energy when they break down food. *Nutrients* are the materials found in food that the body uses for growth and repair. Sugars and starches are types of nutrients called carbohydrates. Other nutrients include proteins, fats and oils, vitamins, and minerals.

Carbohydrates

Carbohydrates are the main source of energy in living organisms. In plants, unused sugars are stored as starches. The plant can break down these starches into food as needed. Animals also break down sugars to get energy. Starches are found in foods such as rice, potatoes, breads, cereal, and pasta. They provide the body with long-lasting energy. Sugars are found in fruits, but the energy from sugars does not last as long as energy from starches.

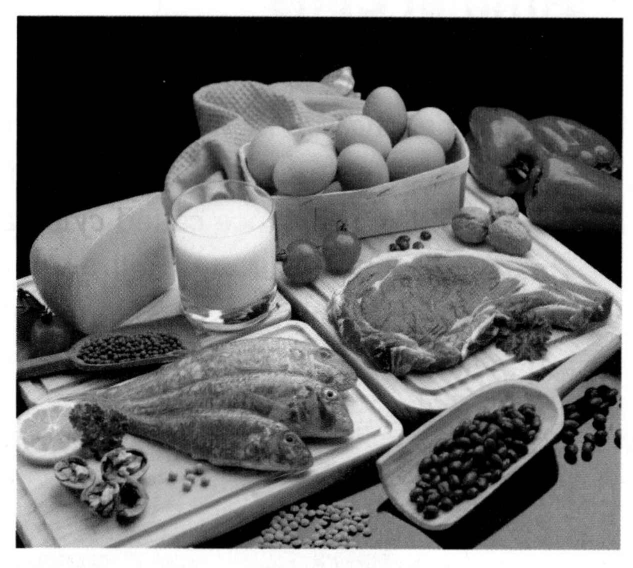

Different nutrients are found in a variety of types of food.

Proteins

Proteins are nutrients that are mostly found in meats, fish, eggs, nuts, and dairy products such as milk, yogurt, and cheese. Legumes are another food source of proteins. These include foods that come from pods, such as beans, lentils, and peanuts. Proteins are essential for growth and repair of the body's cells.

Fats and Oils

The bodies of animals use fats to keep warm. Fats also help the body process other nutrients and absorb vitamins. Sources of fats are ice cream, butter, meats, and oils. Fats are stored in the body and can be used as a source of energy.

Cells to Tissues

What makes your heart different from your skin? What makes the phloem in a plant different from its xylem? In multicellular organisms, the cells are specialized, meaning they perform certain functions. The human body has many different cell types that each perform specific functions. Muscle cells, for example, specialize in movement. Red blood cells carry oxygen to all parts of the body. Plants also have specialized cells. For example, root cells do not make food, but they are specialized to take in water.

muscle cells

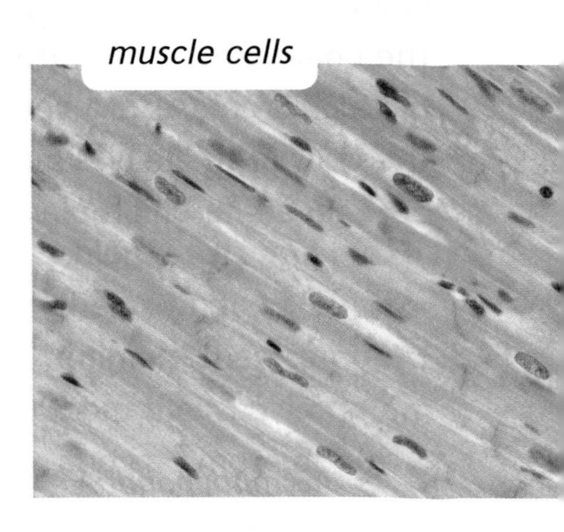

Tissues

Cells are organized into tissues. A **tissue** is a group of similar cells working together at the same job or function. There are four main types of tissues in the human body.

- *Muscle tissue* controls movement both inside and outside the body.
- *Epithelial tissue* is the skin tissue that covers the body. It also lines the internal organs of the body.
- *Connective tissue* holds the body together. It forms ligaments that connect bones with other bones and tendons that connect bones with muscles.
- *Nerve tissue* receives and sends messages between the brain and the rest of the body.

Plant tissues are also specialized. Xylem tissue makes up tubes that move water and minerals through the stem to the leaves. Phloem tissue moves sugars made in the plant's leaves to other parts of the plant.

epithelial cells

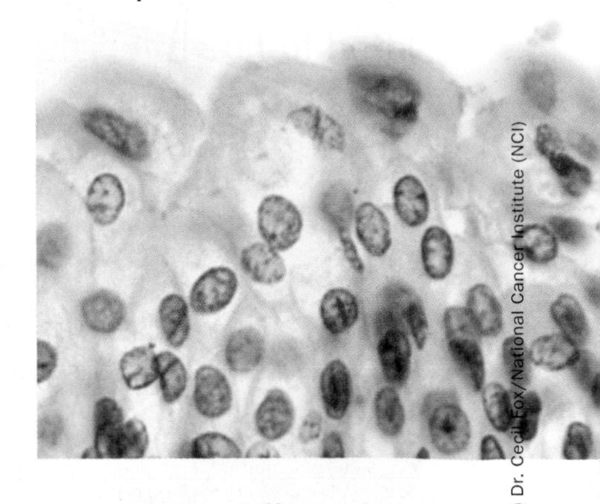

nerve cell

connective cells

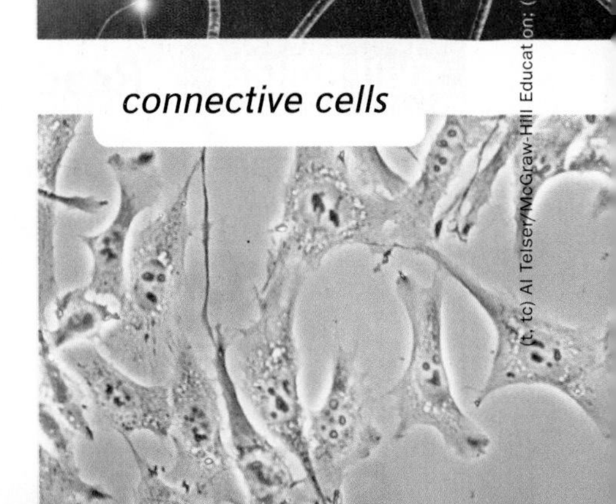

Organs and Organ Systems

An **organ** is a group of tissues that work together to perform a specific function. Your heart, brain, stomach, kidneys, and even your skin are organs. When a group of organs work together to perform a specific function, they make up an *organ system*.

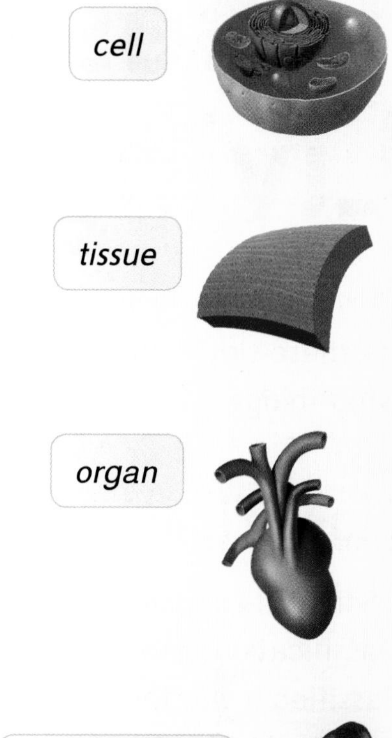

The digestive system performs the functions of breaking down food into smaller and smaller components that the body can absorb. It is made up of the mouth, pharynx, esophagus, stomach, and small and large intestines. These organs are made up of specialized tissues that help them perform a job. For example, the stomach is made up of muscle tissue to help it contract. It is lined with epithelial tissue. The muscle tissues are made up of muscle cells, and the epithelial tissues are made up of epithelial cells.

Many organ systems work together. The digestive system works with the blood cells, blood, veins, arteries, and heart of the circulatory system to transport nutrients to the cells of the body. The muscular and skeletal systems work together to provide both support and movement for the body. The respiratory and circulatory systems work together to transport oxygen from the lungs to the cells of the body.

Make Connections

Jump to the Human Body section to learn about organ systems.

What Could I Be? **Biomedical Engineer**

Interested in engineering, biology, and healthcare? Biomedical engineering might be for you! Some of these scientists are even researching how to engineer new organs! Learn more about how this might be possible by turning to the Careers section.

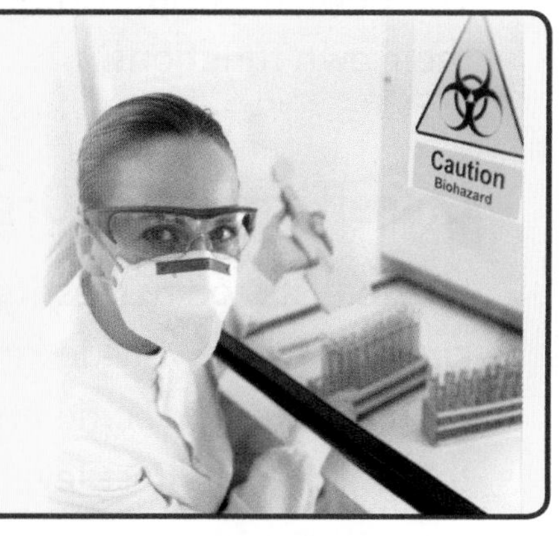

Types of Living Things

Living things have similarities and differences. Scientists use these similarities and differences to group, or classify, living things.

Classification

Scientists classify living things to make them easier to study. Classifications are also useful in naming organisms. Early scientists classified living things based on observable characteristics, such as color. Modern scientists use characteristics such as body structure to classify organisms. They also use information about how an organism obtains food and whether it can move.

Unicellular and Multicellular Organisms

The number of cells in a living thing is one of the characteristics scientists use to classify organisms. *Unicellular* organisms are made up of one cell. Although these living things consist of just one cell, they carry out life processes such as growing, reproducing, responding to the environment, and getting food.

Multicellular organisms are made up of more than one cell. The cells in these living things work together as well as carry out their own functions.

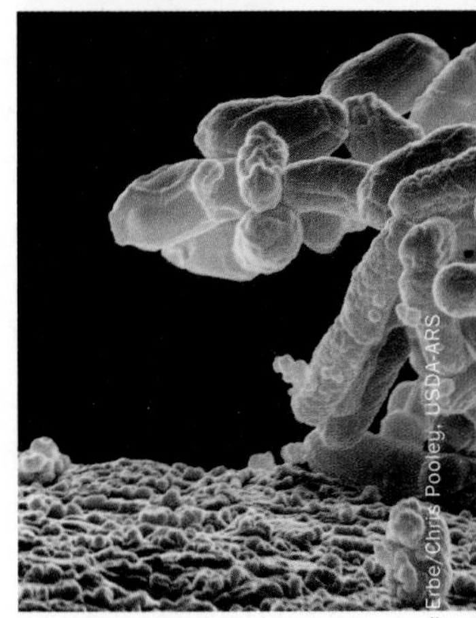

Both the multicellular tree and the unicellular bacteria are living things that carry out life processes.

Classification of Organisms

Scientists today classify living things into six large groups called **kingdoms**. They divide kingdoms into smaller groups. Each smaller group is more specific and contains fewer different types of living things. The smallest level of organization is the species. A **species** is a group of very closely related organisms.

Classification of Horses

kingdom

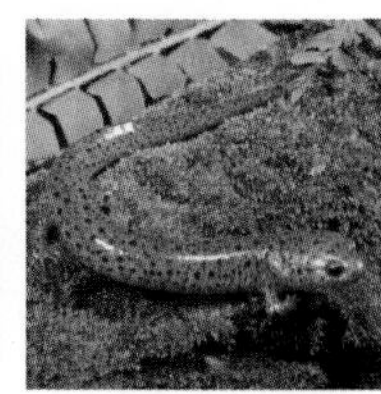

phylum

class

order

family

genus

species

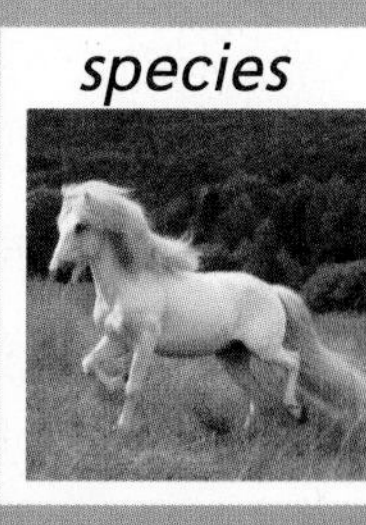

Skill Builder

Read a Chart

This chart shows how a horse is classified. Notice how each level of organization becomes smaller and more specific.

Did You Know?

Scientists estimate that there are as many as 8.7 million kinds of living things on Earth. Of those, 86 percent of land species and 90 percent of ocean species have yet to be named.

Classification - The Six Kingdoms

A kingdom is the largest group into which organisms can be classified. All the members of a kingdom share the same basic traits. The table below shows some of the ways organisms in each kingdom differ.

Kingdom	Archaea	Bacteria	Protists	Fungi	Plants	Animals
Number of Cells	one	one	one or many	one or many	many	many
Can They Make Their own Food?	most	some	some	no	yes	no
Nucleus	no	no	yes	yes	yes	yes
Move from Place to Place	some	some	some	no	no	yes

Archaea Some of the organisms that are classified into the archaea kingdom live in extreme environments. These are places that are too hot or too salty for other living things. Scientists used to classify archaea and bacteria as one kingdom, but research showed many differences between these two types of organisms. They now classify them into separate kingdoms.

Some archea live in very hot water.

Bacteria Some bacteria play important roles in ecosystems, where they are decomposers. Many bacteria affect human health. Some can cause diseases or infection. Other bacteria help keep people healthy.

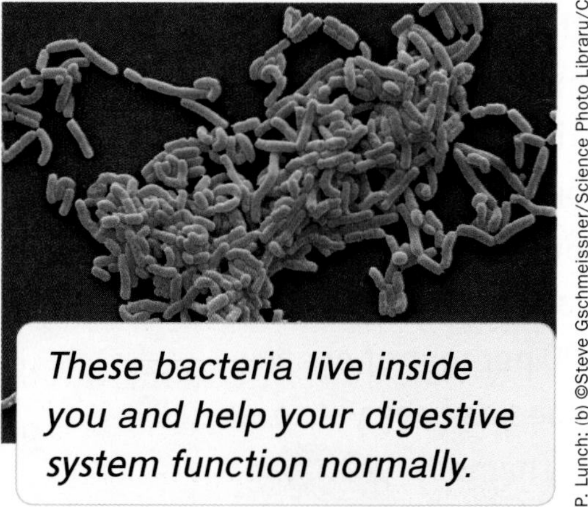
These bacteria live inside you and help your digestive system function normally.

Protists Organisms in the protist kingdom are very different from one another. Some protists are producers; some are consumers; and some are decomposers. Some are unicellular, and some are multicellular. The best way to classify protists is that they are not fungi, plants, or animals.

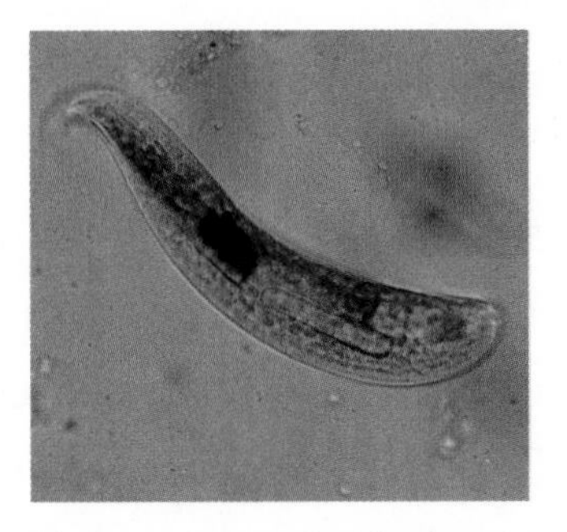

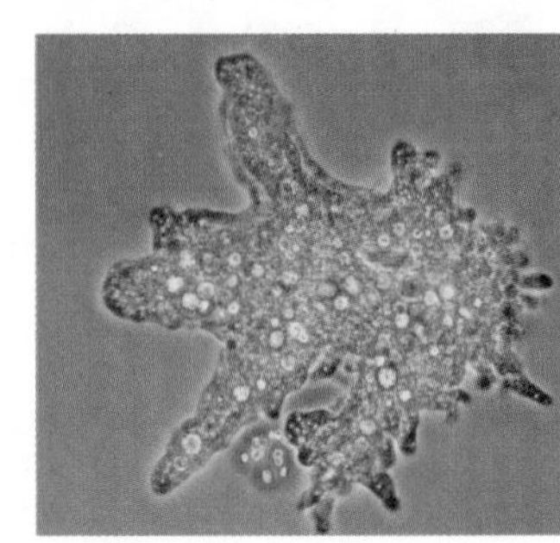

Even though these two organisms look very different, they are both classified as protists.

Fungi Some familiar types of fungi include mushrooms and yeast, which is used to make bread. Fungi cause some human health problems, such as athlete's foot. Fungi are important decomposers in many ecosystems.

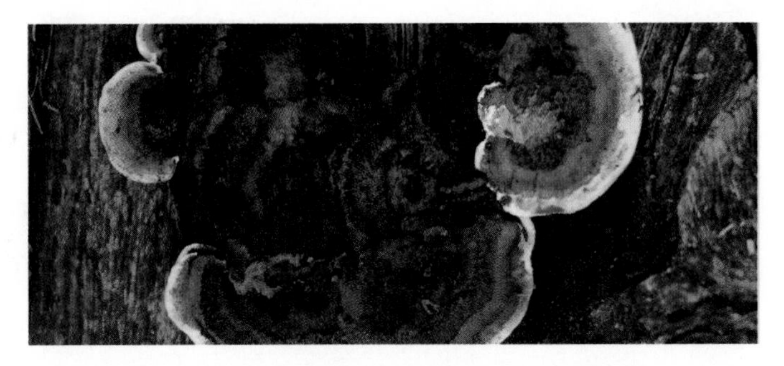

Fungi come in many shapes and sizes.

Plants There are about 400,000 known types of plants. Many are familiar, such as grasses and trees. Most of these plants have roots, stems, and leaves, but some plants do not have these familiar parts. Plants are producers in ecosystems.

Wort plants and mosses do not have typical roots, stems, and leaves.

Animals The animal kingdom contains a wide variety of living things. Sea sponges and humans are both members of the animal kingdom! One way that animals differ from one another is whether or not they have a backbone. Some animals, called *vertebrates*, have a backbone. Other animals, called *invertebrates*, do not have a backbone.

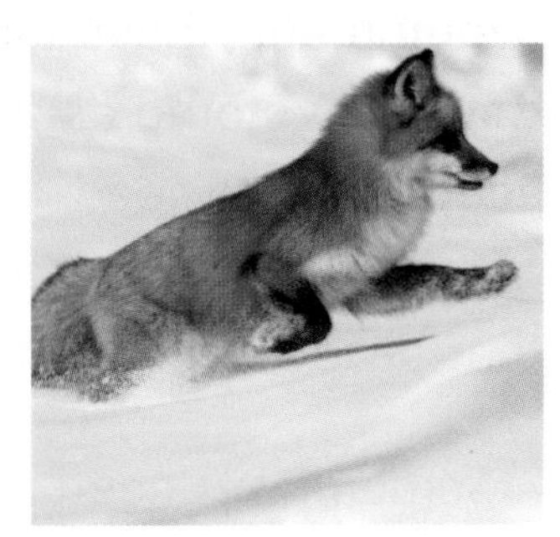

The sea sponge is an invertebrate—an animal with no backbone. The fox is a vertebrate. It has a backbone.

Plants

Plants are an essential part of our lives. All the foods we eat come directly or indirectly from plants. Plants also produce some of the oxygen we breathe. Like all living things, plants have basic needs, a set of typical structures and processes, and life cycles.

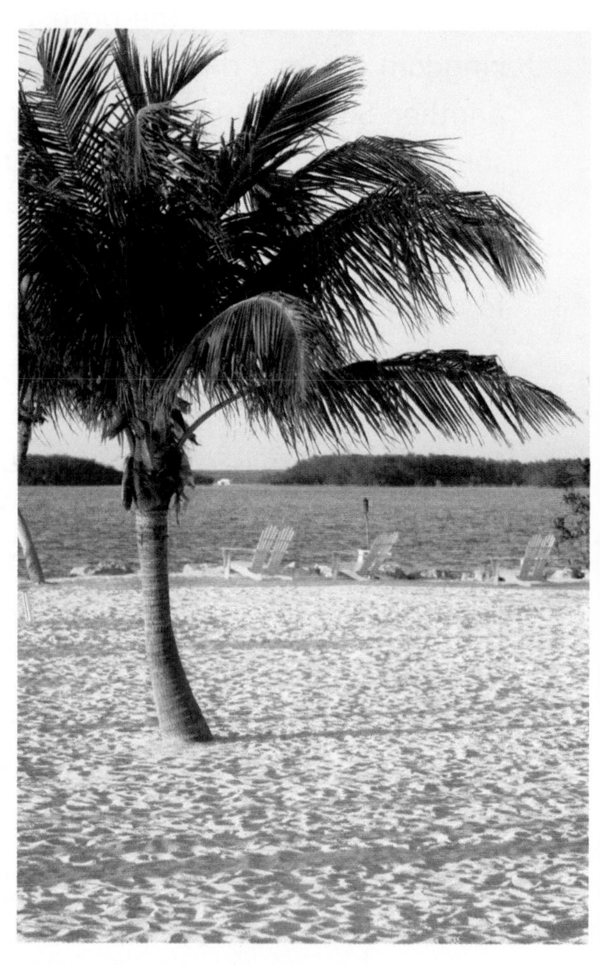

Palm trees require a lot of sunlight. Plants that grow on forest floors or in other shady areas can survive with less light.

Plant Needs

Plants have basic needs—things they need to live and grow. The basic needs of plants are air, sunlight, water, nutrients, and space. Plants must live in an environment where their needs are met.

Gases Plants need gases. The air around Earth is a mixture of gases. Plants need one of these gases, carbon dioxide, to make food in the process of photosynthesis. Plants also need oxygen to break apart food molecules during cellular respiration. Although they do not breathe in and out, plants do take in air from the environment. Stomata in the leaves of plants allow gases to move into and out of plants.

Sunlight Plants need sunlight to make food. When plants make food, the energy in the sunlight is stored as sugars. The sugars provide the energy plants needs to survive. These same sugars pass to your body when you eat plants or animals that have eaten plants.

Some plants need more sunlight than others. Mosses and ferns are plants that can grow in shady areas. Palm trees grow well in places where there is more sunlight.

Make Connections

Jump to the Basics of Life section to review photosynthesis, cellular respiration, and the parts of leaves.

Water Water is a need of all living things, including plants. Water is one of the materials plants need to make food. Water is also used to move materials through a plant—many substances move through a plant dissolved in water. In plants, water also provides support. That is why a plant that does not get enough water starts to droop, or wilt. Most plants take in the water they need through their roots. The water then moves into a system of tubes that distributes the water throughout the plant.

Nutrients Substances that a living thing needs to stay healthy are called nutrients. Plants need nutrients, such as nitrogen. Most plants take in nutrients, which are dissolved in water, through their roots.

Space Plants need enough space to get the air, water, sunlight, and nutrients they need to survive. Remember that plants cannot move around to get what they need. Plants that are crowded close together have a harder time getting the things they need.

Did You Know?

A plant gets most of the material it needs for growth from air and water.

Some plants need more water than others. Cacti can survive in deserts with little rain, while the plants in a rain forest live in a very wet area.

Plant Parts

Most plants have roots, stems, and leaves. These structures help the plant meet its needs and carry out life functions. Plants also have parts used in reproduction, such as flowers, fruits, and seeds.

Roots Plant roots are structures that take in water and dissolved nutrients from the soil. Roots also hold the plant in place. Some roots store food the plant has made. Root hairs help absorb water. The vessels in the center of the root allow materials to move to and from the roots.

Types of Roots	
Description	**Example**
Aerial roots are found in plants that grow on surfaces such as trees and rocks. These roots attach the plant to the surface it grows on and absorb water from the air.	
Fibrous roots are very thin and have many branches. Many grasses have fibrous roots.	
Prop roots extend outward from the bottom of a plant's stem, helping the plant stay upright. Mangrove trees have prop roots.	
A plant with a taproot has one main root that grows deep in the ground with a few smaller side roots. Carrots and dandelions have taproots.	

Stems The stem is the plant structure that supports the plant. It is also a part of a plant's transport system. There are two types of stems: soft stems and woody stems. Soft stems are green and can bend. Woody stems do not bend easily and are often covered in bark. Tree trunks are examples of woody stems. Both types of stems hold up the branches and leaves of the plant and support it.

Stems also allow materials to move inside the plant. The center of plant roots contains a transport system. That transport system extends into the stems. Xylem and phloem are the two types of tissue in this system. Both of these tissues have a structure similar to a tube or drinking straw.

In xylem, materials move up from the roots and through the stem. Xylem carries water and dissolved nutrients. Sugars made in a plant's leaves are transported in phloem. Materials can move both upward and downward in the phloem. Recall that some plants store food in their roots. The food is transported downward through the phloem to the roots. Later, when the stored food is needed, it moves upward from the roots and is carried throughout the plant. The *cambium* is the layer in which xylem and phloem are produced.

Skill Builder

Read a Diagram

Compare how the xylem, phloem, and cambium are arranged in the woody stem and the soft stem.

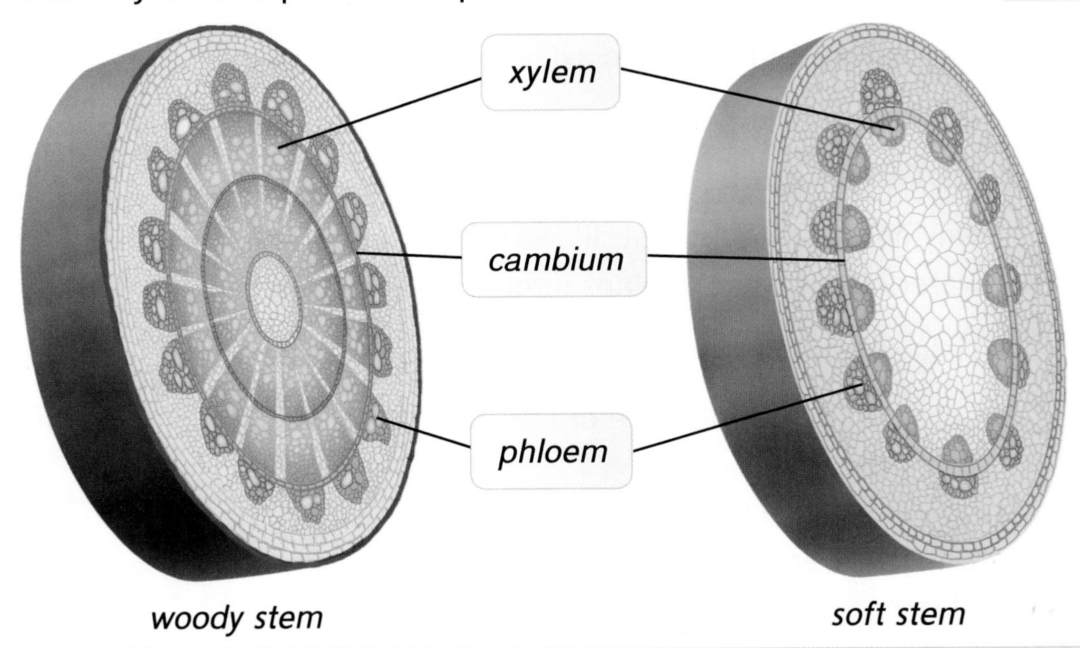

Both soft stems and woody stems contain xylem, phloem, and cambium.

Plant Parts

Leaves Leaves occur in a variety of shapes and sizes. They are the site of photosynthesis (food production) and **transpiration**, the release of water into the air. Transpiration drives the movement of materials throughout a plant. As water evaporates from the leaves, more water is carried from the bottom of the plant to the top. Water moves into the leaf, replacing the water that has evaporated.

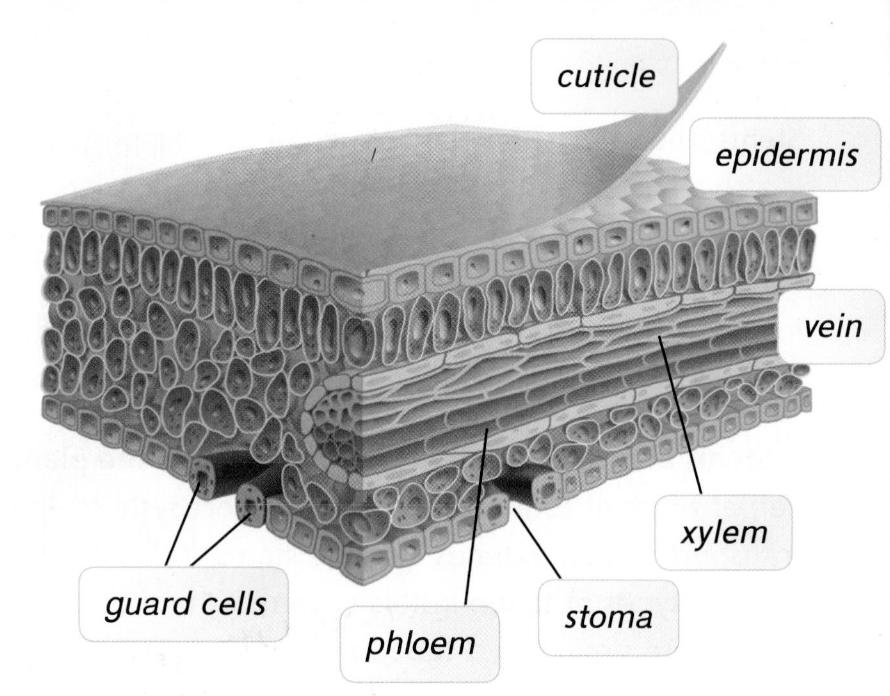

This diagram shows a cross-section of a leaf. Each of the labeled structures has an important function.

Leaf Structures and Their Functions	
Structure	**Function**
Cuticle	A waxy coating that keeps water from evaporating
Epidermis	Contains the cells that are the main site of food production
Vein	A structure that holds the xylem and phloem
Xylem	Tissue that transports water and dissolved nutrients throughout the plant
Stoma (plural, *stomata*)	Tiny holes, or pores, that allow gases to enter and leave the plant; water also leaves the plant through the stomata.
Phloem	Tissue that transports sugars throughout the plant
Guard cells	Cells that control the size of the stomata to keep the plant from losing too much water

Flowers In most seed plants, flowers are the structure used for reproduction. Most flowers contain male and female parts within the same flower. The male part is the stamen. Pollen, which contains the male sex cells, is produced in the anther. The pistil is the plant part that makes the female sex cells. The ovary stores these cells. In the process of pollination, pollen is moved into the pistil. In some plants, pollen is moved by the wind. Animals such as bees and birds move the pollen of other plants. After pollination, the pollen moves to the ovary. Fertilization occurs when the male and female sex cells join.

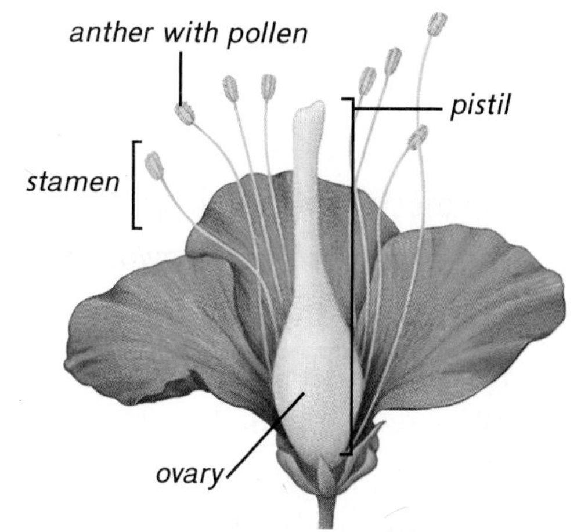

Although flowers come in many colors and shapes, they all contain structures used in reproduction.

Fruits Fertilization occurs in the plant ovary, resulting in a seed. The ovary then develops into a fruit, which protects the seed. Some fruits appeal to animals, which eat the fruits and spread the plant's seeds in their droppings.

Fruits develop from the ovary in a plant's flower.

Cones Some seed plants reproduce with cones instead of flowers. These plants, called *conifers*, usually produce both male and female cones. The male cones produce pollen that is released into the wind. The female cones produce a sticky liquid that captures the wind-blown pollen. Pollination and fertilization occur in the female cone.

Pollen is produced in the smaller male cones. Seeds develop in the larger female cones.

Plant Life Cycles

Like all living things, plants have life cycles. Each stage of development a living thing goes through in its life is a part of the organism's *life cycle*. The life cycles of four different kinds of plants are shown on these pages.

The life cycle of a moss includes an asexual stage, in which spores form a new plant. The new plant that grows produces male and female sex cells, and sexual reproduction occurs. The life cycle alternates between sexual and asexual stages.

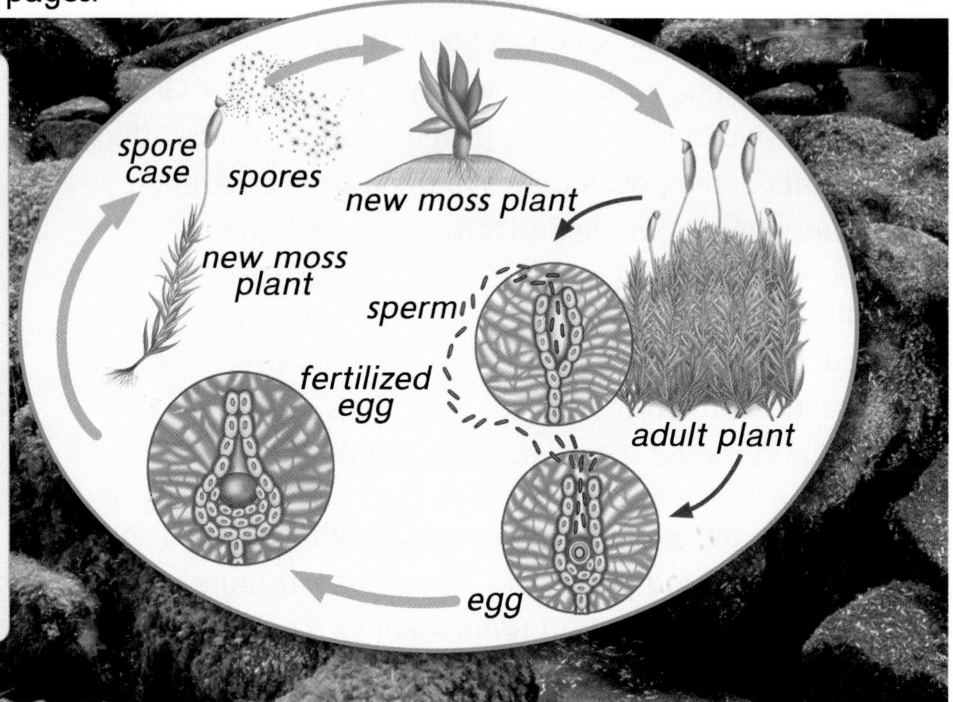

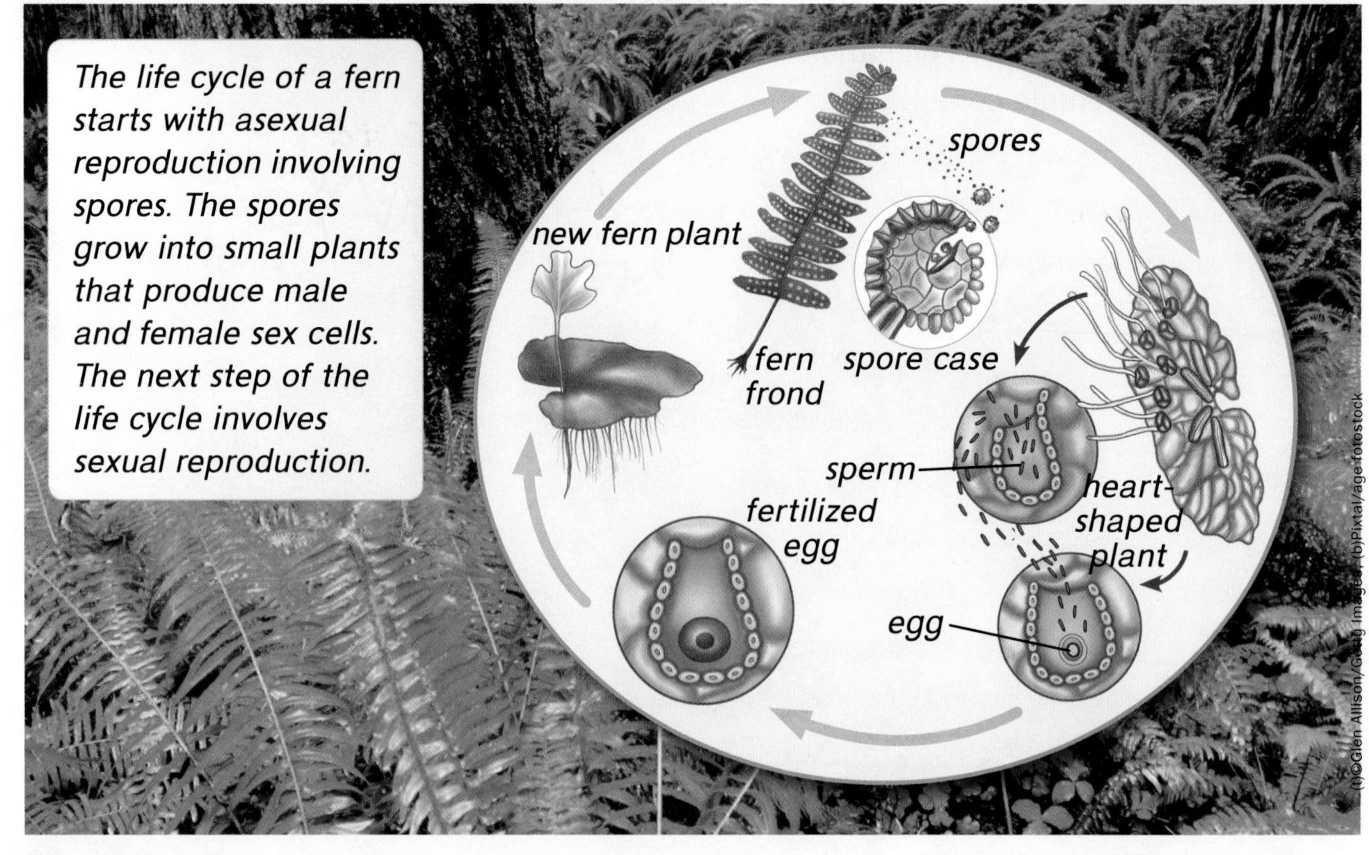

The life cycle of a fern starts with asexual reproduction involving spores. The spores grow into small plants that produce male and female sex cells. The next step of the life cycle involves sexual reproduction.

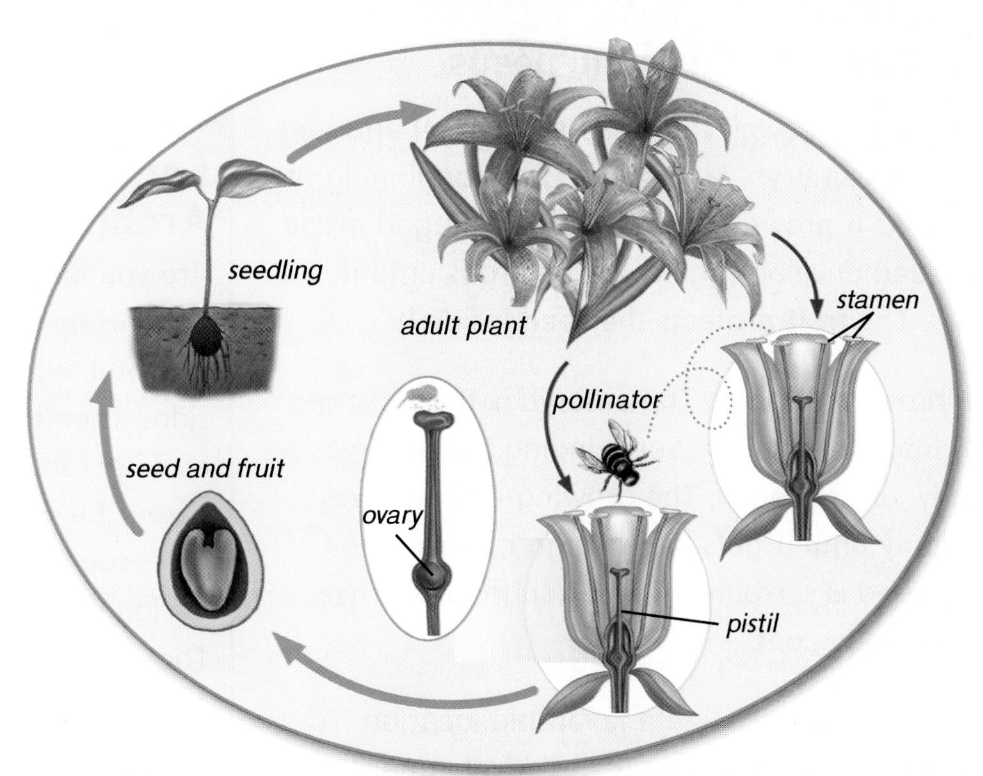

This diagram shows the life cycle of a flowering plant. The seeds are encased in fruit. The seeds germinate, or begin to grow, if conditions are right. A seedling then grows into a plant that produces flowers. The asexual portion of the flowering plant life cycle is not shown in this diagram.

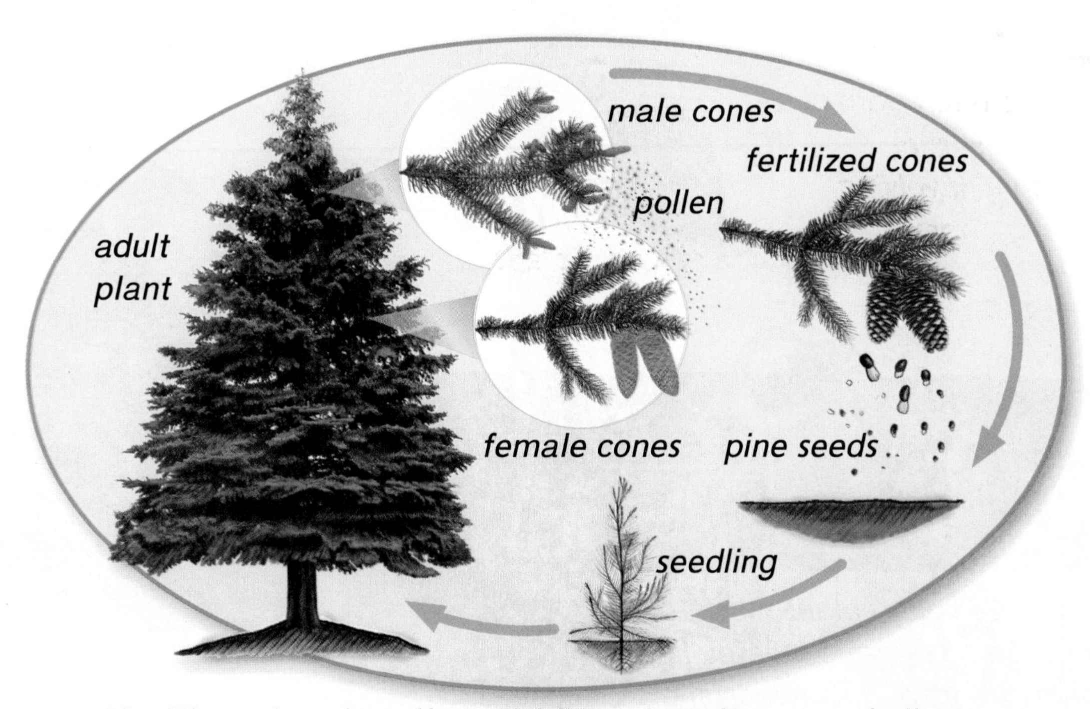

The life cycles of conifers and flowering plants are similar. Recall that conifers produce seeds in cones rather than in flowers. The asexual portion of the conifer life cycle is not shown in this diagram.

Plant Life Cycles - Growing from Seeds

Inside the ovary of a flower, the fused sperm cell and egg cell develop into an embryo. An embryo is the beginning of a new offspring. As it grows, the embryo is packaged inside a seed. As the seed develops, the ovary enlarges until it becomes a fruit. The fruit protects the seed inside it.

A seed has three main parts. The embryo is the offspring that can grow into a new plant. Surrounding the embryo is its food supply, or cotyledon. The growing embryo lives off this food supply until it gets big enough to make food on its own. The seed is surrounded by a tough, protective covering called a seed coat.

Once formed, seeds must fall in a favorable location in order to germinate. Germination is the development of a seed into a new plant. Seeds will not sprout unless conditions are right. Favorable conditions include enough water and space to grow. Seeds do not need sunlight to germinate, but growing seedlings need sunlight to produce food after the cotyledons are used up.

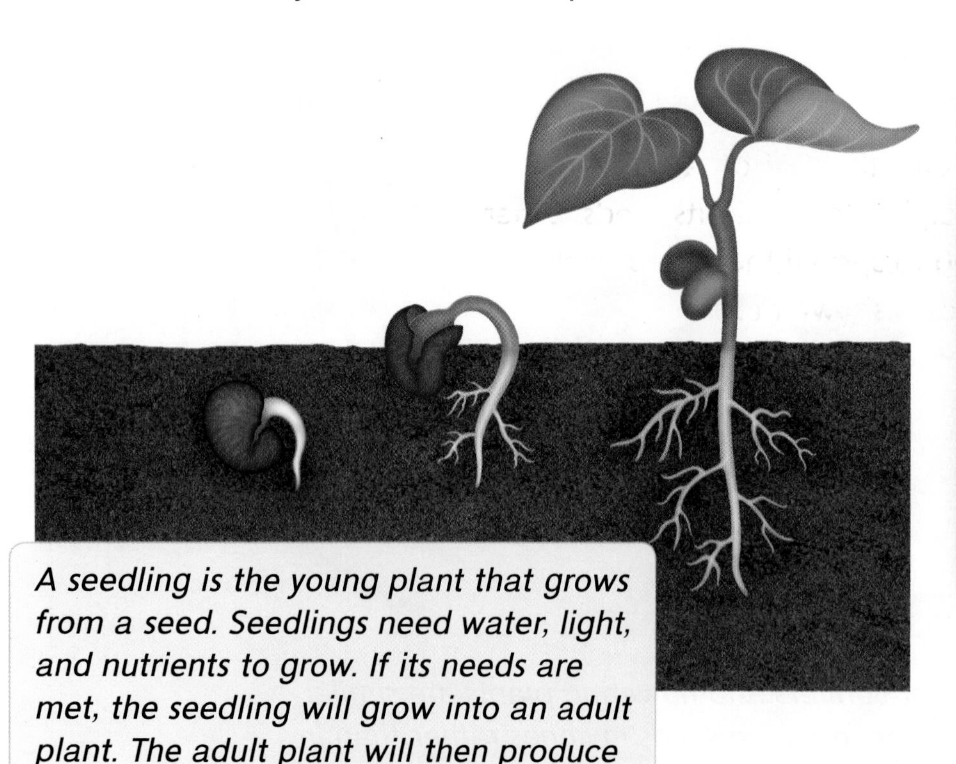

A seedling is the young plant that grows from a seed. Seedlings need water, light, and nutrients to grow. If its needs are met, the seedling will grow into an adult plant. The adult plant will then produce seeds, and the life cycle continues.

What Could I Be?

Landscape Architect

Are you fascinated by the variety of plants? Do you have an artistic side? Then landscape architecture may be the career for you! These professionals plan and design parks and gardens. They also restore natural areas disturbed by humans. To find out more about what landscape architects do, turn to the Careers section.

How Seeds Spread

Seed dispersal helps ensure that the seeds will have room to grow. Seeds are spread, or dispersed, in many ways.

Animals Animals disperse seeds in many ways. Some seeds, like the one shown here, have hooks that catch on the fur of passing animals. The seeds drop as the animal walks away. Seeds that develop in fleshy fruits are eaten by animals and dispersed in their droppings. Other seeds, such as acorns, are buried by animals preparing for winter. Forgotten seeds can grow into new plants.

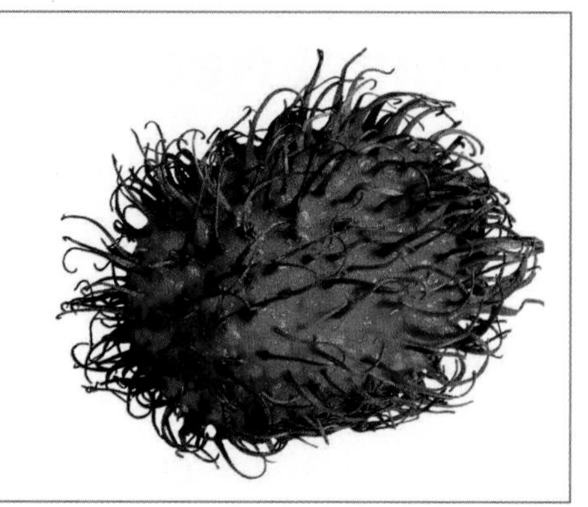

Explosions These Himalayan balsam flowers forcefully eject their seeds when the seedpods are touched.

Wind Have you ever blown the fluff off a dandelion? If you have, you were helping to spread its seeds. Other plants that use the wind to spread their seeds include maple trees and milkweed, shown here.

Water Many coconut palms use nearby water sources to disperse their seeds.

(t) lynx/iconotec.com/Glow Images; (tc) Design Pics/David Chapman; (bc) Bear Dancer Studios/Mark Dierker; (b) PeskyMonkey/E+/Getty Images

Animals

Animal Needs

Animals need food, water, space, and shelter to survive. They also need *oxygen*, which is a gas found in air and water. While animals all have the same needs, they meet their needs in a variety of ways and places.

Food

Food provides energy for animals. Unlike plants, which produce their own food, animals eat other organisms. Scientists classify animals based on the kind of food they eat. *Herbivores* are animals that eat plants. Deer and squirrels are examples of herbivores. *Carnivores*, such as owls and other birds of prey, eat other animals. *Omnivores* are animals that eat plants, fungi, eggs, and other animals. Raccoons, humans, and bears are omnivores.

Butterflies have a body part called a proboscis that helps them get food from flowers and fruits.

Water

The human body is made mostly of water. So are most other living things. Water keeps the cells, tissues, and organs in an animal's body working correctly. It helps animals' bodies break down the food they eat and dispose of waste. Water also helps some animals regulate and control their body temperature.

An abundant water supply is important for all living things.

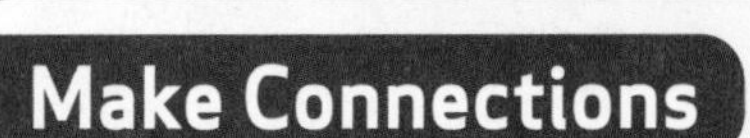

Jump to the Plants section to learn about the needs of plants.

Space

Animals need space, or room, so that they can move from place to place, hunt for food, escape from predators, and build homes. Large animals need a great deal of space. Elephants, for example, roam wide-open plains and large, dense forests. Smaller animals do not need much space at all. Squirrels, mice, and birds have plenty of room right in your backyard.

Orangutans move by swinging from tree to tree. They need plenty of space to find shelter and food.

Shelter

A shelter is where an animal makes its home. Shelters come in a variety of shapes and sizes and provide protection for the animals that live in them. They also provide a place for animals to give birth to and raise their young. Some animals, such as bears, wood frogs, and deer mice, spend many months of the year in a state of inactivity called hibernation. In order to hibernate, these animals need a safe and dry place to rest.

Robins build nests that are high off the ground so that they can safely lay eggs and raise their young.

Oxygen

Oxygen is a gas that is found in both air and water. Every time a land animal breathes, it takes oxygen into its lungs. Some animals that live in water need to come to the surface to breathe air. Most animals that live in water take in oxygen from the water, though. Fish do so through their gills. Oxygen helps an animal's cells process food and turn it into energy.

Fish take in oxygen from the water through their gills.

Arthropods

Arthropods are the largest animal group on Earth. In fact, some scientists estimate that tens of millions of them exist. Spiders, crabs, and insects are just a few examples. Arthropods are invertebrates, which means that they do not have a backbone. This trait is just one of many traits that set arthropods apart from other animals.

Exoskeleton An **exoskeleton** is a hard skeleton on the outside of an arthropod's body. It provides protection, strength, and support. As an arthropod grows, it sheds its exoskeleton and grows a new, larger one.

Body Segments All arthropods have bodies that are divided into segments. Some, like spiders, have two body segments. Insects have three. Centipedes and millipedes have numerous body segments.

Jointed Limbs Arthropods have sets of limbs on both sides of their bodies. These limbs can be legs, wings, or even claws. Because arthropods cannot bend their bodies like humans and other animals, their limbs have joints that help them move around more easily.

Body Symmetry If you were to draw an imaginary line down the middle of an arthropod's body, its two sides would look exactly the same. This trait is called *bilateral symmetry*. In this way, arthropods are like humans and many other animals.

A ladybug's exoskeleton protects its fragile wings.

A millipede has anywhere from 40 to 400 jointed legs.

Fact Checker

Spiders are not insects. They are actually part of a subgroup of arthropods called arachnids.

Types of Arthropods

There are five main arthropod subgroups.

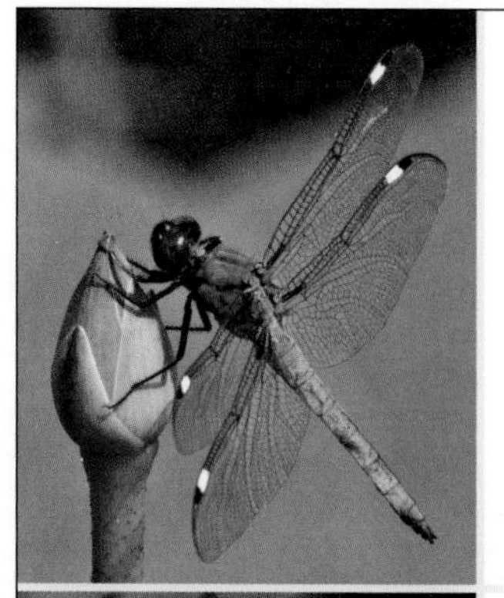

Insects The largest subgroup of arthropods are insects. Scientists estimate that there are over one million different species of insects on Earth. Insects have three body segments: a head, a thorax, and an abdomen. They also have three pairs of legs and two antennae. Adult insects may or may not have wings. Ants, bees, dragonflies, and cicadas are insects.

Arachnids While different from insects, arachnids are also arthropods. They have only two body segments and no antennae. They also have eight legs instead of six. Some examples of arachnids are spiders, ticks, and mites.

Crustaceans Although these arthropods mostly live in water, a few crustaceans live on land. Crustaceans have two body segments and at least five pairs of legs. They also have two sets of antennae. Crabs, shrimp, and lobsters are crustaceans.

Chilopods The chilopod group is made up of many species of centipedes. Most of them are small in size, although a few species can grow to 25 centimeters (10 inches) in length. Centipedes have numerous body segments that each have one pair of legs. Centipedes also have venomous fangs that they use to bite and kill prey.

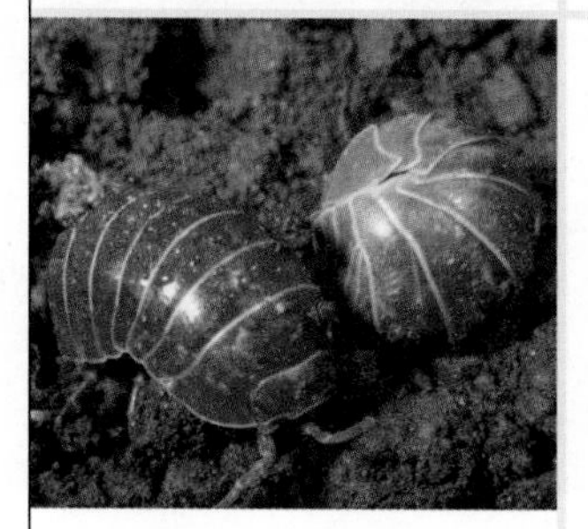

Diplopods Millipedes are diplopods. Some may look similar to centipedes, but they are quite different. Only the first four body segments have one pair of legs each; the rest have two pairs of legs. Also, diplopods do not have fangs. To defend themselves, these animals may roll into a ball or emit a foul smell.

Mammals

Another large animal group is the mammal group. There are over 4,000 different species of mammals, and they are found in habitats all over the world. Mammals come in a variety of shapes and sizes. The smallest known mammal is the hog-nosed bat, which weighs just 1.5 grams (about 0.05 ounces). The largest mammal, the blue whale, weighs over 135 metric tons (150 tons). All mammals are vertebrates, which means that they have backbones.

All mammals have fur or hair, but the amount of body coverage varies from species to species. Because mammals are warm-blooded, their bodies stay at a constant temperature. One function of fur and hair is to help mammals keep their bodies warm or cool.

Another characteristic of mammals is that they produce milk and feed it to their young. Because their young rely on them for food, adult mammals spend more time caring for their offspring than most animals in other groups.

Cats are mammals that have a great deal of fur.

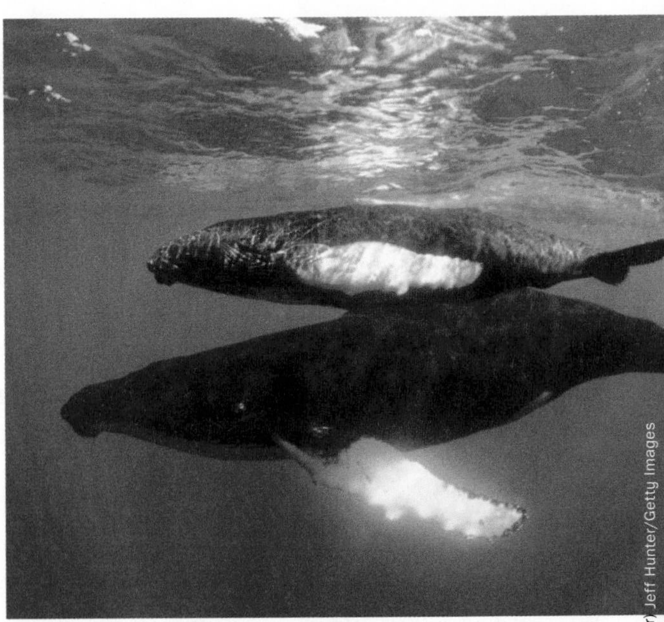

Like humans, whales have very little body hair.

Types of Mammals

Similar to arthropods, mammals are classified into smaller groups based on certain characteristics. Mammals are grouped according to how they bear young. The three groups are marsupials, monotremes, and placental mammals.

Baby kangaroos are called joeys. Kangaroos are marsupials.

Marsupials are mammals that have pouches. When a marsupial gives birth, the baby is small and undeveloped. It is not ready to live on its own. Instead, it lives and feeds in its mother's pouch until it is mature. This process can take weeks or even months. Kangaroos, koalas, and wombats are examples of marsupials.

Monotremes are mammals that lay eggs. There are only five living species of monotremes: the duck-billed platypus and four kinds of spiny anteaters. Monotremes are different from other egg-laying animals because they have fur instead of feathers and they produce milk.

Placental mammals have young that develop inside their mothers. The young mammal is attached to a placenta that provides nutrients and helps get rid of waste. Most mammal species are placental mammals, including humans, rodents, bats, dogs, cats, and whales. Young born to placental mammals are more developed than those born to marsupials and monotremes.

Did You Know?

The spiny anteater has not changed for thousands of years. It has continued to survive while other species have become extinct.

A duck-billed platypus lays its eggs in a burrow dug near the water's edge. This unique animal is a montreme.

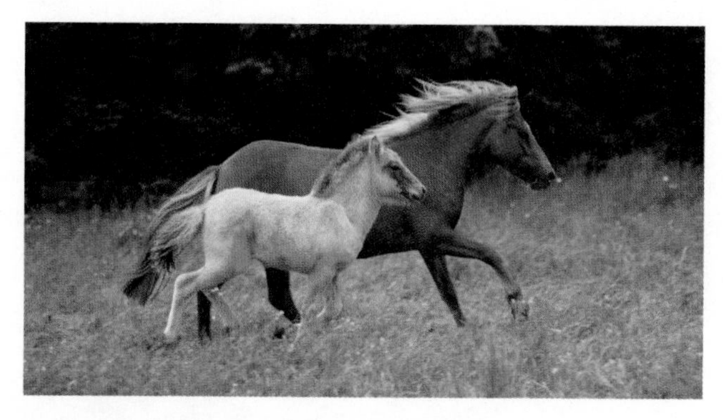

Horses are placental mammals. A foal usually begins walking very soon after it is born.

Reptiles

Like mammals, all reptiles are vertebrates. However, unlike mammals, reptiles are cold-blooded. This term means that their body temperature changes depending on the environment. In a cold habitat, a reptile's temperature will drop. These animals rely on sunlight to warm their bodies. In a hot environment, a reptile's temperature will rise. The animal will seek shade, mud, or water to cool itself.

The micra chameleon may be the smallest reptile in the world. Adults grow to a length of only 2.5 cm (1 in.).

Reptiles have scales covering part or all of their bodies. Scales keep reptiles from losing water through their skin. The eggs that reptiles lay are also covered in a hard, scaly material so that the young reptiles inside the eggs will not dry out.

Reptiles are found all over the world, but are most plentiful in warm environments. Most live and reproduce on land, although a few reptiles, such as sea turtles, live a great deal of the time in water. Even sea turtles lay their eggs on land, though, and must come to the surface to breathe. All reptiles have lungs rather than gills.

An American alligator's muscular tail helps propel it through water.

Types of Reptiles

Reptiles are organized into four main groups: lizards and snakes, turtles and tortoises, alligators and crocodiles, and tuatara.

Lizards and Snakes Lizards and snakes live in a variety of different habitats. Both kinds of animals have loosely connected jawbones, which allow them to eat large prey. Some members of this group lay eggs while others give birth to live young.

Turtles and Tortoises Turtles live mainly in the water. They have webbed feet or flippers that help them swim. Tortoises live mainly on land and have round, clawed feet. Both lay eggs. Turtles and tortoises cannot hear well, but they can sense vibration and have a good sense of smell.

Alligators and Crocodiles In addition to alligators and crocodiles, this group also includes caimans and gharials. Crocodiles and alligators spend time both on land and in water. These animals use their excellent hearing and eyesight to hunt for prey as they move powerfully through the water.

Tuatara Tuataras are a small group of reptiles found only in New Zealand. Tuataras have a row of spikes down their backs and clawed feet. They thrive in cooler weather and feed mainly on insects. Adult tuataras lay eggs only every two to five years.

Birds

Like mammals, birds are warm-blooded vertebrates. Birds are closely related to reptiles. Their bodies are covered in feathers, but their legs are covered in scales like the skin of their reptilian ancestors. All birds lay eggs.

There are about 10,500 species of birds. Because birds are warm-blooded, they are able to live in almost any climate. Penguins, some species of albatrosses, and a few other birds make their homes in Antarctica. The Sahara Desert, a very hot and dry climate in Africa, is home to ostriches, some species of crows, and a variety of other birds.

Birds vary in size. The largest bird in the world is the ostrich. An adult male ostrich can grow up to almost 3 meters (9 feet) tall and weigh nearly 140 kilograms (about 300 pounds). Ostriches cannot fly, but they can run extremely fast. An ostrich uses its wings for balance and stability as it runs.

The smallest bird in the world is the bee hummingbird. Females, which are larger than males, are usually less than 6 cm (2.5 in.) in length. Their wingspan is more than half the length of their bodies.

The brightly colored scarlet macaw lives in the rain forests of Mexico, Central America, and South America.

Ostriches lay an average of 60 eggs per year.

The Parts of Birds

Most birds' bodies are built for flight. Hollow bones keep their bodies light. The shape of their wings helps them glide, soar, or flap. Birds also have other physical characteristics that help them meet their needs.

Eyes Birds have excellent vision. This feature is important for finding food, choosing mates, and avoiding collisions during flight. Not only can birds see an array of colors, but they can also detect ultraviolet light that reflects off of prey and other birds. Nocturnal birds, such as owls, have very large eyes, which allow more light to enter.

Beak All birds have a beak or a bill. The shape and size of a bird's beak depends on what the bird eats. Some beaks are short, cone-shaped, and are good for cracking seeds. An owl's hooked beak helps it grab food and tear it into smaller pieces. A duck's bill has fringes around the edge that filter seeds, plants, and small insects from the water.

Legs and Feet Since birds do not have hands or paws, they use their feet for gripping. Most birds have four clawed toes. Birds of prey, like the owl, have long, sharp claws called talons that help them catch food. Other birds' toes are arranged so that they can easily grip branches. Ducks and other aquatic birds have webbed feet that help them move through water.

What Could I Be?

Ornithologist

Fascinated by birds? Ornithologists study every aspect of bird life. Learn more about a career in ornithology in the Careers section.

(t) Design Pics/Natural Selecti Curtin; (b) Erica Simone Leeds

Amphibians

Amphibians are another large animal class. There are over 6,000 different species of these cold-blooded vertebrates. Amphibians are organized into three subclasses: frogs and toads, salamanders, newts and mudpuppies, and caecilians. Caecilians are large, worm-like amphibians.

The green salamander prefers cool, moist, shady habitats.

Amphibians take in oxygen in two ways: with their lungs and through their skin. An amphibian's skin is covered in a thin layer of mucus and must always be moist. If it dries out, the amphibian will not survive. Some species of amphibians, such as toads, spend more time on land than others.

Word Study

The word *amphibian* comes from the Greek word *amphibios,* which means, "having two lives." The term refers to most amphibians having two life stages, the larval stage and the adult stage.

Many species of amphibians grow and change considerably from the time they hatch from eggs through a process called metamorphosis. For example, when a frog larva hatches, it has gills instead of lungs. It also has a tail, but no legs. As the frog larva grows and changes, it develops lungs, loses its tail, and grows legs. It is then suited for life on land.

Amphibians are found in many habitats, from rain forests to ponds. In climates where there is a cold winter season, amphibians may bury themselves in mud underneath the water. Most amphibians are carnivores and eat insects, slugs, worms, snails, and other amphibians.

Make Connections

To learn more, jump to the Animal Life Cycles section.

A Mantella's bright colors warn predators that it is dangerous.

Fish

Fish have body parts, such as fins and gills, that are different from most other animals. One class of fish, the bony fish, is the largest group of vertebrates in the world. Bony fish have skeletons that are made of bone. Their bodies have scales, paired fins, one set of gills, and one pair of nostrils. Goldfish, clownfish, sunfish, and marlins are a few examples of bony fish.

bony fish

There are two other groups of fish: cartilaginous fish and jawless fish. Cartilaginous fish have skeletons that are made of cartilage. Cartilage is a hard substance but is softer than bone. Sharks, rays, and skates are cartilaginous fish.

jawless fish

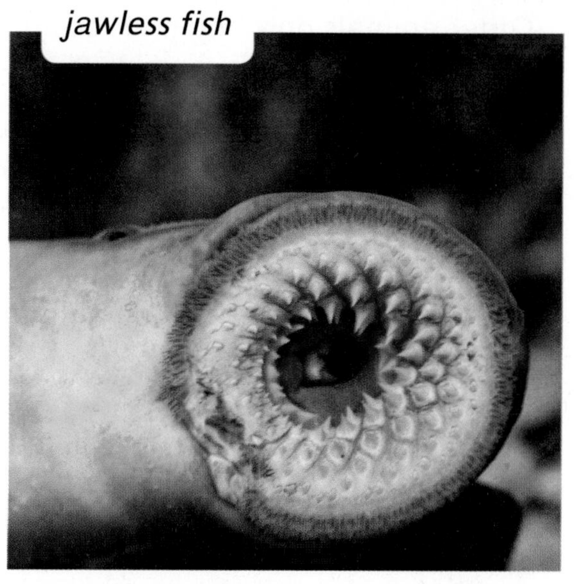

Scientists think that jawless fish might be the earliest vertebrates. They have mouths that are similar to suction cups. One species of jawless fish called lampreys attach themselves to other fish and suck out fluids and tissue.

Like reptiles and amphibians, most fish are cold-blooded. They live in salt water, freshwater, and brackish water, which is a mixture of both. They live in warm water, extremely cold water, and everything in between. Some fish lay eggs, while others give birth to live young.

cartilaginous fish

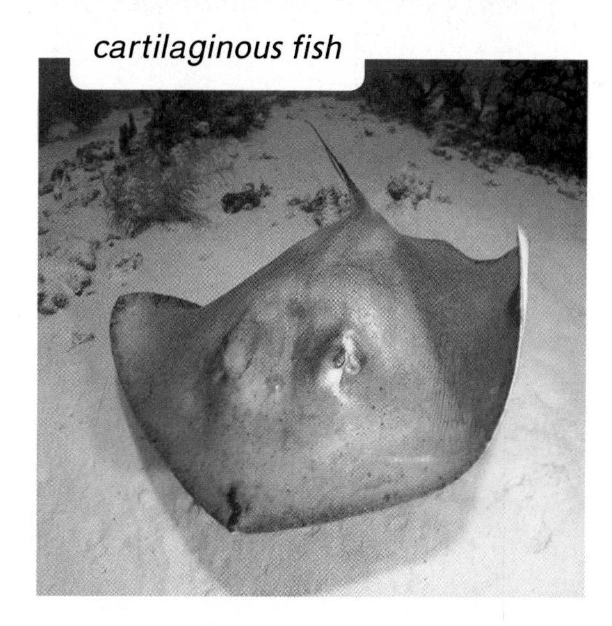

Animal Life Cycles

All living things have life cycles. A life cycle is how an organism changes from birth as it grows into an adult. Life cycle stages include birth, growth, reproduction, and death.

Make Connections

To learn about plant life cycles, jump to the Plants Section.

Life cycles vary from organism to organism. Many animals are born looking similar to their parents. For example, when a foal is born, its body continues getting bigger until it grows into an adult horse. Nothing changes except its size and sometimes the color of its hair. Its overall body plan stays the same.

Other animals only begin to resemble their parents as they grow. A frog hatches as a tadpole, which has a tail, no legs, and gills. It cannot survive out of the water. As the tadpole grows, its tail disappears. It grows legs, develops lungs for breathing, and increases in size. It is then able to move between water and land.

On the following pages, you will learn more about different kinds of animal life cycles.

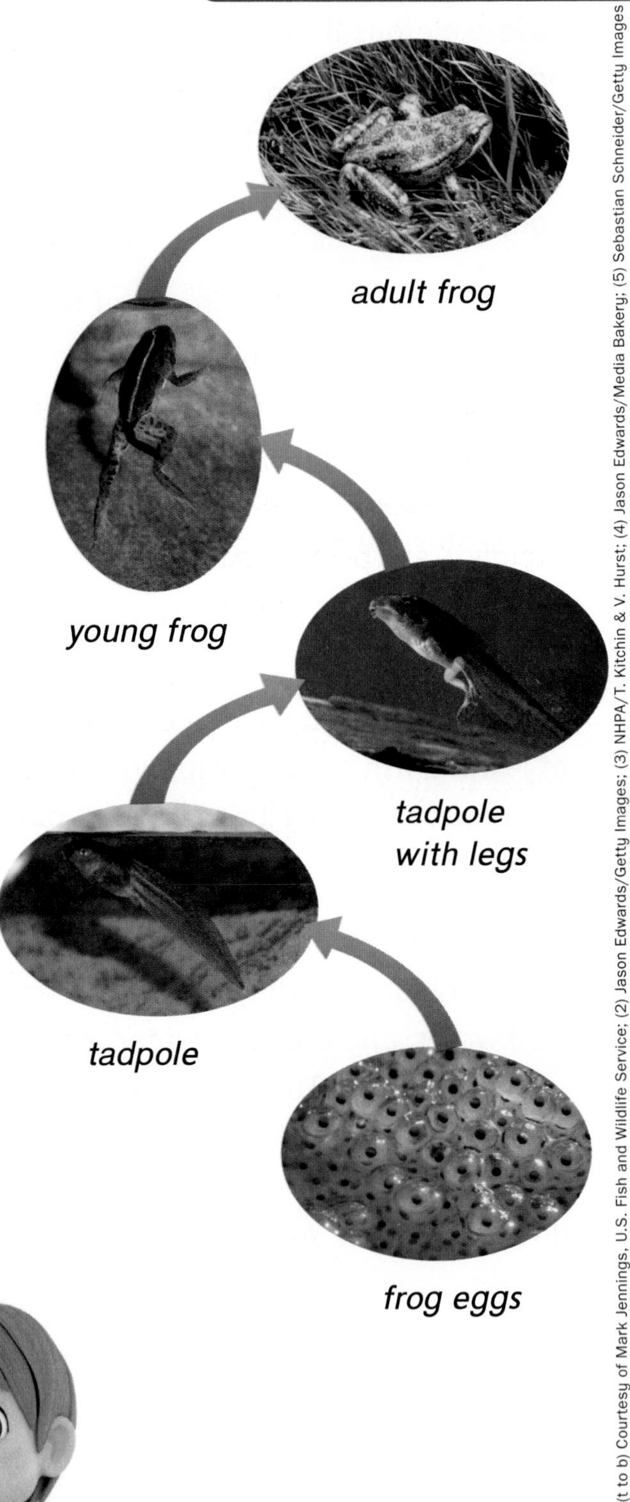

Skill Builder **Read a Diagram**

Follow the arrows to understand the stages in the frog's life cycle.

Complete Metamorphosis

Some animals, such as butterflies, moths, frogs, and salamanders, develop through a life cycle called *complete metamorphosis.* In complete metamorphosis, there are four different life stages. The animal looks different during each of the stages.

In the insect group, a primary stage of complete metamorphosis is the larval stage. A larva is a worm-like form with no wings. It hatches from an egg and then spends much of its time eating. In fact, a larva can eat several times its body weight each day. Mealworms and caterpillars are types of larvae.

The next stage is the pupa stage. During this stage, the animal develops a hard shell or case around itself and does not move or eat. Butterfly caterpillars make chrysalises and moth caterpillars make cocoons. Inside these shells, the pupa's body changes rapidly. Caterpillars become butterflies or moths, and mealworms become darkling beetles.

The final stage of complete metamorphosis is the adult stage. The insect emerges from its shell. It has wings, antennae, legs, and other body parts that it did not have during the other stages. The adult insect lays new eggs, and the cycle begins again.

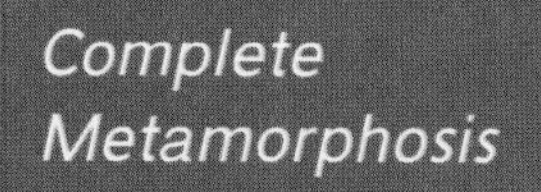

monarch butterfly egg

monarch butterfly larva

monarch butterfly pupa

adult monarch butterfly

Animal Life Cycles - Incomplete Metamorphosis

Grasshoppers, crickets, praying mantises, damselflies, dragonflies, and cockroaches are some insects that go through incomplete metamorphosis. During *incomplete metamorphosis,* an insect develops in three stages instead of four. These stages are the egg stage, the nymph stage, and the adult stage.

During the first stage, the insect hatches from an egg. When it hatches, it is a nymph. A nymph is a small version of an adult insect. However, a nymph does not have wings or parts that help with reproduction. The nymph stage is very gradual; during this stage a nymph spends most of its time eating and growing. As it grows, it sheds its skin. This process is called molting.

The final stage of incomplete metamorphosis is the adult stage. During this stage, the nymph reaches its adult size. It may grow wings and its reproductive systems have finally developed. Female adults mate and lay eggs, beginning the life cycle once again.

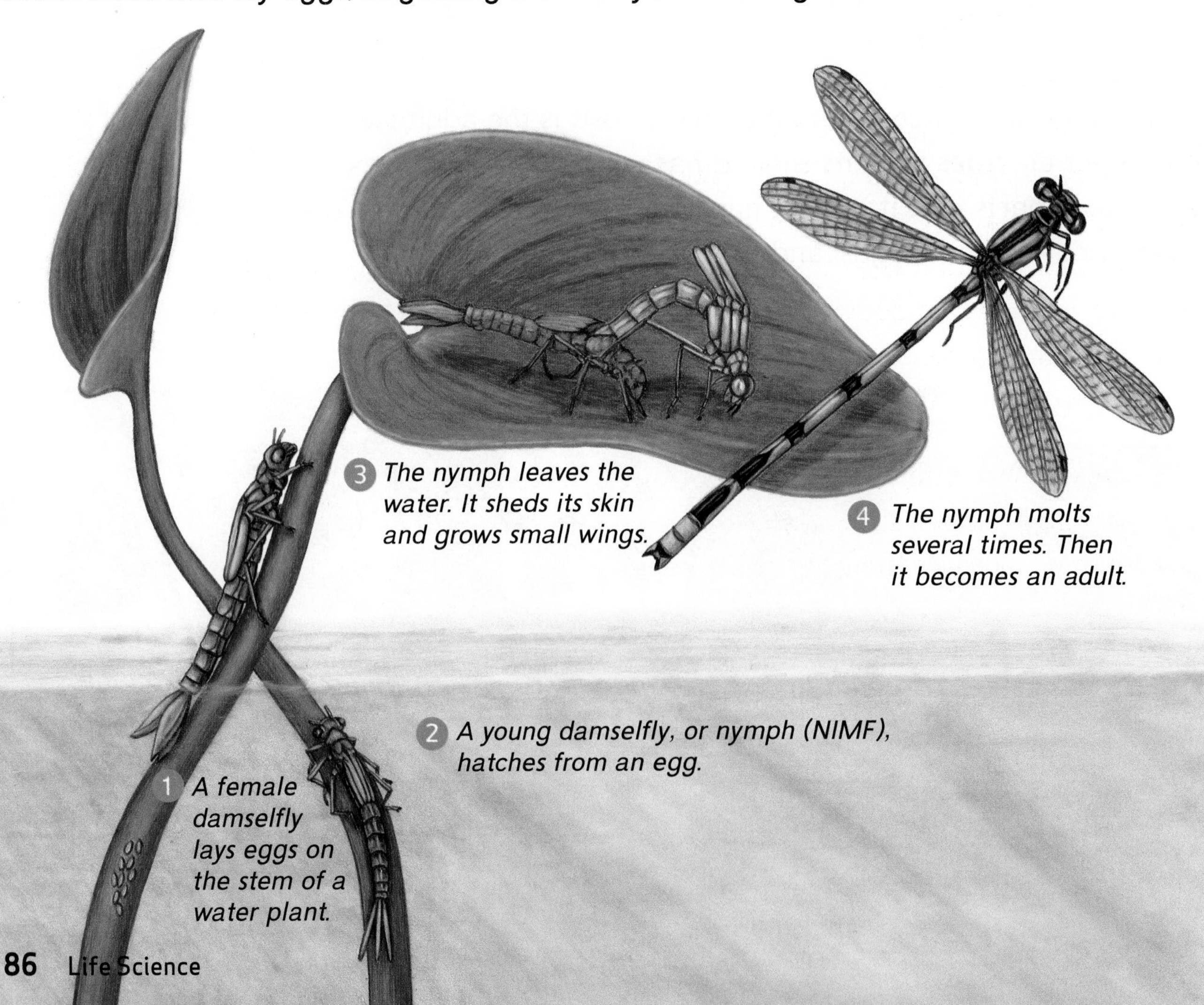

Animal Reproduction

Reproduction occurs when adult organisms make offspring. There are two ways in which animals can reproduce. *Asexual reproduction* involves only one parent. *Sexual reproduction* involves two parents.

One way that animals reproduce asexually is through *budding*. During budding, a growth forms on an adult animal's body. This growth is called a bud. The bud gradually grows and develops into a new animal. The nutrients that the parent takes in from food help support the bud as it matures. Freshwater hydra and some species of jellyfish reproduce this way.

A second kind of asexual reproduction is *regeneration*. Regeneration occurs when a completely new animal grows from a body part of the parent animal. In both budding and regeneration, the offspring is an exact copy of its parent. Sea stars and some other echinoderm species reproduce this way.

Sexual reproduction requires cells from two parents. The cell that comes from the female parent is the egg, and the cell from the male parent is the sperm. The egg and the sperm must join in order to produce offspring. This process is called *fertilization*. Once the egg is fertilized, it develops into an embryo. Unlike offspring that come from budding or regeneration, embryos inherit traits from both parents.

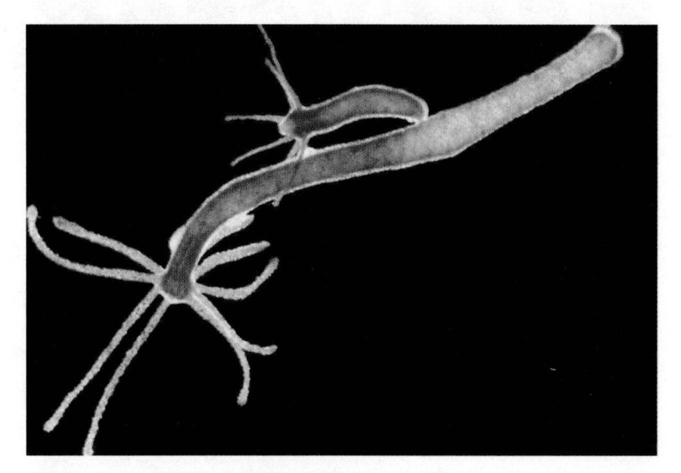

The hydra reproduces through budding from one parent.

Sea stars reproduce through regeneration from one parent.

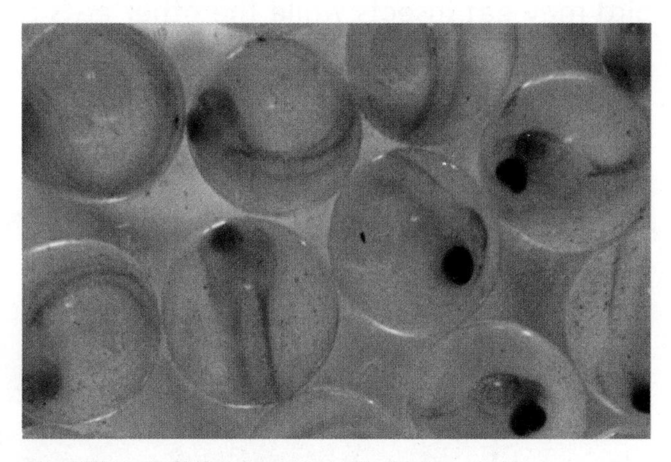

Fertilized fish eggs contain genetic information from two parents.

(t) NHPA/M. I. Walker; (c) Barry Barker/McGraw-Hill Education; (b) USFWS/Peter Steenstra

Ecosystems

An *ecosystem* is made up of all living and nonliving things in an environment. All of the living things in an environment are called **biotic** factors. Plants, animals, fungi, and bacteria are all biotic factors. **Abiotic** factors are the nonliving things in the environment. Air, water, soil, rocks, and light are abiotic factors. Biotic and abiotic factors in an ecosystem interact with each other. They supply the resources that living things need. For example, plants need air, water, and sunlight to survive. Plants then provide food for many animals that live in that ecosystem.

Habitats and Niches

Ecosystems can be small, like a single log or a pond, or very large, like a forest or desert. Each organism must have its own space. The place in an ecosystem where an organism lives is its *habitat*. Habitats vary depending on the type of ecosystem. The logs in a forest ecosystem provide a habitat for insects, fungi, and moss. Trees and bushes provide a habitat for birds and squirrels. Each living thing has its own **niche**, a special role that an organism plays in the ecosystem. For example, two birds may share the same habitat in a rainforest canopy but eat different foods. One bird may eat insects while the other eats plants. The birds have different niches in the canopy ecosystem.

Species, Populations, and Communities

The living things in any ecosystem can be organized into groups. The most specific group an organism can be classified into is a *species*. For example, there are many different kinds of cats, but the lion is one particular species of cat. In any ecosystem, a *population* includes all members of a single species in an area at a given time. In a desert ecosystem, all of the saguaro cactus plants form a population. The sidewinder, diamondback, and tiger rattlesnakes are different species that all form separate populations of rattlesnakes in the desert ecosystem.

Together, all of the populations in an ecosystem make up a *community*. A pond community may have populations of ducks, bullfrogs, fish, water lilies, and dragonflies. In addition to plants and animals, a community has populations of bacteria, protists, and fungi such as mushrooms. For example, a forest community has huge populations of mold and bacteria living in its soil. The size of a community depends on its abiotic factors such as shelter, water and light, as well as the amount of food that is available. A living community may cover a huge area and include thousands of populations.

This fallen log is part of a tiny ecosystem that includes populations of fungi, moss, and bacteria.

Skill Builder **Read a Diagram**

Look carefully to identify the species, populations, and community shown in this pond ecosystem.

Types of Ecosystems

Ecosystems vary around the world. Many factors such as temperature, location of water sources, type of soil, and amount of sunlight contribute to the differences in ecosystems. A **biome** is one of Earth's major land ecosystems with its own characteristic animals, plants, soil, and climate. A **climate** is the average weather pattern for a region over a period of time.

There are six major land biomes: desert, grassland, tundra, taiga, rain forest, and deciduous forest. Near the equator are tropical rain forests rich with thousands of species of plants and animals. Closer to the poles are cold-weather taiga and tundra biomes. Fewer living organisms are found in these colder environments. Some biomes, such as the desert biome in Northern Africa, can stretch across an entire continent. A taiga biome covers the distance across Northern Asia for a length of about 10,000 kilometers (6,000 miles). The map below shows the location of the world's different biomes.

Skill Builder

Read a Map

This map uses color to indicate the different biomes. Study the key to determine what each color means.

Global Biomes

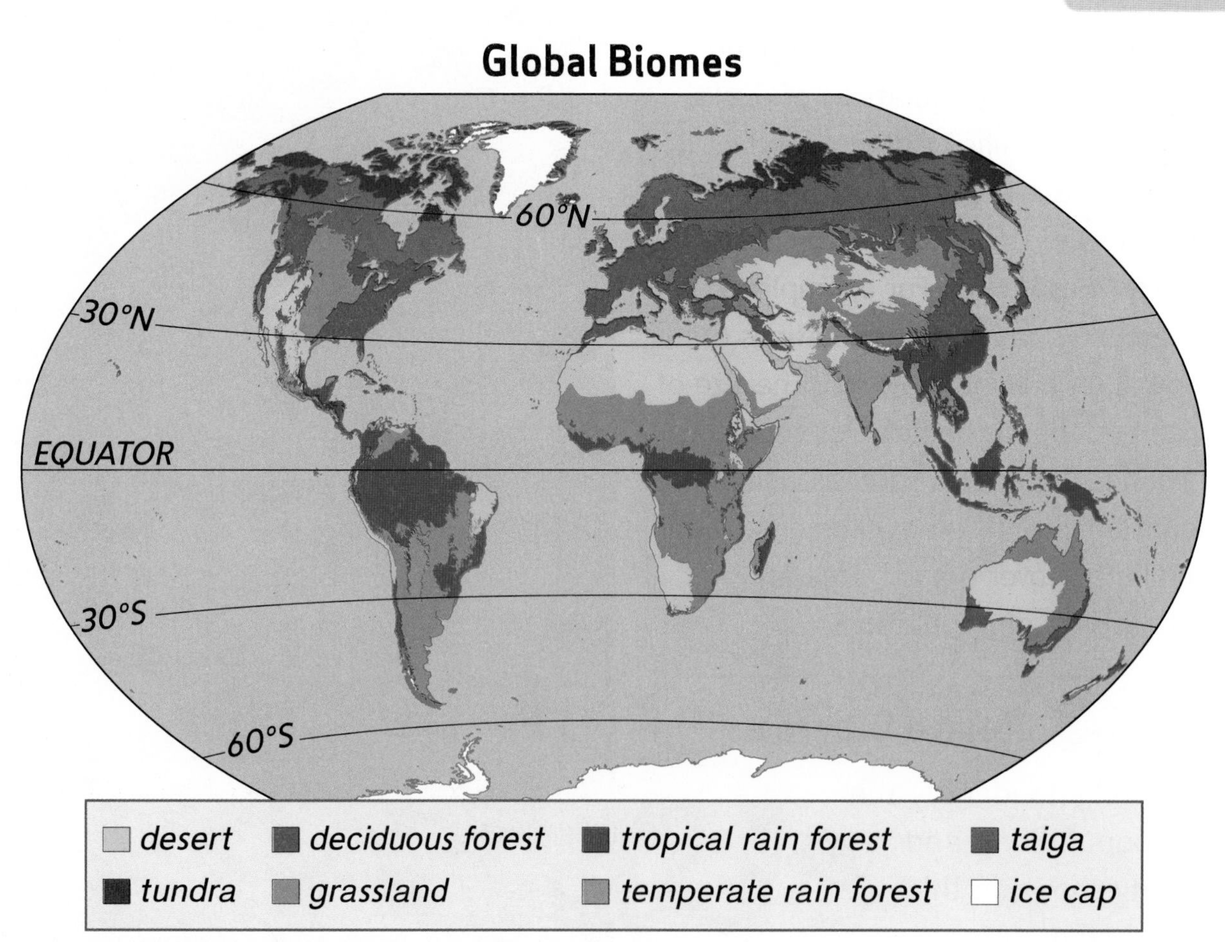

Desert

A *desert* is a sandy or rocky biome with little precipitation or plant life. There is at least one desert on each continent. Desert biomes can be hot or cold. Temperatures in hot deserts can range from over 38°Celsius (100°Fahrenheit) in the daytime and as low as -4°C (25°F) during the night. Few organisms can live in such an extreme environment. Plants and animals need special adaptations to survive. Cactus plants have thick, waxy cuticles that prevent stored water from evaporating in the hot Sun. Many cold-blooded animals, such as reptiles, use the Sun's heat to warm their bodies. However, most desert animals are active at night, and they find shelter or stay underground during the heat of the day.

A kit fox's large ears allow it to lower its body temperature to stay cool.

Grasslands

In a *grassland* biome, grasses are the main plant life. A temperate grassland has cold winters and hot, dry summers. There is enough rainfall to support some plant life, but not enough for many trees to grow. American prairies are one kind of grassland with a mild climate. During hot, dry summers, the grasses may burn, giving the prairies rich soil that is good for farming. Prairie life includes wildflowers, worms, insects, spiders, mice, prairie dogs, and snakes.

An African grassland is called a savanna. It has shrubs and a few trees and stays warm all year long.

Types of Ecosystems

Tundra

The *tundra* is a large biome where the ground is frozen all year. Winters are long and cold, and summers are short and cool. A layer of permafrost, or permanently frozen soil, prevents trees from growing. Plants grow close to the ground. Their shallow roots help them get water from melting snow. Tundra plants include wildflowers, mosses, grasses, and shrubs. A small number of animals have adapted to live in the tundra. Mammals such as caribou, polar bears, and musk ox have an extra layer of fat to stay warm. Arctic hares and foxes have fur that changes color with the seasons to help them blend into the environment.

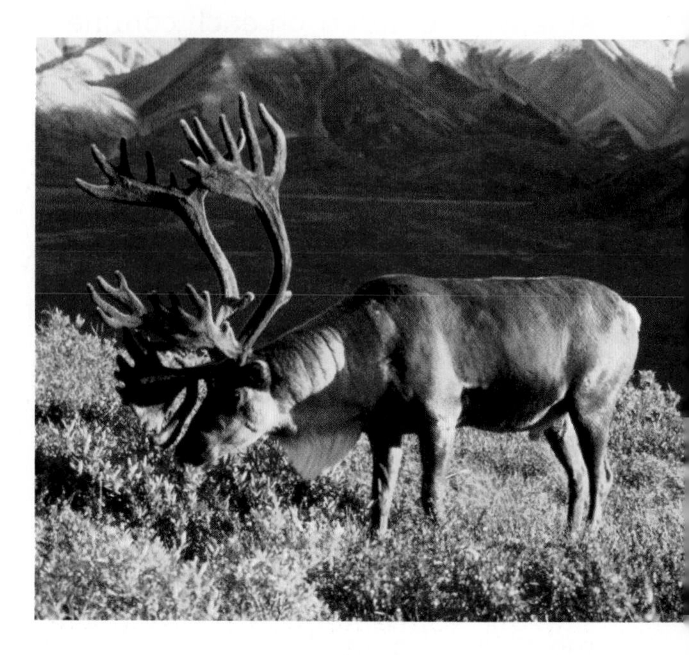

This caribou feeds on grass in a tundra biome.

Taiga

The *taiga* is a cool forest biome found in Earth's upper northern regions. It is the largest biome in the world. Just south of the tundra, the taiga is full of coniferous evergreen trees such as pines, firs, and spruce. Temperatures are cold, but not as cold as those in the tundra. The taiga also contains lakes and ponds that were formed thousands of years ago by glaciers. Taiga winters are long and cold. Mammals such as black bears and snowshoe hares have thick coats of fur and layers of fat for extra warmth. Many taiga animals hibernate during the cold winter months.

The Siberian tiger is the largest cat in the world. It is native to the Russian taiga.

(t) U.S. Fish & Wildlife Service/Dean Biggens; (b) Kyslynskyy/iStock/Getty Images

Rain Forest

A *tropical rain forest* is a hot, humid biome near the equator, with heavy rainfall and a great variety of life. Most tropical rain forests have distinct layers. A *canopy* of treetops spreads over the top of the forest. The canopy is full of animal life such as monkeys, toucans, tree frogs, and snakes. Below the canopy, an *understory* is made up of a dense layer of tree trunks, shrubs, and vines. The *forest floor* is dark and dim. Few plants grow there because the thick leaves of the canopy prevent sunlight from entering the lower levels. *Temperate rain forests* are found in cooler climates. They contain large evergreen trees and a variety of animals such as cougars, bobcats, and owls.

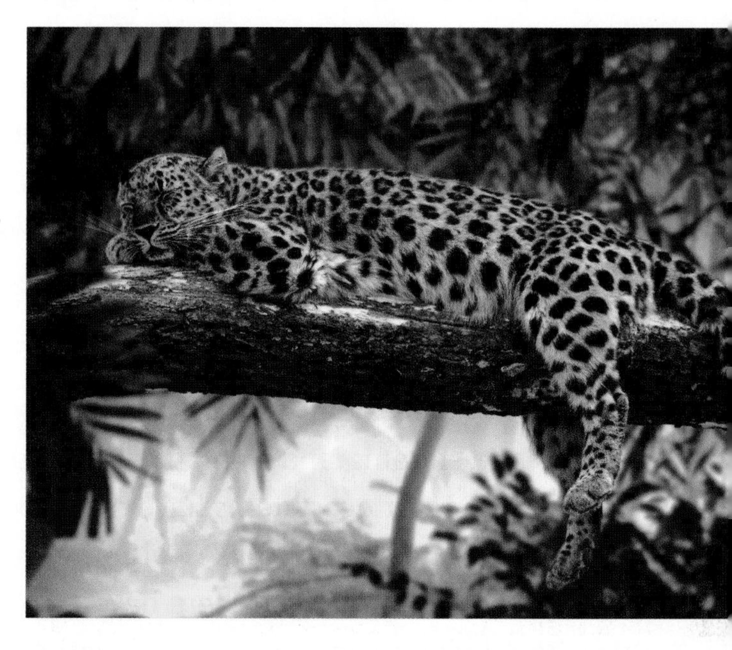

Leopards and other large mammals live in the understory of the rain forest.

Deciduous Forest

A *deciduous forest* biome undergoes distinct changes during the four seasons. Deciduous trees change color in fall and lose their leaves in winter. Common types of deciduous trees include oaks, beech, hickory, and maples. Deciduous forests have cold winters and warm summers. In winter, these forests are snowy and bare. Birds may migrate to warmer climates, and animals may hibernate. In spring, temperatures warm, and the forest trees, ferns, shrubs, flowers, and saplings grow new leaves. The forest becomes active again as migrating birds return and animals come out of hibernation.

Not all birds migrate. Cardinals are a common sight in deciduous forests in winter.

Water Ecosystems

Freshwater and saltwater ecosystems vary in location and in the kinds of plants and animals that live there. About 97 percent of the world's water is found in oceans as salt water. Salt water is also found in other water ecosystems such as marshes and swamps. Fresh water is found in rivers, lakes, and streams. Abiotic factors such as the amount of light, dissolved salt, and dissolved oxygen determine which organisms can survive in water ecosystems.

Freshwater Ecosystems

Freshwater ecosystems include running water ecosystems such as rivers and streams, and standing water, such as ponds and lakes. Fast running water ecosystems have more oxygen than slower running water because air mixes with the water as it flows. The plants and animals that live there have adaptations that prevent them from being swept away by the swift current. Some organisms attach themselves to rocks. Fish such as salmon have powerful muscles that allow them to swim against strong currents.

In standing water ecosystems, most organisms live in the shallow zone along the shore. Algae and plankton float near the surface of open-water zones. Few organisms can survive in the deepwater bottom zone where there is little sunlight and little oxygen. Some invertebrates such as worms and mollusks are found here.

Skill Builder

Read a Diagram

Use the labels on the diagram to help you identify the different zones in this standing-water ecosystem.

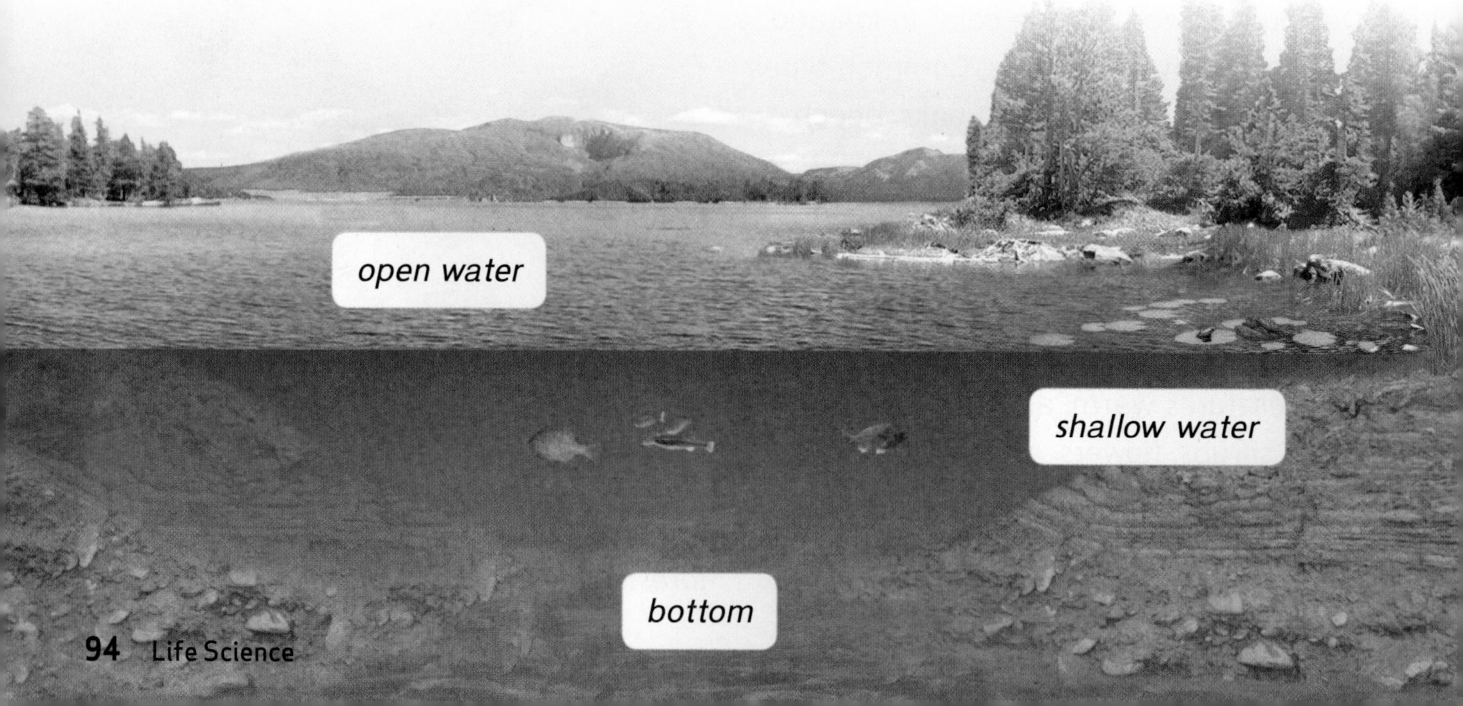

Wetlands and Estuaries

A *wetland* is an ecosystem that is wet during some or all of the year. Wetlands are found between land and water and include marshes, bogs, and swamps. Grasslike plants, moss, and some shrubs have adapted to live in the water-soaked soil. Beavers, bobcats, muskrats, otters, herons, raccoons, turtles, and fish are found in wetlands.

Estuaries are areas where fresh water feeds into salt water. Estuaries change with the tide. When the tide is high, estuary water has a higher *salinity*, which means it becomes saltier. When the tide goes out, the estuary has a lower salinity, or more fresh water. Salt marshes are found near estuaries. These areas are covered with grasses and other marsh plants and are filled with large shrimp, oysters, and clams. Insect larvae, young fish, and small invertebrates such as crabs begin their lives in the protected waters of a salt marsh. Larger organisms feed on these smaller animals. Mangrove swamps, found in coastal tropical or subtropical areas, are dominated by mangrove trees. Their roots are specialized to grow in the changing water level of the swamp. Manatees, American alligators, and red-tailed hawks are some of the animals that live in a mangrove swamp.

The bobcat is a top predator in estuary ecosystems.

Mangrove trees have roots that rise out of the water. These roots can reach the air when the area is flooded at high tide.

Water Ecosystems - Ocean Ecosystems

Ocean ecosystems are saltwater ecosystems that are divided into zones. The shallowest part of the ocean is the *intertidal zone,* where the ocean meets the shore. Here, tides rise and fall in regular intervals throughout the day. Organisms such as crabs, clams, sea stars, and barnacles are adapted to the daily changes in water levels.

Beyond the intertidal zone are the neritic and the oceanic zones. The *neritic zone* has abundant sunlight, which allows producers such as algae to grow. Coral reefs and kelp forests provide habitats and food for other living organisms in this zone.

The *oceanic zone* is divided into two layers based on the amount of sunlight. The *bathyal zone* is home to many consumers, but few producers. Marine mammals also live here, but must stay close to the surface to breathe.

Little, if any, light reaches the *abyssal zone* of the open ocean. This deep layer is dark and cold. Sunlight is completely blocked by the water above. Because of the lack of sunlight, organisms that carry out photosynthesis cannot survive here. Producers that do live here get energy from the heat inside Earth instead of from the Sun. Organisms in this zone tend to be scavengers and decomposers that feed on nutrients that sink down from other zones.

Skill Builder

Read a Diagram

Examine the diagram carefully to determine which zones have the most and the fewest organisms.

Resources in Ecosystems

An ecosystem has a limited supply of resources. These resources can be abiotic or biotic. The organisms in an ecosystem compete for resources such as space, food, and water. The fight for limited resources is called *competition*. Plants compete for space and sunlight in an ecosystem. Animals compete for food and water. Sometimes organisms within a population compete with one another for resources. For example, all the individuals in a population of foxes compete for the same food resource—rabbits. Different populations might compete with each other as well. Foxes and hawks both feed on rabbits, so these populations compete for this limited source of food. Rabbits feed on plants, so they must compete with other plant-eating organisms for food resources.

The survival of populations in an ecosystem depends on their resources. A **limiting factor** is any resource that restricts the growth of populations. Abiotic limiting factors include water, temperature, weather, soil type, space to grow, shelter, and sunlight. Changes in these abiotic factors affect the sizes of populations. For example, an ecosystem that is experiencing winter has fewer resources than in summer. Food is scarce in the winter so the ecosystem cannot support as many organisms as in summer.

Make Connections

To find out more about how biotic and abiotic factors affect population sizes, go to the Changes in Ecosystems section.

In winter, bison must search harder to find food.

Energy in Ecosystems - The Oxygen-Carbon Dioxide Cycle

Earth's early atmosphere contained no breathable oxygen. Life as we know it could not have existed then. Early producers used water, carbon dioxide, and energy from the Sun to make their own food. Over time, the oxygen that is released as a waste product of photosynthesis gradually built up in the atmosphere. Today, producers continue to add oxygen to the air.

When you breathe, you take in oxygen and release carbon dioxide. The carbon dioxide is a waste product made through cellular respiration as your cells burn energy. Activities such as burning fossil fuels also release carbon dioxide. Plants take in some of the carbon dioxide from the air to use during photosynthesis.

Plants and animals are both part of the oxygen-carbon dioxide cycle. Plants take in carbon dioxide and give off oxygen. Animals take in oxygen and give off carbon dioxide.

The Water Cycle

Water is also cycled through ecosystems. The continuous movement of water between Earth's surface and the air, changing from liquid to gas to liquid, is the **water cycle**. Water on Earth's surface evaporates, or rises into the atmosphere and condenses, or forms droplets, and then falls back to the surface as precipitation. Water also cycles through living things. When you exhale, your breath releases water vapor. Plants release water too. Water exits the stomata of leaves during transpiration.

All living things need water to survive. Water falling as precipitation is just one part of the water cycle.

Make Connections

Jump to the Earth's Changing Surface section to learn more about the water cycle.

Roles of Organisms in Ecosystems

The plants and animals in an ecosystem have different roles based on how they feed. All ecosystems have organisms that are producers, consumers, or decomposers.

Producers Every organism in an ecosystem relies on producers. A *producer* is an organism that uses the energy of sunlight to make sugar and oxygen in a process called photosynthesis. Producers include green land plants such as grasses and trees. In lakes and oceans, the main producers are algae.

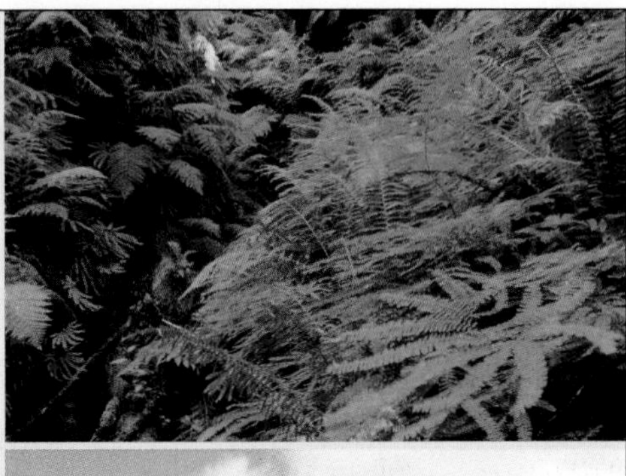

Consumers Organisms that cannot make their own food are called *consumers*. Consumers can be classified by the kind of food they eat. *Herbivores* eat only producers. Rabbits, cattle, and some birds and insects are herbivores. Bobcats, sharks, and hawks are *carnivores*, consumers that eat only other animals. Animals that eat both plants and other animals are *omnivores*. Raccoons, some mice, woodpeckers, and bears are omnivores.

Decomposers Organisms that break down dead or decaying plant and animal material are *decomposers.* They turn this material into simpler substances. Bacteria and fungi are decomposers.

(t) Pixtal/age fotostock; (c) Ingram Publishing; (b) Tinke Hamming/Ingram Publishing

Food Chains

The path that energy and nutrients follow in an ecosystem is called a **food chain**. Food chains model a series of feeding relationships among organisms in an ecosystem. Energy flows only in one direction. The source of all the energy in a food chain is the Sun. This energy is stored in the food made by producers through the process of photosynthesis. Producers are at the base of all food chains. The energy moves from producers to consumers.

An example of a food chain is shown in the diagram. Algae or green plants capture the Sun's energy and store it in the sugars they produce. When an animal such as a mayfly eats the producer, the stored energy moves into the animal's body. When a carnivore such as a sunfish eats the mayfly, the sunfish gets some of the Sun's energy as it passes from the mayfly. A blue heron may then eat the sunfish. The energy of the Sun continues along the food chain from the sunfish to the blue heron.

All food chains end with decomposers. All organisms, after they die, become food for decomposers. Decomposers, such as bacteria and fungi, break down the dead plant and animal matter into simple nutrients. The cycle repeats itself as nutrients go back into the soil to once again be used by producers.

Pond Food Chain

Skill Builder **Read a Diagram**

Follow the arrows to see how energy travels in a pond food chain.

Food Webs

In an ecosystem there are many food chains that overlap. A **food web** is a model that shows how the food chains in an ecosystem are linked together. As with food chains, arrows represent the energy flow from one organism to another. The food web shows the predators and prey in an ecosystem. A *predator* is an organism that hunts for its food. The organisms that it hunts are its *prey*. Predators are important in any ecosystem. They limit the size of prey populations. When the number of prey is reduced in an ecosystem, producers and other resources are less likely to run out.

Fact Checker

The arrows in a food web always point in the direction that energy flows. An arrow begins with a food source and points to the organism that eats it.

In any food web, a predator may feed upon more than one type of prey. Likewise, a type of prey might have more than one predator. In the land food web below, the mountain lion, hawk, and snake all feed upon the mouse. The three predators all compete for the same food source. Remember, the struggle for resources such as food, water, and other needs is called competition.

Changes in Ecosystems

Changes in an ecosystem can affect the sizes of the populations that live there. Changes can be either biotic or abiotic. That is, the changes can occur in the living or nonliving parts of the ecosystem. Some changes can make it difficult for organisms to survive.

Biotic and Abiotic Factors

Together, biotic and abiotic factors determine the carrying capacity for the populations in a community of organisms. The *carrying capacity* is the greatest number of individuals within a population that an ecosystem can support. For example, a rain-forest ecosystem can only support a certain number of jaguars. If the jaguar population increases, it becomes more difficult for the jaguars to find food. The number of animals that jaguars prey upon may decrease. Some of the jaguars may die, and the population returns to its former level.

Abiotic Factors Abiotic factors, such as water, temperature, weather, soil type, space to grow, shelter, and sunlight can change in an ecosystem. Some ecosystems undergo regular changes in abiotic factors, such as when seasons change. In winter, food resources may become more difficult to find in a forest ecosystem. In summer, there is more rainfall and sunlight. Temperatures are warmer, so the forest can support more organisms than it can in winter. In this case, temperature, sunlight, and rainfall are the limiting factors, or resources that restrict the growth of populations.

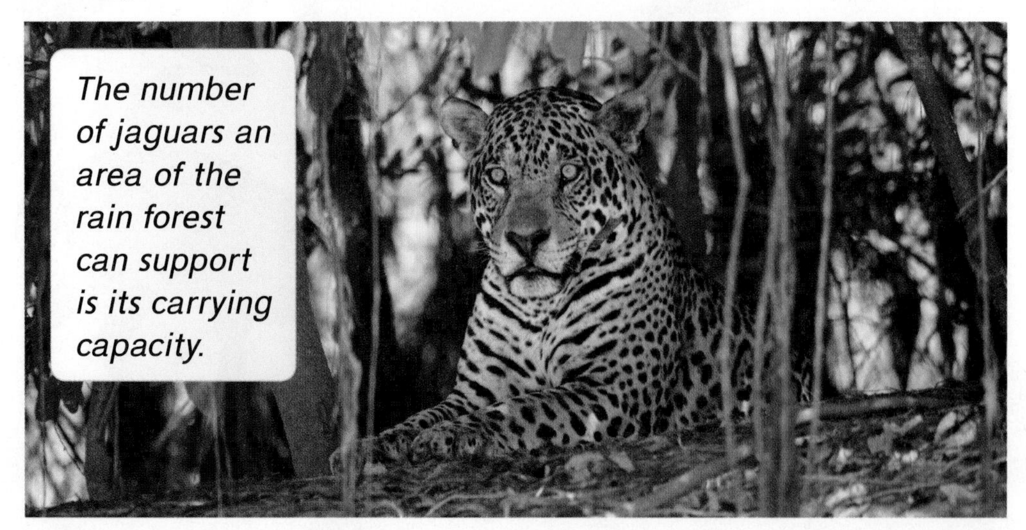

The number of jaguars an area of the rain forest can support is its carrying capacity.

Images & Stories/Alamy

Population Movement Changes in abiotic factors may result in movement of populations. In winter, when it is colder in certain ecosystems, many animals migrate to warmer climates to find food sources. Movement of populations in and out of ecosystems can upset the balance of nature as predator-prey relationships change. For example, if most of the bobcats left a forest ecosystem, the populations of birds, mice, and raccoons would increase. Soon there would be less grass, shrubs, and other producers to support these organisms.

Biotic Factors Biotic factors can also limit ecosystems. A prairie ecosystem has more producers than a desert ecosystem. As a result, the prairie can support more herbivores, which in turn, can support more carnivores. In this case, the amount of available food is the biotic limiting factor for the desert ecosystem. With more available food, the prairie ecosystem can support more populations of different types of consumers.

The movement of a population into an ecosystem can also affect it. For example, in small numbers, locusts pose little threat to an ecosystem. But a swarm of millions of locusts can clear out the plants in an ecosystem, leaving consumers without food.

Disease is another limiting factor that can wipe out a population of plants. The consumers that depend on the plants for food may die or leave the ecosystem.

A swarm of locusts can devastate an ecosystem.

Structural Adaptations

Structural adaptations are adjustments to internal or external body parts. Fur color, long limbs, strong jaws, and the ability to run fast are structural adaptations. Some structural adaptations help organisms survive in certain environments. For example, squirrels have claws that help them cling to forest trees. Moles have large claws that help them dig through the ground as they search for insects. Other structural adaptations protect prey from predators or enable predators to hunt more successfully. Turtles have hard shells that protect them from predators. The spines of cacti keep many organisms from taking a bite. Long legs and strong muscles help cheetahs catch their prey.

What Could I Be?

Biomimicry Engineer

Interested in studying the structures of living things and applying what you learn to solve human problems? A career in biomimicry may be for you! Biomimicry engineers studied gecko feet to determine how to make robots climb walls! Learn more about a career in biomimicry in the Careers section.

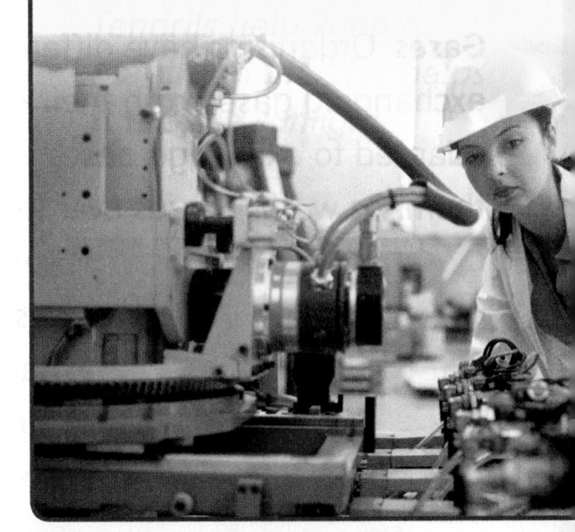

Structural Adaptations in Animals

Animals have many structural adaptations that help them survive in their environments.

Feet Environments and needs vary, and so do animals' feet. Camels live in the sandy desert. They have wide hooves that allow them to walk on sand without sinking. Bat's feet have long, curved claws that are adapted to clinging to the roof of a cave or the inside of a tree. Many animals that spend time in the water, including ducks, frogs, and polar bears, have webbing between their toes to help them swim.

Feet are not used solely for movement. Owls, hawks, and eagles have feet that are adapted for getting food. Their sharp claws allow them to catch and hold prey while they fly.

(t) Don Mennig/Alamy; (b) Pixtal/age fotostock

Camouflage *Camouflage* is any coloring, shape, or pattern that allows an organism to blend in with its environment. Predators with camouflage can sneak up on prey. Camouflage also helps prey animals hide from predators. Protective coloration is a type of camouflage in which the color of an animal helps it blend in with its background. The arctic fox and arctic hare change fur color with the seasons. During the snowy season, these animals have white fur. During the warmer months, they have brown fur. These color changes help both predator and prey blend with the environment.

Both lions and the zebras they hunt as prey have protective coloration. The zebra's stripes provide protection for individual animals, especially their young, when they are with the herd. The stripes all run together, making it harder for a lion to identify one individual. The lion's light brown fur helps it blend into its surroundings, making it easier for the animal to surprise its prey.

Mimicry *Mimicry* is an adaptation in which an animal is protected against predators by its resemblance to a different animal or object. For example, different kinds of hoverflies resemble stinging bees or wasps. Their appearance helps protect the flies from predators. The back of a hawk moth caterpillar looks like a snake's head. This shape frightens away most predators.

The coral snake is poisonous. Predators recognize its coloring and avoid it. Although the milk snake is harmless, predators avoid it because it looks like a coral snake.

coral snake

milk snake

This stick bug avoids predators by looking just like bark, leaves, or twigs.

Structural Adaptations - Animal Senses

Animals use their sense organs to gather information about their surroundings. Some of their senses are similar to yours. Others are quite unusual. Sense adaptations help living things survive in their environments.

Bats can see, but their eyes are small and weak. Bats use echolocation to hunt for food. Bats send out high-pitched pulses of sound. The sound hits the prey and bounces back to the bat. The bat can then judge the direction and distance to its next meal.

The duck-billed platypus does not send out signals, but it can detect them. When a platypus hunts for invertebrates at the bottom of a river, it keep its eyes, nostrils, and ears closed to keep water out. Its bill, which has thousands of sensory cells, can detect weak electrical fields put out by animals as they move. The bill can also detect movement in the water. Using this information, the platypus quickly locates its prey.

Pit vipers and some other snakes can detect infrared light given off by their warm-blooded prey. The light enters a small pit organ, which is located between the snake's eye and nostril. A heat-sensitive membrane in the organ sends a message to the brain, and the snake strikes.

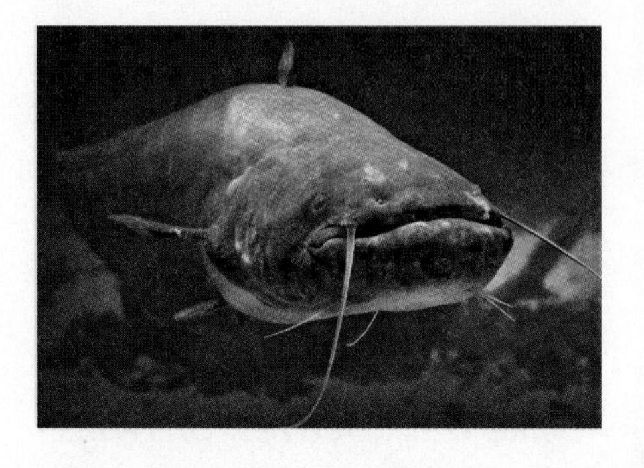

Many animals' sense of taste is localized in their nose and mouth. The body of a catfish, however, is covered with taste-sensitive cells. These cells help catfish locate distant prey by tasting the water they swim through.

Mouthparts All animals need to take in energy to survive. The mouthparts of various animals are adapted to help them get the food they need.

Many animals have teeth that are used to chew food. Chewing helps start the process by which the animal's body digests food. Animals that eat different foods have teeth that are adapted to their diet. Herbivores, such as giraffes, have flat teeth that can grind up plant matter. Carnivores, such as lions and tigers, have sharp teeth that can hold prey and tear meat.

Birds' beaks are adapted to the food the birds eat. A bird that eats large seeds and nuts will have a strong beak. A bird that eats smaller seeds will have a thinner, more delicate beak.

Insects have a wide variety of mouthparts. Grasshoppers, cockroaches, and termites have mouthparts that are adapted for chewing. Their powerful jaws work sideways, like a pair of scissors. This movement allows the insects to both cut food and chew it. Butterflies and moths have mouthparts that look like slender drinking tubes. The tubes uncurl to allow the insects to suck nectar from flowers. Some insects, such as bedbugs and cinch bugs, have mouthparts that look like long, grooved beaks. Four sharp needles in the groove pierce plants or animals and then suck liquids from them.

Did You Know?

Like other omnivores, humans have both sharp, pointed teeth and flat teeth.

Teeth and beaks are important adaptations for getting the food animals need.

Insects have mouthparts that are adapted to the food they eat.

Structural Adaptations -
Structural Adaptations in Plants

Plants, like animals, have adaptations that help them survive in their environments. Flowering plants have scented flowers that attract certain pollinators. They have leaves that catch sunlight and roots that soak up water. These and other adaptations help plants survive.

Some plants have specific structural adaptations for different environments. For example, rain forest plants, such as orchids, have adaptations that help them survive wet conditions and hot temperatures. Orchid stems have water storage organs called pseudobulbs. An orchid's aerial roots help secure the plant high in a tree. These roots also absorb water from the moist air. Like many rain-forest plants, orchids have drip-tip leaves. These leaves are adapted to the constant wet conditions in a rain forest. Their tips drain excess water from the leaves.

Drip tips allow water to run off plant leaves. This helps discourage the growth of mold on the leaves.

Growing in Harsh Conditions Many plants have adaptations that allow them to grow in harsh conditions. The stem of a cactus can store enough water from one rainfall to survive years of drought. Plants that live in temperate forests, such as oak trees, lose their leaves in the winter. This adaptation helps prevent water loss. Cold climate plants, such as mosses, are able to complete their life cycle in a shortened growing season.

Carnivorous plants are "meat-eating" plants. Plants such as Venus flytraps, sundews, and pitcher plants grow in nitrogen-poor soil. To make up for the lack of this nutrient, these plants have parts adapted to trapping and digesting small animals. Most of the animals these plants trap are insects, but they also trap small vertebrates including lizards, mice, frogs, and rats.

Cup-shaped pitcher plants are full of a sweet-smelling liquid. Insects attracted by the liquid fall into the cup and are trapped. Compounds in the juice then digest the insect.

Defense Many plants have adaptations that defend them against herbivores. Some plants produce chemicals that give them a bad taste. Most herbivores that eat these bad-tasting leaves will stop after a few bites. Other plants, such as milkweed, produce chemicals that are poisonous to most animals. Roses have thorns that discourage animals from eating them.

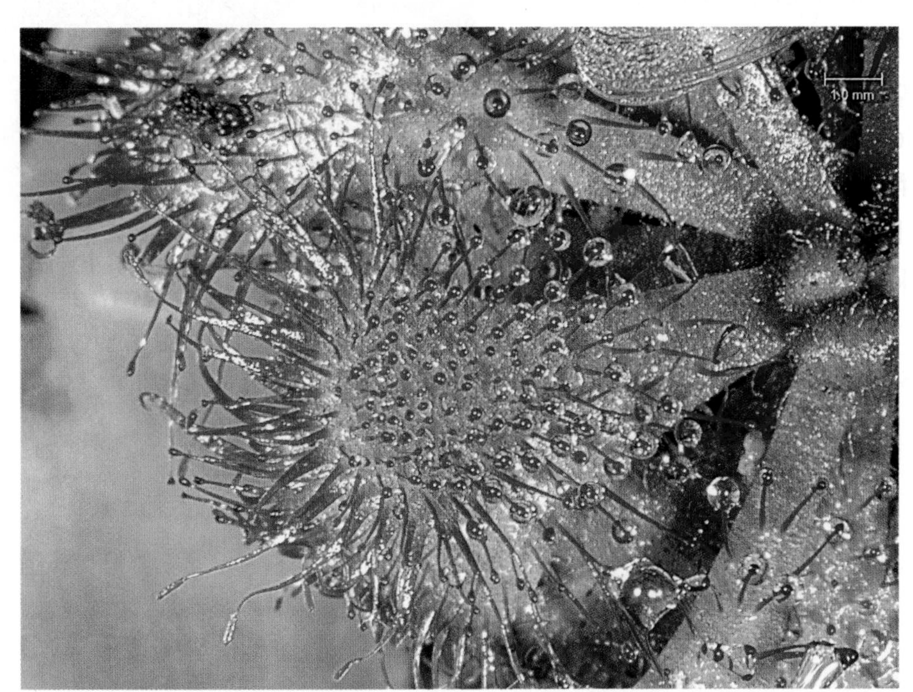

Sundews trap insects using a sticky, nectar-like substance. Insects land on their leaves and stick. Then compounds in the nectar digest the insect.

Behavioral Adaptations - Behavioral Adaptations in Animals

Attracting Mates Some behaviors help animals attract mates. Some birds perform courtship "dances" that show off their bright colors to attract mates. Crickets chirp by rubbing their wings together. Male elk fight by tangling their antlers. Fireflies emit light. The light also helps the insects defend their territories and warn away predators.

Male peacocks display tail feathers to attract mates.

Behaviors and Seasonal Changes

Some behaviors are related to seasonal changes. White-tailed deer change their diet with the seasons. In warm months, they eat shoots and leaves. In autumn, their diet includes acorns and other tree nuts as well as dried leaves and grasses. In winter, a deer's body chemistry changes so that it can digest twigs and stems of small trees and shrubs. As spring approaches, the deer switch to eating buds of flowers and young leaves.

Dormice escape the cold of winter by hibernating.

When temperatures drop and days become shorter, some animals, such as chipmunks, bats, snakes, turtles, frogs, and ground squirrels, hibernate to escape the cold. *Hibernation* is defined as a period of inactivity. During hibernation, body temperature drops and other life functions slow. When temperatures warm, the animals become active again.

Many animals, such as birds, butterflies, and fish, migrate. To *migrate* is to move from one place to another. Some animals migrate seasonally in response to their environments. A change in climate, availability of food, or changing habitat are also factors that can cause animals to migrate.

These monarch butterflies migrate to avoid cold weather.

Symbiosis

Symbiosis is a relationship between two or more kinds of organisms that lasts over time. Symbiotic relationships are also behavioral adaptations.

Mutualism A symbiotic relationship that benefits both organisms is called **mutualism**. The relationship between a pollinator and a flowering plant is an example of mutualism. The pollinator, usually an insect or a bird, gets nectar from the flower. The plant gets its pollen transported to the pistil of another flower. Both organisms rely on the relationship for survival.

Commensalism Remoras are fish that attach themselves to the bodies of rays and sharks. The remora gets food scraps, transportation, and protection from the ray. While the remora does not hurt the ray in any way, it does not help the ray either. A symbiotic relationship that benefits one organism without harming the other is called **commensalism**.

Parasitism Some partnerships are harmful to one of the individuals in the relationship. **Parasitism** is a symbiotic relationship in which one organism benefits and the other is harmed. The organism that benefits from the relationship is called a parasite. The organism that is harmed is called a host. Ticks are parasites on mammals. A tick uses its host's body for a home and a food source. The tick attaches itself to a host and then harms the host by taking its blood. The host gets no benefit from this relationship.

Ants and acacia trees have a mutualistic relationship. The tree provides food and shelter for the ants. The ants defend the tree from other insects.

Barnacles growing on the backs of whales are commensal. The barnacles gain a home. The whales are not hurt by the barnacles.

A wasp laid its eggs under this tomato hornworm's skin. The larvae of the wasps fed on the hornworm and then surfaced and made these pupas.

Heredity

Have you ever stopped to wonder why a dalmatian has spots or a plant has pink flowers? These organisms got these traits from their parents. The passing of traits, or characteristics, from parent to offspring is called *heredity*.

Heredity applies to all organisms. In plants, for example, flower color and plant height are *inherited traits*. An inherited trait is a trait that an offspring receives from its parents. Some inherited human traits include dimples, facial features, and hair and eye color.

Heredity can also affect an organism's behavior. In animals, inherited behaviors are called instincts. An instinct is a way of acting or behaving that an animal is born with and does not have to learn. For example, spiders do not have to learn how to spin a web. This information is an inherited instinct. Just as puppies do not learn how to breathe, spiders do not learn how to build webs.

Similarly, birds are born with the instinct to build a nest. Different types of birds build different types of nests. For example, weaver birds always build elaborate hanging nests out of small plant pieces. Other birds, such as penguins, build nests out of pebbles. Young birds instinctively build nests that are like those of their parents.

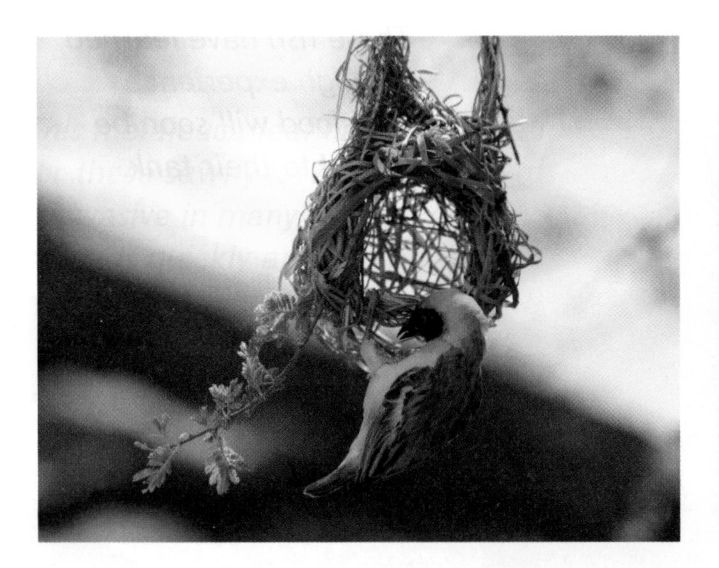

A weaver bird instinctively knows how to build this type of nest.

Web building is an instinct in spiders.

How Traits Are Inherited

The first person to study how traits are inherited was an Austrian monk named Gregor Mendel. In 1856 he began experimenting with garden pea plants. He crossed plants that had different traits and observed how those traits were passed on.

Mendel spent years conducting experiments. He determined that inherited traits are passed from parents to offspring during reproduction. He thought that each trait is controlled by two factors. The offspring receive one of these factors from each parent. Today, scientists refer to these factors as genes. A *gene* contains chemical instructions the body uses to produce inherited traits. Genes are stored on structures called *chromosomes,* which are found in the nucleus of each cell.

Mendel found that for each trait he studied, one form of the trait could mask the other. For example, pea plants can have purple or white flowers. When Mendel crossed a purple pea with a white pea, all the offspring had purple flowers. When he crossed two of the purple offspring, the trait for white flowers reappeared. The trait for white flowers had been hidden.

Mendel concluded that for every trait there is a dominant form and a recessive form. The *dominant trait* is one that masks another form of the trait. A *recessive trait* is one that is masked by the dominant form. The traits an individual inherits depend upon the combination of genes received from the individual's parents.

In pea plants, purple flowers are dominant and white flowers are recessive.

dominant trait

recessive trait

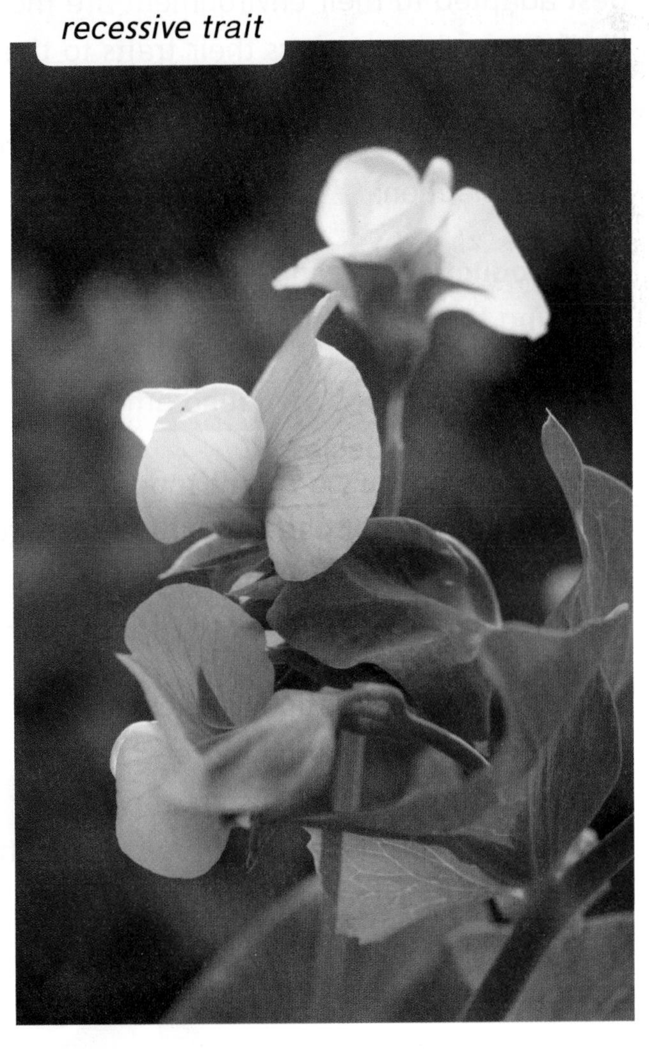

Humans and the Environment

Earth's Natural Resources

A **natural resource** is a material that people can use that is found in nature. Natural resources can be living or nonliving and include air, water, sunlight, soil, rocks, minerals, plants, and animals. They are classified into two groups: renewable resources and nonrenewable resources.

Nonrenewable Resources

Nonrenewable resources are natural resources that are used up much faster than they are made.

Metals, such as copper, are nonrenewable resources that people mine from beneath Earth's surface. These materials are used for building homes and roads and manufacturing cars, computers, and appliances. Once these metals are used up, they will be gone forever.

Many of the fuels we use are nonrenewable. Some of our electricity comes from nuclear energy. Uranium is used in nuclear power plants to produce this energy. Uranium is a nonrenewable resource.

A much larger portion of the energy we use to make electricity and drive vehicles comes from burning fossil fuels. A **fossil fuel** is an energy source that formed millions of years ago from the remains of plants and animals. Layers of sediment built up on top of the dead organisms. The diagrams show how these nonrenewable resources form.

Dead organisms in a swamp form peat.

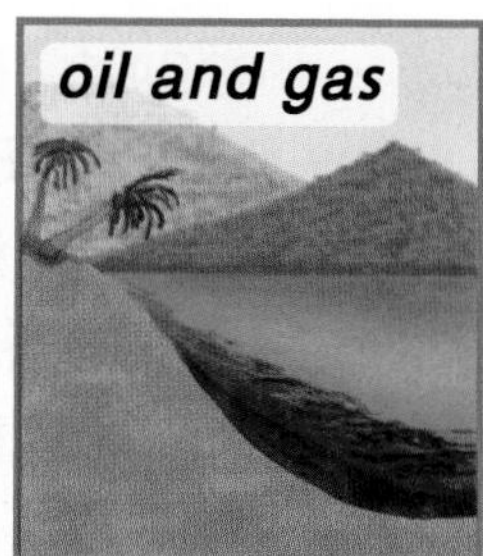

Dead organisms fall to the ocean floor.

Peat is covered with layers of sediment.

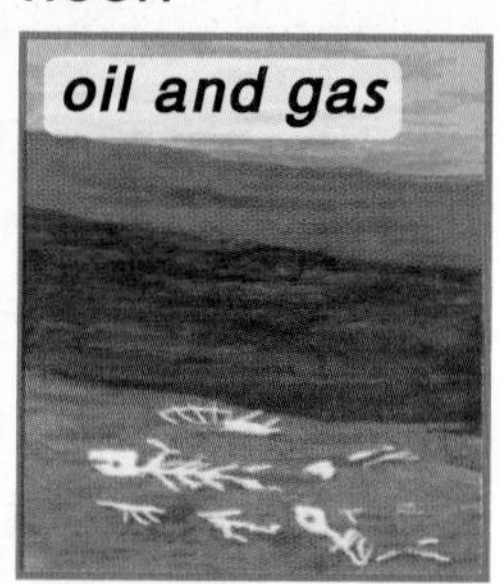

The dead organisms are buried in sediment.

Pressure turns peat into coal.

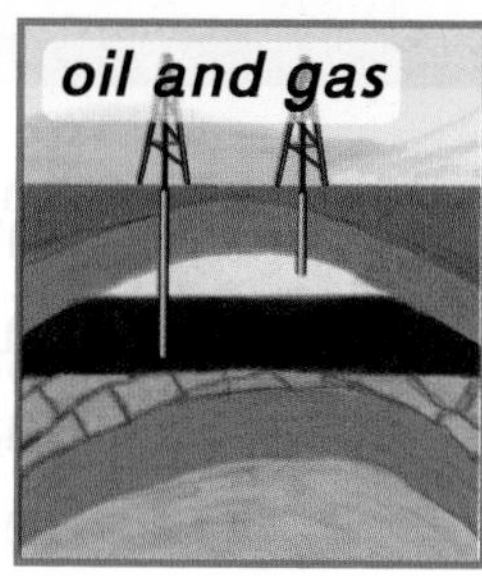

Pressure forms oil and gas.

There are three main types of fossil fuels.

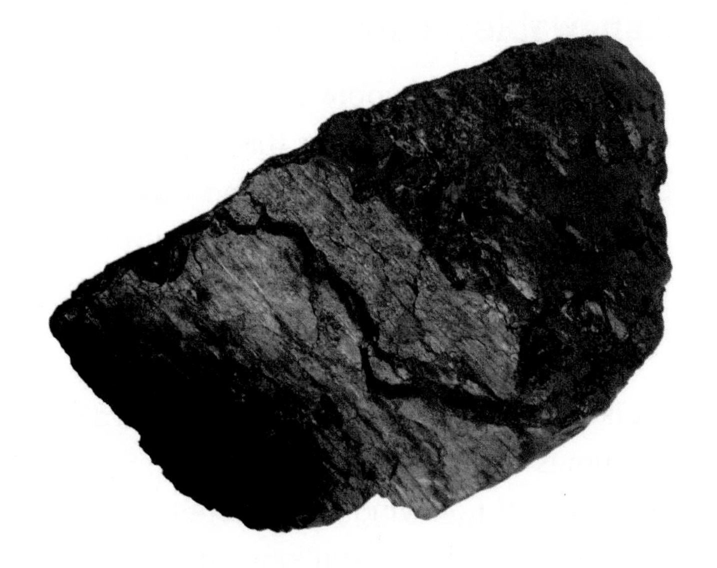

Coal Coal is a rock that is found underground. People mine coal by working in mines or by using giant shovels to dig away the layers of rocks and soil above the coal. Coal is the most plentiful fossil fuel. It is used mainly to generate electricity. In fact, nearly half of all of the electricity that people use in the United States is produced by burning coal.

Oil Oil is a thick, black substance that is also called *petroleum*. Like coal, oil is located in beneath Earth's surface in the spaces inside rocks. People drill into the rocks to find oil and then pump it to the surface. Once the oil comes out, it is delivered to refineries where it is turned into gasoline and other kinds of fuel. It can also be burned in power plants to generate electricity. Today, oil is drilled in 31 states and many other countries.

Natural Gas Where people find oil, they usually also find natural gas. Natural gas is colorless and odorless. It is pumped out of the ground through pipelines and then stored until it is needed. Natural gas is an especially important fossil fuel in homes because it can provide energy for cooking and heating.

Earth's Natural Resources - Renewable Resources

Eventually, our supply of nonrenewable resources will run out. For this reason, people are working hard to find ways to use resources that can be replaced in nature or that simply never run out. These are called **renewable resources**.

Like nonrenewable resources, renewable resources can be living or nonliving. Some examples of nonliving renewable resources are water, wind, and sunlight. Some of these resources, such as moving water in ocean waves, are always available. Trees and other plants, as well as animals, are considered living renewable resources.

It may seem strange that plants and animals can help provide energy. Remember that plants absorb energy from the Sun during photosynthesis. This energy is passed on to animals that eat plants, and then to animals that eat other animals. Wood, crops, and animal waste all contain this energy. When these materials are burned, they release the energy as heat. These materials can also be changed into other forms of energy, such as gas and fuel.

Sources of Electricity

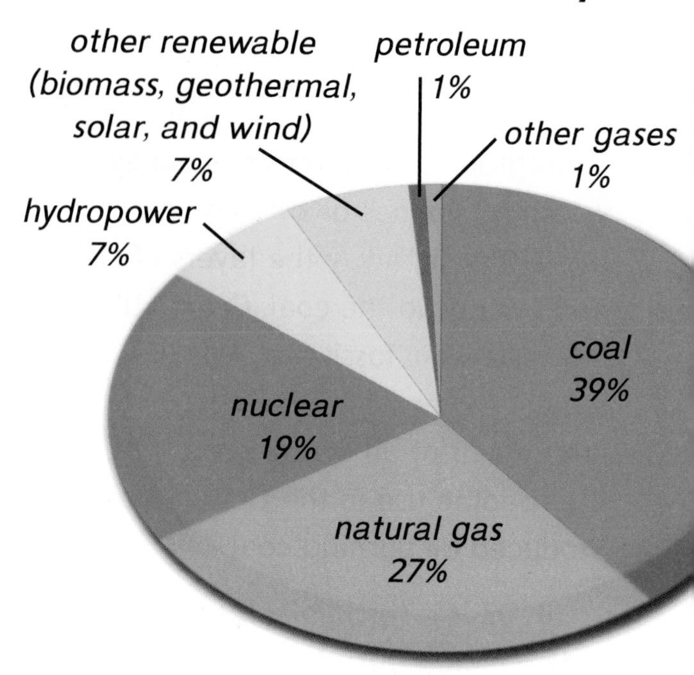

Skill Builder **Read a Circle Graph**

A circle graph shows how parts relate to a whole. In this graph, the smallest slices represent the resources that were used the least as a source of electricity. The largest slices show which resources were used the most.

Wood was once a main source of energy. At one time, it supplied more than 90% of the energy needs in the United States.

Borut Trdina/Getty Images

Alternative Energy Sources

Wind, moving water, and sunlight are **alternative energy sources**. Alternative energy sources are not fossil fuels. In the future, humans will need alternative energy sources to fulfill most of our energy needs.

Wind You may have seen fields dotted with large wind turbines like the one in the photograph. Wind moves a turbine's blades. The blades are connected to gears and shafts that are connected to a generator, which transforms the wind energy into electricity. Thirty-nine states currently have operating wind turbines. Wind turbines work well only in areas where the wind blows consistently.

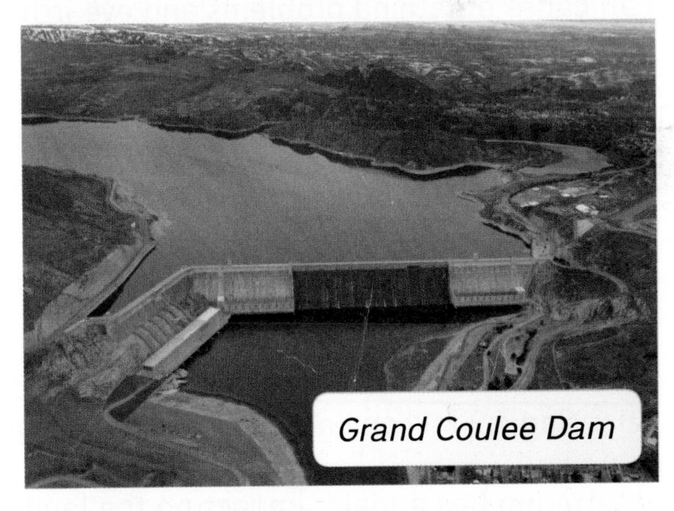

Grand Coulee Dam

Water Energy from moving water is called *hydropower*. Hydropower generates more electricity than any other renewable resource. Hydropower plants are built near sources of moving water, such as rivers and waterfalls. Dams are built and collect water and store it in reservoirs for later use. The Grand Coulee Dam in Washington State is the largest source of hydroelectric power in the United States.

Solar Energy Energy that comes from the Sun is called *solar energy*. Solar energy heats water and land. Solar energy will last as long as the Sun shines and is available almost everywhere, although it works best in places where sunny days are common. Solar energy can be used to generate electricity, heat, and light. It is estimated that in just one hour, and with the right tools, the Sun could provide one year's worth of energy to Earth!

(t) Image Source/Getty Images; (c) Aric Vyhmeister/iStock/Getty Images; (b) Design Pics/Ken Welsh

People Change Environments

You and the living and nonliving things around you form an environment. In order to survive, all of the living organisms in an environment need clean air, water, and soil.

People impact the environment every day. Sometimes these impacts are negative and can harm the environment. One example is pollution. *Pollution* is any harmful substance that affects Earth's land, air, and water. One of the main causes of air pollution is particles that are released during the burning of fossil fuels. Other causes include dust from farming, construction sites, and mines. Sometimes these harmful particles produce smog, which can cause breathing problems and eye irritation.

Air is not the only natural resource that can become polluted. Heavy rains can wash fertilizers used in farming into lakes, rivers, and streams. These fertilizers can negatively impact water quality and cause a toxic kind of algae to grow. Oil spills are another cause of water pollution. In 2010, the worst oil spill in United States history occurred when an oil rig in the Gulf of Mexico exploded, releasing 4.9 million barrels of oil into the gulf.

Littering has a major impact on the land. Sometimes people dump trash, chemicals, and other waste on the ground instead of disposing of them properly. This dumping destroys habitats and can kill plants and animals.

What Could I Be?

Environmental Engineer

Are you interested in preventing or solving problems caused by pollution? Environmental engineers use chemistry, biology, and mathematics to solve environmental problems. Learn more about a career in environmental engineering in the Careers Section.

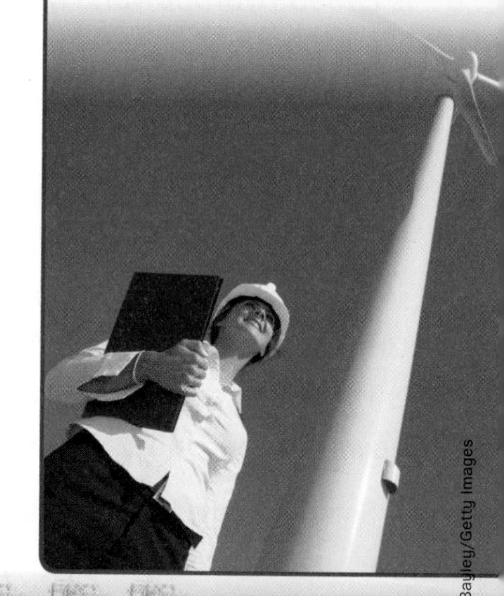

Smog is a problem in major cities, where fumes given off by cars, trucks, and buses pollute the air.

Protecting Our Resources

Even though some human activities harm natural resources, there are many people who work hard to protect them. In 1963, the U.S. government passed the first Clean Air Act. It gave scientists money to study different pollutants and how they affect the environment. In 1970, a new Clean Air Act was passed and the Environmental Protection Agency (EPA) was formed. The EPA helps states and communities reduce pollution. The agency also makes sure that chemical plants and other companies are not releasing harmful pollutants into the air.

Planting trees can decrease levels of harmful gases and reduce air pollution.

In 1974, the government passed a law to help protect our water. The Safe Drinking Water Act protects America's drinking water. It also protects water sources, such as rivers, lakes, and reservoirs. Farmers are now using safer, more natural ways of controlling pests and providing nutrients to plants.

There are many simple ways to help protect resources.

- Organize a group to pick up trash. Dispose of all trash properly.
- Plant new trees, bushes, and flowers.
- Compost garbage, grass, and leaves. Use the compost to feed plants instead of using chemical fertilizers.
- Ride a bike, walk, or take mass transit instead of riding in a car.

Cleaning up litter helps protect natural resources.

Conservation

Earth does not have an unlimited supply of natural resources. Many resources are being used more quickly than nature can replace them. Humans can help slow the use of natural resources through conservation. **Conservation** is the practice of using resources wisely.

People conserve resources in many ways. Turning off lights when leaving a room conserves energy. Planting trees helps conserve soil. Taking shorter showers helps conserve water. The "three Rs" also guide people in conserving resources. The 3 Rs are *reduce, reuse,* and *recycle.*

Reduce To reduce means to lessen the amount something is used. The chart on this page shows how people can reduce water use in the home. Paper is made from material that comes from trees. People can conserve trees by writing or printing on both sides of each sheet of paper. Running many errands at once instead of taking multiple trips helps reduce the amount of fuel used. Installing low-energy light bulbs instead of standard ones helps reduce electricity used.

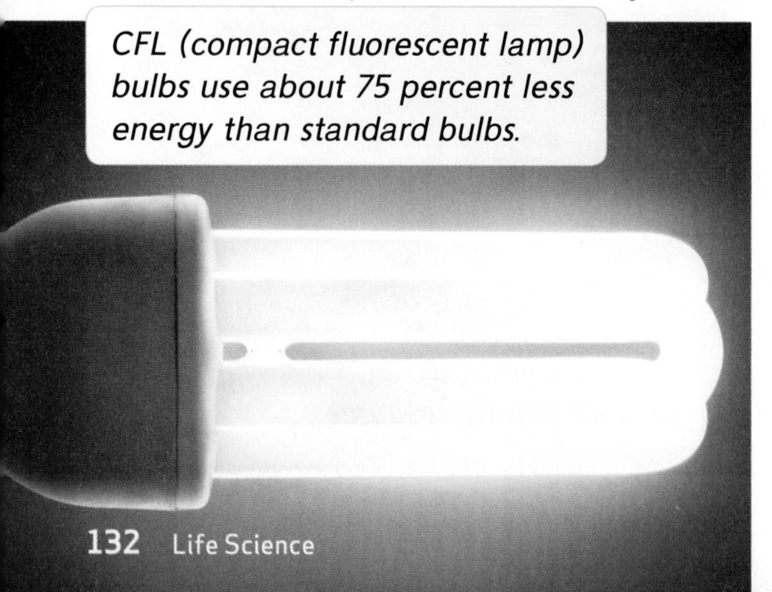

CFL (compact fluorescent lamp) bulbs use about 75 percent less energy than standard bulbs.

Use water-conserving showerheads and take shorter showers.

Do not leave water running when you are not using it.

Wash dishes by hand. If you use a dishwasher, use a water-saving model and do not run it unless it is full.

Fix leaking pipes or faucets.

Use a water-saving washing machine and wash full loads of clothes.

Grow plants that do not require frequent watering and water your plants after dark so the water does not evaporate.

Reuse To reuse means to use something more than once. Cloth grocery bags are reusable. They can be used many times. These bags replace new plastic or paper bags. Old T-shirts can be torn into strips and used as rags or to tie up plants in the garden. Scrap paper can be saved and used for future art projects. Food scraps can be composted. Compost can provide nutrients for soil and plants.

Did You Know?

In 2012, Americans recycled nearly 79 million metric tons (87 million tons) of trash!

Recycle When an item is recycled, it is made into a new product. Many communities encourage people to put recyclable materials, including metals, glass, plastic, and paper, in separate bins rather than in the trash. These items are picked up and taken to a facility where they are cleaned and prepared for recycling. For example, old tires are sometimes recycled into soft playing surfaces for playgrounds. Plastic bottles can be recycled into fleece clothing. Recycling products keeps waste out of landfills and helps keep soil and water clean.

Recycling helps conserve natural resources.

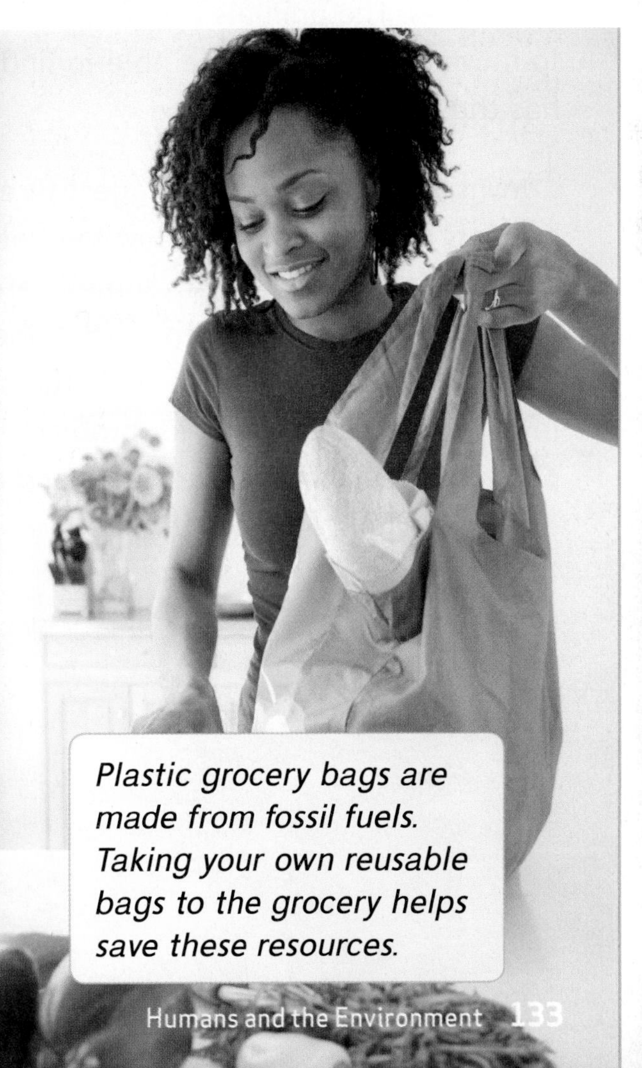

Plastic grocery bags are made from fossil fuels. Taking your own reusable bags to the grocery helps save these resources.

Protecting Plants and Wildlife

Threats to Ecosystems

There are thousands of species that are considered *endangered*. This term means that they could soon become extinct, or die out. Human activities are partly to blame for this.

Pollution affects people as well as plants and animals. Acid rain, which falls to the ground when air pollution mixes with water vapor, falls into soil and fresh water. It can kill plants and make animals very sick. When people throw litter on the ground, animals may try to eat it.

Another threat is *deforestation*. Deforestation is the removal of large areas of trees. The trees are used to make goods or land is cleared to build farms, homes, and businesses. The animals that live in the forests are forced to move. They may not be able to find another habitat that has the resources they need.

Japanese honeysuckle is an invasive species. It takes up space and steals nutrients that other plants need.

Invasive species are the biggest threat to plants and animals around the world. Invasive species are plants or animals that come from somewhere else. They may arrive by ship or in trees that have been chopped down. Sometimes they are accidentally released into nature. Invasive species are a major problem because they can pass on diseases to native plants and animals. They can also steal vital resources from native living things.

The emerald ash borer is an insect that came to the United States in a cargo shipment from Asia. It has killed ash trees in 21 states and is still spreading.

Helping Living Things

Some humans cause problems in ecosystems, but others work hard to protect them. In 1916, President Woodrow Wilson established the National Park Service. The National Park Service protects and cares for 84 million acres of land as well as the plants and animals that live there. Rangers in the service help teach people about living things and ways in which we can protect them in our own neighborhoods and communities.

The United States Fish and Wildlife Service (USFWS) was formed in 1940. This service manages areas called wildlife refuges. A *wildlife refuge* is a place where plants and animals, including endangered species, can live in a healthy and protected environment. The USFWS also enforces the Endangered Species Act (ESA), which was passed in 1973. The ESA protects endangered species. It also helps populations of endangered species recover.

People are working to protect living things right in their own backyards. They are using compost instead of fertilizers to provide nutrients for plants. They are putting trash where it belongs and recycling whenever possible. They are removing invasive plants from their gardens and natural areas. They are planting trees to provide habitats for many living things.

This grizzly bear's habitat is part of Yellowstone National Park, which is America's first national park.

Did You Know?

There are over 560 national wildlife refuges in the United States. They help provide habitats for more than 2,000 different species of animals. They also protect 380 endangered species of animals and plants.

What Could I Be?

Park Ranger

Do you love being outside? Do you like interacting with a lot of different people? Park rangers fill a variety of roles in our national parks and other government installations. Learn more about a career as a park ranger in the Careers Section.

Future Career

Ocean Engineer Have you ever wondered what lies on the ocean floor? An ocean engineer studies this mysterious part of Earth. They develop vehicles that explore parts of the ocean floor that are dangerous for humans to go to. Ocean engineers identify the effect of the ocean on the shore and restore beaches that have worn away. They also examine coastal ecosystems for changes. These engineers are looking for safe ways to drill for oil and natural gas on the ocean floor.

Earth and Space Science

Earth's Structures

Earth has air, water, land, and living things. Energy constantly flows through these different parts. This interaction of matter and energy makes up all of the structures on Earth. The two main sources of energy on Earth are energy from the Sun and heat from inside Earth.

Earth has four main systems. These are the atmosphere, geosphere, hydrosphere, and biosphere.

Earth's Systems

Scientists have organized the parts that make up Earth into four main systems. They are the atmosphere, the hydrosphere, the geosphere, and the biosphere. These systems all interact.

Atmosphere The **atmosphere** is the layer of gases surrounding Earth. Made up mostly of nitrogen and oxygen, the atmosphere also contains water vapor, carbon dioxide, and other gases.

Geosphere The **geosphere** includes the solid and melted rock inside Earth. It also includes the soil, rock pieces, and land features at Earth's surface. Hills, mountains, erupting volcanoes, and other landforms are all part of the geosphere.

Hydrosphere All of Earth's liquid and solid water, including oceans, lakes, rivers, glaciers, and ice caps, makes up the **hydrosphere.** The hydrosphere covers more than 70 percent of Earth's surface. It exists in two basic forms: salt water and fresh water. Most of Earth's fresh water exists as ice. Most of Earth's salt water is in the ocean.

Biosphere The **biosphere** is the collection of all of earth's living things. Organisms that make up the biosphere are found from the lower atmosphere to the depths of the ocean floor. All living things are part of the biosphere.

Word Study

The word *sphere* comes from the Greek word for "ball."

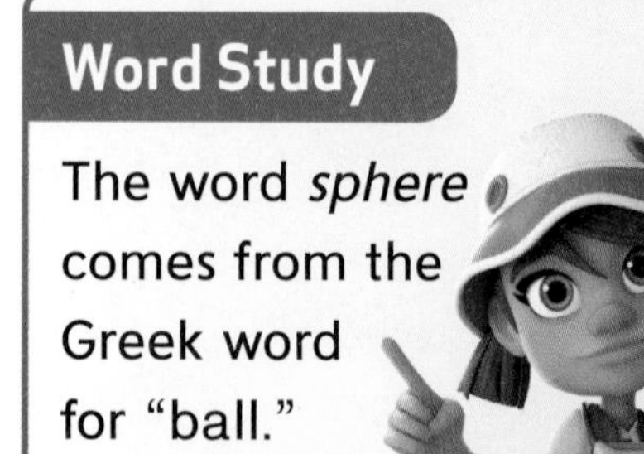

Earth's Systems Interact

The biosphere, atmosphere, hydrosphere, and geosphere are interconnected. For example, the atmosphere contains carbon dioxide, a gas that plants—part of the biosphere—need in order to make food. Volcanic eruptions in the geosphere produce carbon dioxide. The atmosphere also contains oxygen, the gas that animals need in order to breathe. Earth's upper atmosphere contains a layer called the ozone layer. This layer protects living things in the biosphere from the Sun's damaging ultraviolet rays.

Weather takes place in the lowest layer of the atmosphere. Evaporation of water from Earth's oceans—part of the hydrosphere—drives most of Earth's weather. The geosphere influences these weather patterns. For example, a mountain range can affect patterns of rainfall. As warm, moist air moves up a mountain, the air cools. Water vapor condenses, and rain falls on the windward side of the mountain. The dry side of the mountain is called a rain shadow. Different habitats and organisms are found on the wet and dry sides of the mountain.

Skill Builder

Read a Diagram

Examine the labels to understand how Earth's four major systems interact with each other.

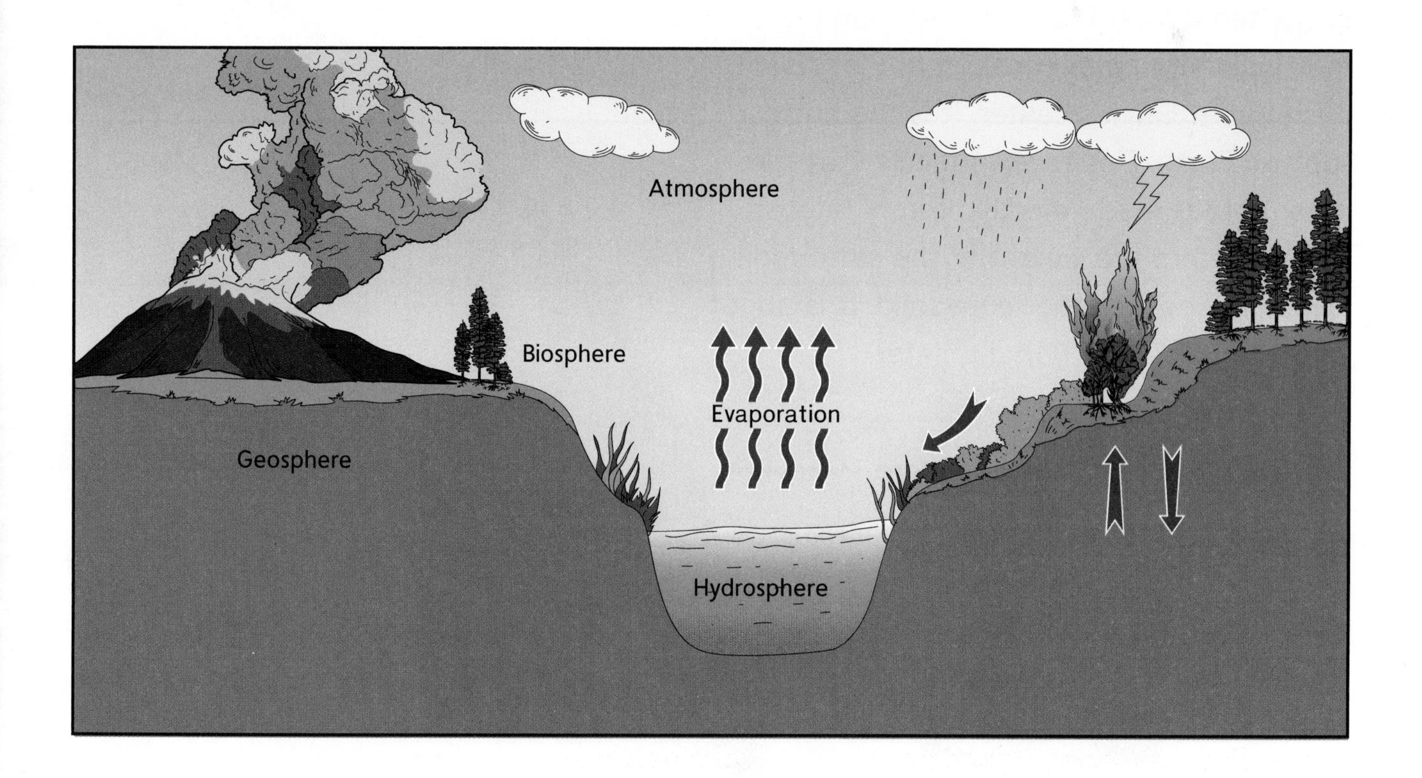

Earth's Layers

The geosphere includes both Earth's surface features and its interior. Scientists divide Earth into several layers.

Crust The **crust** is the rocky, outermost layer of Earth's surface. It varies in thickness, from 6 kilometers (4 miles) to 70 km (40 mi). The crust includes the continents and the ocean floor. All of Earth's land and underwater features are found here. The crust is brittle and can crack easily.

Mantle The **mantle** is the layer of Earth's interior below the crust. It is about 2,900 km (1,800 mi) thick. It is divided into the upper and lower mantle. The top of the upper mantle is solid rock. Together, the crust and the top of the upper mantle form the *lithosphere.* The rest of the upper mantle is almost melted rock. This part of the upper mantle is called the *asthenosphere.* The rock here is under great pressure. It can flow slowly like thick putty. The lower mantle is a thick, slow moving liquid.

Core The innermost layer of Earth is the **core**. The *outer core* is a liquid layer made mostly of melted iron. It is about 2,300 km (1,400 mi) thick. The *inner core* is a ball of solid material at Earth's center. It is the hottest spot on Earth. Evidence suggests that the inner core is made mostly of iron. Intense pressure from the weight of the surrounding layers makes the inner core solid. It is about 2,400 km (1,500 mi) in diameter.

Did You Know?

The crust under the continents is thicker than the crust on the ocean floor.

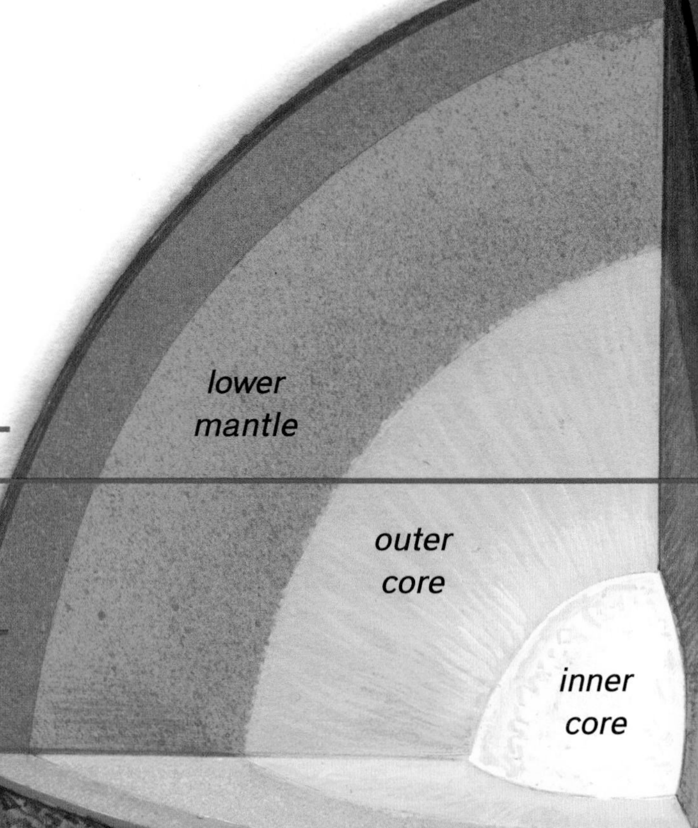

Landforms

A **landform** is a physical feature on Earth's surface. Landforms vary greatly in shape and size. They include features such as level plains, rounded hills, and jagged mountains.

Skill Builder

Read a Diagram

Match the numbers on the map to the numbers on the photographs.

Landforms of the Continental United States

If you were to travel across the continental United States, you would observe a variety of landforms.

The Continental United States

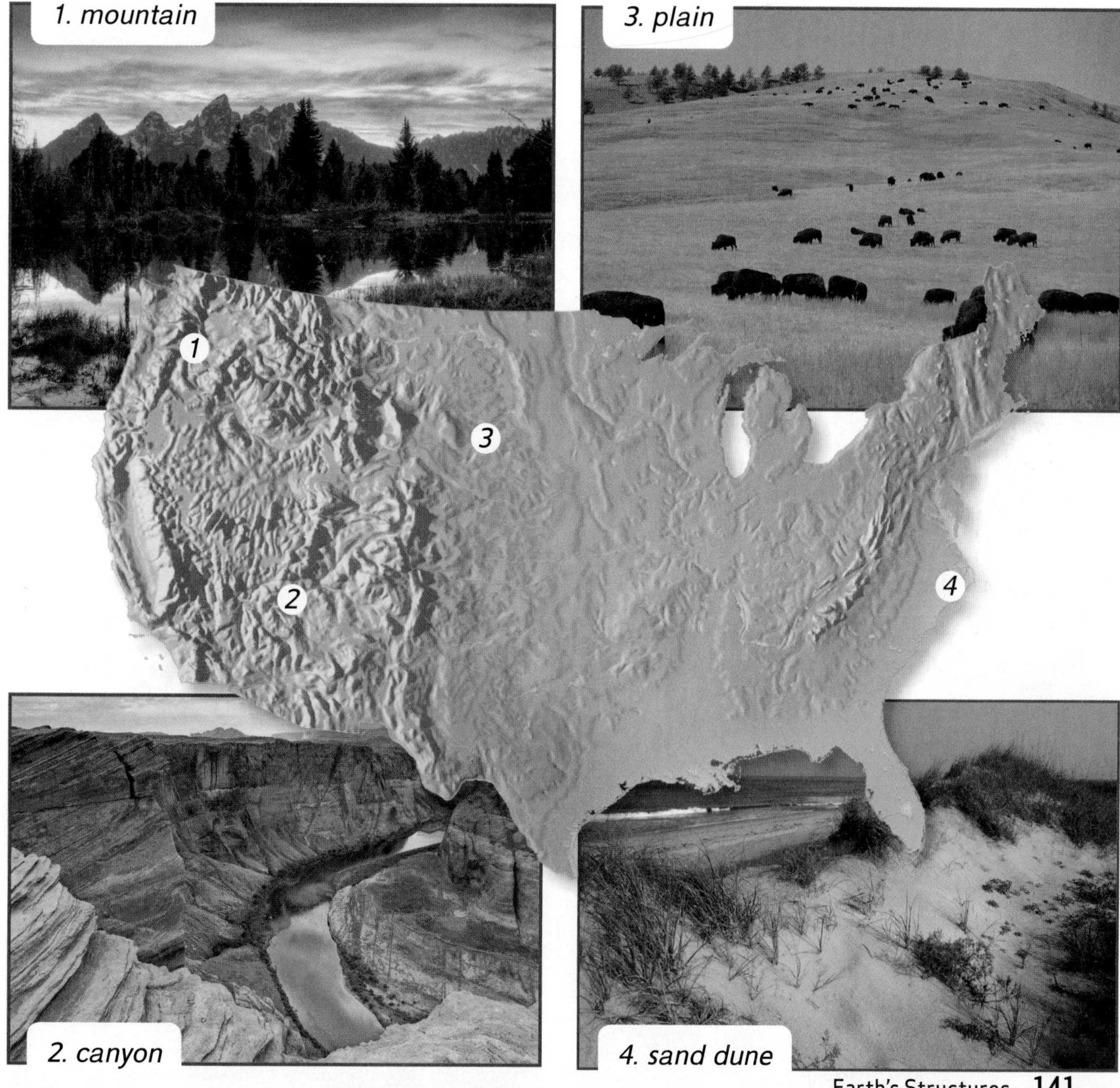

Landforms

Types of Landforms

From outer space, Earth's land might seem flat. On Earth, you can see many landforms on it's surface. Each landform has specific characteristics, and each forms in a different way. Use the labels on the image below and descriptions on the next page to learn more about common landforms.

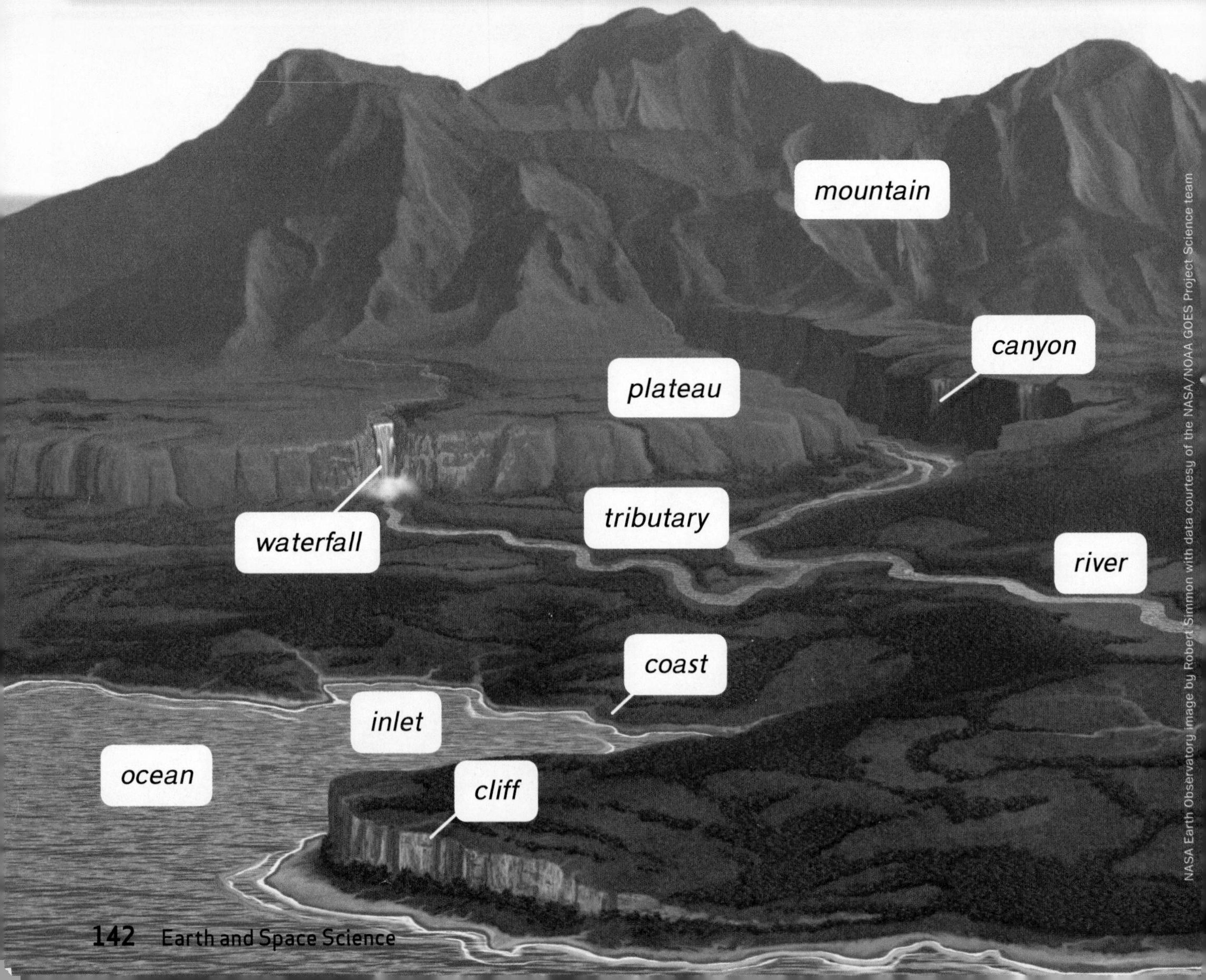

NASA Earth Observatory image by Robert Simmon with data courtesy of the NASA/NOAA GOES Project Science team

Earth's Land Features

Mountain A mountain rises high above the ground.

Hill A hill is lower and rounder than a mountain.

Valley A valley is the low land between hills or mountains.

Canyon A canyon is a deep valley with high, steep sides.

Cliff A cliff is a high, steep section of rock or soil.

Plain A plain is a wide, flat area.

Plateau A plateau is flat land that is higher than the land around it.

Desert A desert is an area with very little precipitation.

Beach A beach is the land along the edge of a body of water.

Dune A dune is a mound or ridge of sand.

Earth's Water Features

Ocean An ocean is a large body of salt water.

Coast A coast is where a body of water meets land.

Tributary A tributary is a small river or stream.

River A river is a natural body of moving water.

Waterfall A waterfall is a natural stream of water falling from a high place.

Lake A lake is a body of water surrounded by land.

Estuary An estuary is where river water and ocean water meet.

Delta A delta is the mass of land that forms at the mouth of a river.

Inlet An inlet is a narrow body of water off a larger body of water.

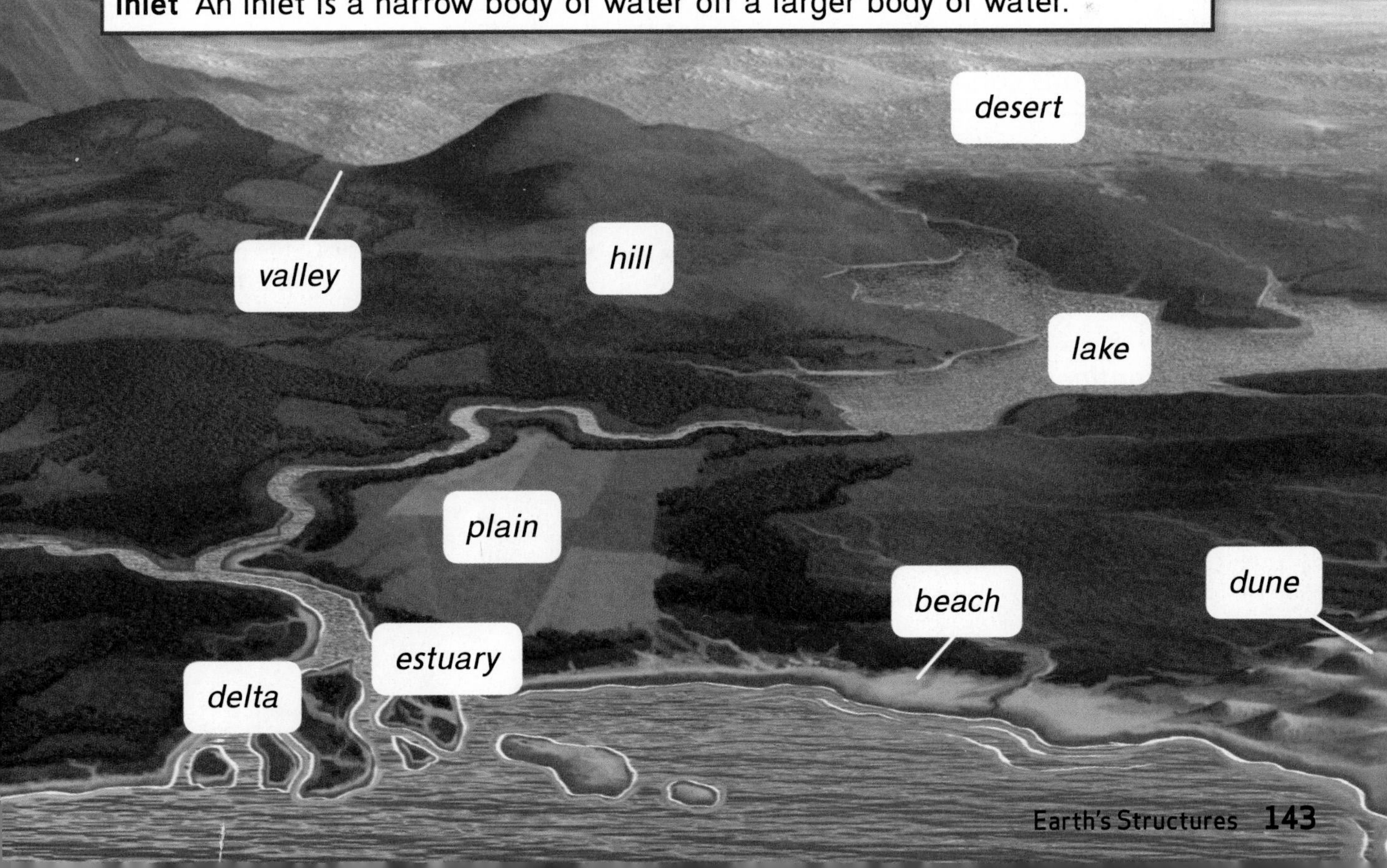

Earth's Ocean Features

Oceans are large bodies of salt water. Earth has five oceans, which are all connected. Together, they cover more than 70 percent of Earth's surface. Ocean water plays an important role in the water cycle and contains nutrients that support a variety of living things in the biosphere.

Ocean Floor

Although waves move over the surface of the ocean, the surface of the ocean is mostly flat. However, if you could travel deep below the ocean's surface, you would find features on the ocean floor that look like mountains, plains, and valleys.

An *ocean basin* is a large underwater area between continents. Along the coast of a continent, the ocean floor is called the *continental shelf.* Here the ocean floor is covered by shallow water and gradually slopes down. The continental shelf can stretch seaward for miles. A continental shelf ends at the point where a sharp downward slope begins. This sharp slope is called the *continental slope.* This land is the steeper part of the continent that slopes down toward the ocean floor. Underwater canyons can form on the continental slope.

Did You Know?

At 11,304 meters deep (36,201 feet), the Marianas Trench is the deepest part of the ocean.

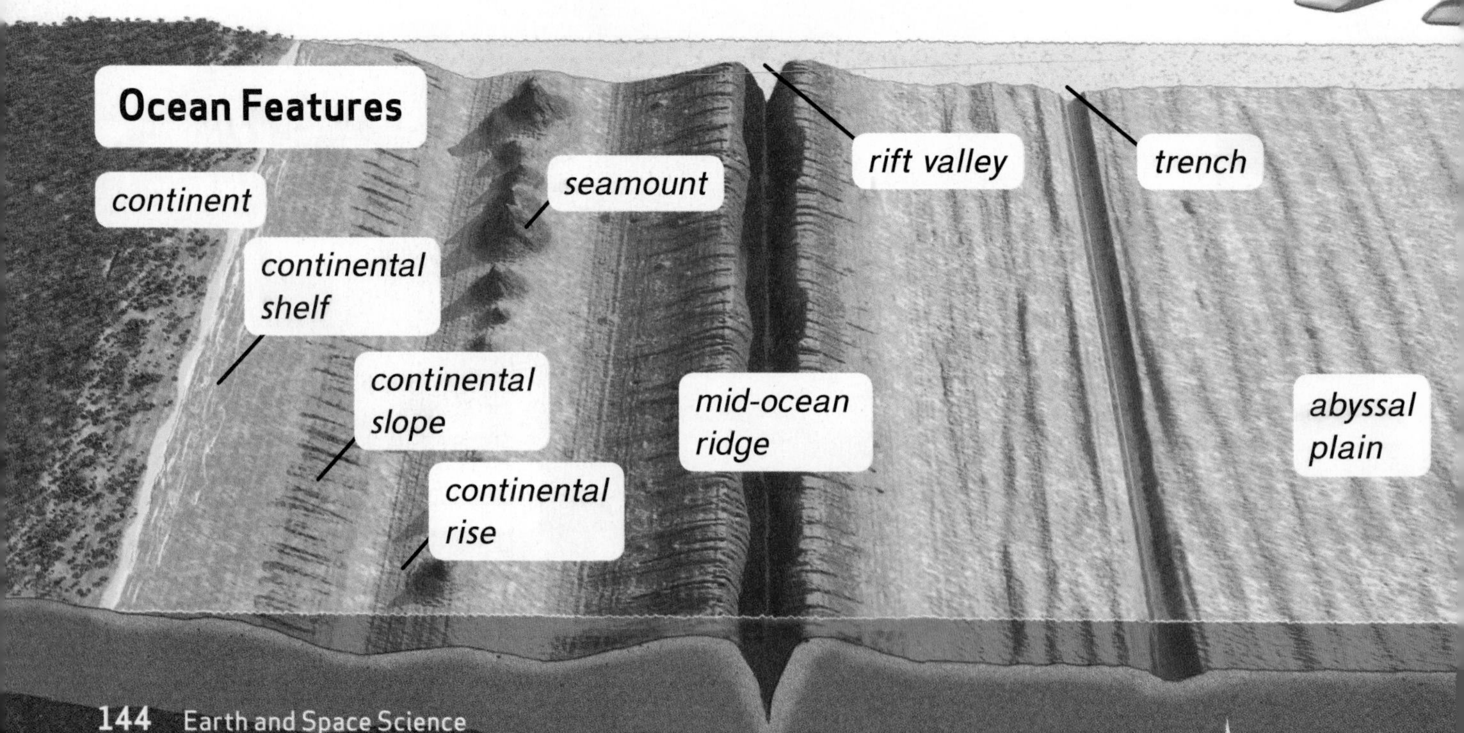

At the base of the continental slope is the continental rise. The *continental rise* connects the continent with the ocean floor. Most of the ocean floor is flat and without features. These wide, flat areas are called the *abyssal plains.* They cover about 40 percent of the ocean depths.

Long mountain ranges stretch through the middle of some oceans. These mountain ranges are called *mid-ocean ridges.* The valley down the center of a mid-ocean ridge is called a *rift valley.*

Other ocean floor features include trenches and seamounts. *Trenches* are the deepest parts of the ocean floor. They are usually long and narrow. A *seamount* is an underwater mountain that rises from the ocean floor but stops before it reaches the surface of the ocean.

Scientists use underwater vehicles to observe the ocean floor. These vehicles may carry cameras, instruments to measure the underwater environment, or mechanical arms to gather samples. Scientists also measure the depth of the ocean floor. They send sounds into the ocean and wait for the echo to come back.

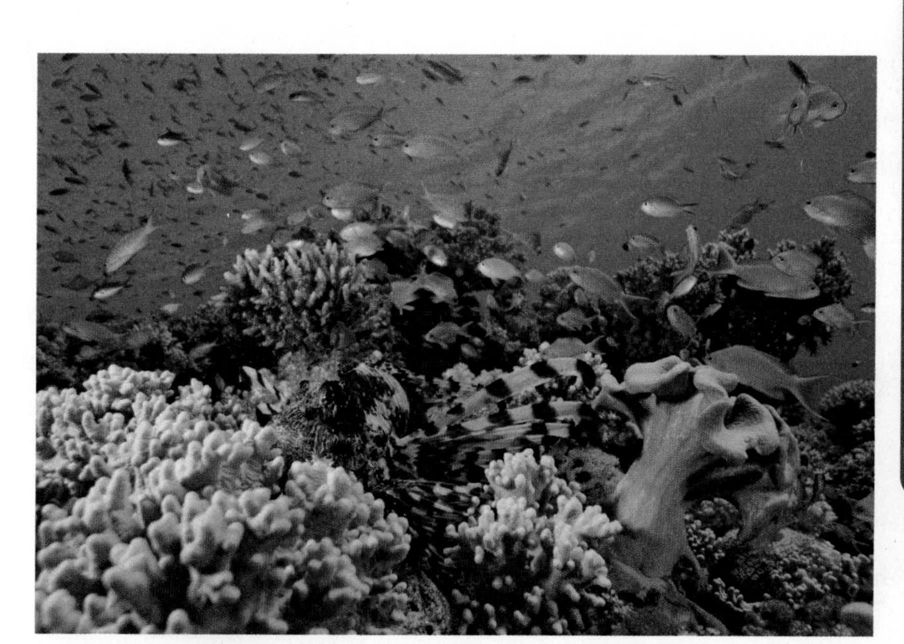

This coral reef formed on the shallow continental shelf.

What Could I Be?

Oceanographer

Interested in exploring the depths of the ocean? Oceanographers apply their knowledge of chemistry, biology, geology, engineering, and math to answer questions about the processes of the oceans. Learn more about a career in oceanography in the Careers section.

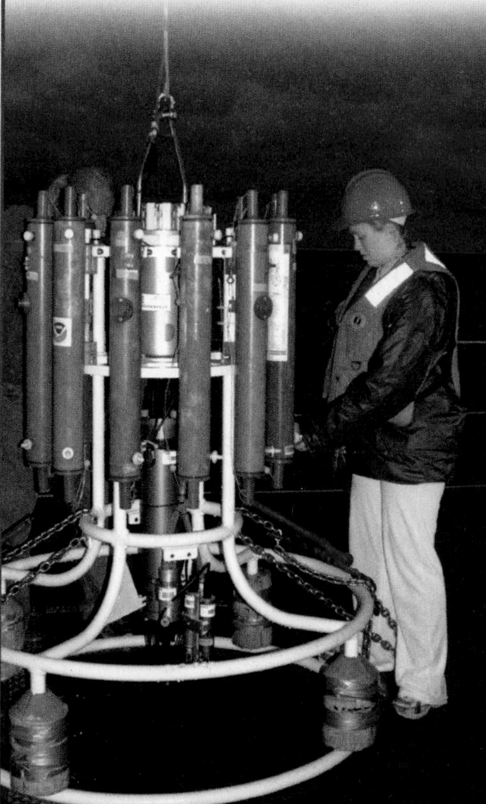

Minerals

A *mineral* is a solid, nonliving substance found in nature. Minerals are found underground, in soil, and even on the ocean floor.

Minerals are made of elements. An *element* is a pure substance that cannot be broken down into a simpler substance. Gold is an element, as are aluminum, oxygen, hydrogen, and iron. Some minerals are made of a single element. Other minerals are made of two or more elements.

Properties of Minerals

There are about 3,800 known minerals. Each mineral has its own properties. You can use the properties of minerals to tell them apart. Some mineral properties include color, streak, luster, hardness, and shape.

The *color* of a mineral is the color that shows on the mineral's surface. Samples of the same mineral can have different colors. Hematite, for example, includes red, black, brown, and silver gray varieties. *Streak* is the color of the powder left when a mineral is rubbed on a rough surface. All varieties of hematite leave a red streak. For this reason, streak is generally a better identifying property than color alone.

Luster is the way a mineral reflects light from its surface. There are two general kinds of luster: metallic and nonmetallic. Minerals with a metallic luster look shiny. Minerals with a nonmetallic luster look dull. Minerals with a nonmetallic luster may look waxy, pearly, earthy, oily, or silky.

Pyrite, or fool's gold, has a yellow color, a shiny metallic luster, and a greenish-black streak.

Ken Karp/McGraw-Hill Education

Mohs' Hardness Scale		
Hardness	**Mineral**	**Can Be Scratched By**
1	talc	
2	gypsum	fingernail
3	calcite	penny
4	fluorite	
5	apatite	steel file
6	feldspar	streak plate
7	quartz	
8	topaz	
9	corundum	
10	diamond	

diamond

Diamond is a 10 on Mohs' hardness scale.

This broken piece of smoky quartz has rough surfaces. It shows fracture.

Hardness is a measure of how well a mineral resists scratching. A hard mineral resists scratching better than a softer mineral. On Mohs' hardness scale, minerals are ranked from 1, which is the softest, to 10, which is the hardest. A mineral with a higher number will scratch a mineral with a lower number. By scratching an unknown mineral with a mineral that has a known hardness, you can find the hardness of the unknown mineral.

Mica breaks in flat sheets. It has cleavage in one direction.

The appearance of the surfaces of a broken mineral can help identify it. When broken surfaces are flat, the mineral shows cleavage. Cleavage is described by the number of surfaces along which the mineral breaks. Minerals that break along uneven or rough surfaces show fracture.

Minerals have a crystal *shape*. A crystal is a solid whose shape forms a fixed pattern. Different minerals have different crystal shapes. Sometimes the larger shape of the mineral shows the same shape as the crystal structure. For example, salt crystals look like tiny cubes. In other minerals, the crystal structure can be seen only with a microscope.

Amethyst has a hexagonal crystal shape.

Classifying Rocks

Rocks are nonliving materials made of one or more minerals. To a person who studies rocks, a rock's *texture* is how its grains look. Some rocks have large grains and a coarse texture. Other rocks have small grains and a fine texture. You can use a hand lens to observe a rock's texture in detail.

Rocks are classified by how they form. There are three main kinds of rocks: igneous, sedimentary, and metamorphic.

Word Study

Igneous is a Latin word that means "fire."

Igneous Rock

Igneous rock forms when melted rock cools and hardens. Melted rock below Earth's surface is called *magma.* When melted rock reaches the surface, it is called *lava*.

Igneous rocks that form from magma inside Earth are called *intrusive igneous rocks.* The magma that forms these rocks cools much more slowly than magma that reaches the surface. This slow cooling occurs because the rocks that surround the cooling magma hold in the heat. When melted rock cools slowly, mineral crystals have more time to form. Because of this, intrusive igneous rocks have large mineral grains and a coarse texture.

Igneous rocks that form from lava on the surface are called *extrusive igneous rocks.* When lava flows from a volcano, it cools and hardens much more quickly than magma underground. Crystals have little or no time to form. For this reason, extrusive igneous rocks have small mineral grains and a fine texture. Some lava cools so quickly that grains do not have time to form at all.

Granite is an intrusive igneous rock. It has large grains that form as magma cools slowly beneath Earth's surface.

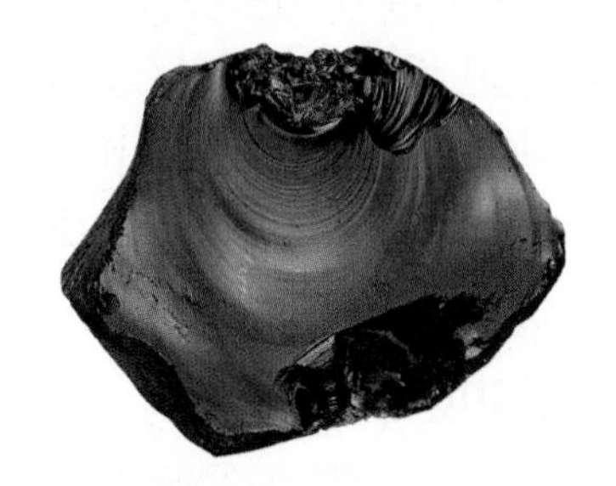

Obsidian is an extrusive igneous rock. It is sometimes called volcanic glass. Obsidian hardens so quickly that grains do not have time to form.

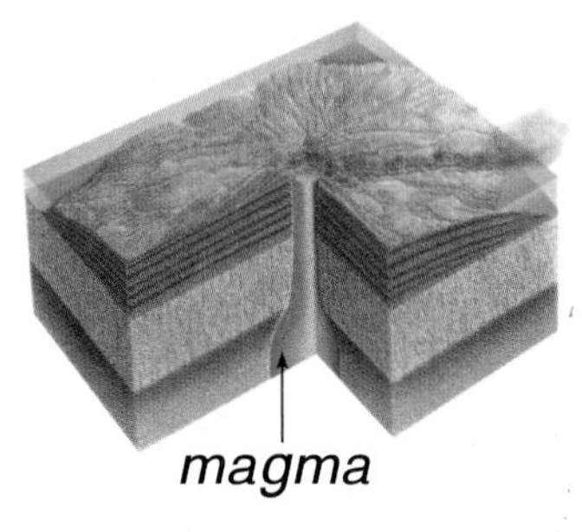

Sedimentary Rock

Sedimentary rock forms from sediments that are cemented or pressed together. *Sediments* are tiny bits of soil and rock. Wind and water deposit most of the sediments. Over time, new sediments are deposited on top of older layers.

Some sedimentary rocks contain minerals that were once dissolved in water. The minerals formed crystals among the sediments that came together to form the rock. Other sedimentary rocks are formed by the weight of the top layers pressing the sediment together. It can take millions of years for sediment to become rock.

The layers in sedimentary rock are stacked in order of their relative ages. *Relative age* is the age of one thing compared with another. The older the relative age of a rock layer, the lower it is found. Relative age also applies to any fossils in the rock layers. **Fossils**, the traces or remains of living things, are preserved in sedimentary rocks.

Conglomerates are sedimentary rocks with a coarse texture. The grains of sediment are very large in this kind of rock.

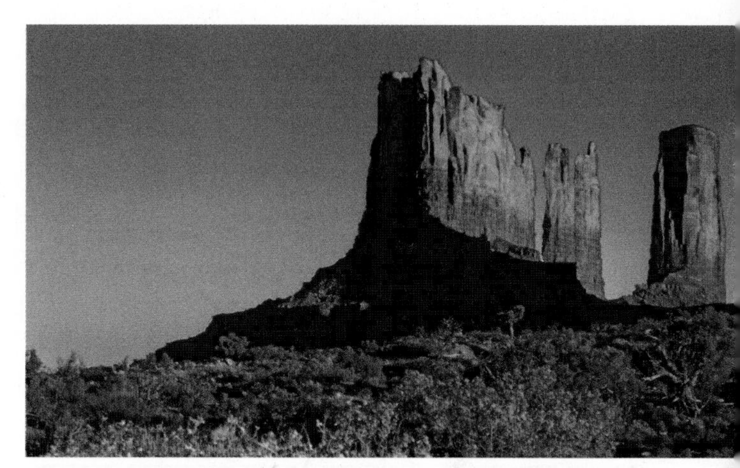

Sandstone is a sedimentary rock that forms from layers of sand.

Fossils are common in limestone rock formations.

Classifying Rocks - Metamorphic Rock

A third kind of rock is *metamorphic rock.* This kind of rock was once a different kind of rock. Metamorphic rocks have been changed by heat and pressure. They can form from igneous, sedimentary, or even other metamorphic rocks.

Temperatures and pressure levels below Earth's surface can be very high. When rocks are under so much heat and pressure, their physical properties can change. The properties of metamorphic rocks depend on the heat and pressure during their formation.

In some metamorphic rocks, the minerals get rearranged and pressed into thin layers, or bands. Bands may be straight or wavy. Gneiss is a metamorphic rock that starts out as granite, an igneous rock. When granite is put under heat and pressure, gneiss forms. The minerals in granite are rearranged to form the banding in gneiss.

Some metamorphic rocks form from sedimentary rocks. Marble, for example, forms from limestone. Unlike gneiss, marble does not have bands. Quartzite is another metamorphic rock that forms from sedimentary rock. Quartzite forms from sandstone and, like marble, lacks bands.

Before (Original Rock)	After (Metamorphic Rock)
granite (igneous rock)	*gneiss*
limestone (sedimentary rock)	*marble*
slate (metamorphic rock)	*phyllite*

Metamorphic rocks are found in many mountain ranges.

The Rock Cycle

All rock comes from other rock. The changing of rocks over time from one type to another is called the *rock cycle*. Over long periods of time, rocks are broken apart, melted, and put under heat and pressure. Through this cycle, any kind of rock can become another kind of rock.

The rock cycle has many paths. Magma cools, hardens, and becomes igneous rock. Rocks are broken down into sediments that collect in layers. These layers are pressed and cemented together, forming sedimentary rock. Under heat and pressure, igneous or sedimentary rocks become metamorphic rock. Any rock that is pushed deep underground can melt and turn into magma once more.

Skill Builder

Read a Diagram

Examine the key closely to understand what each color represents. Then follow the arrows to see how rocks can change in the rock cycle.

Melting

Heat and pressure (metamorphism)

Weathering, erosion, and deposition

Cementing and compacting

Cooling and hardening

Make Connections

Jump to the Earth's Changing Surface section to learn about weathering, erosion, and deposition.

Earth's Water

The water found on Earth makes up the hydrosphere. About 97 percent of Earth's surface water is salt water found in oceans. We cannot drink salt water or use it to grow crops. For those activities, we need fresh water.

Where Fresh Water Is Found

Only about three percent of Earth's water is fresh water. Most of this fresh water is frozen in the form of permanent snow cover, glaciers, and ice caps. A giant ice cap covers Antarctica—the continent at the South Pole. This frozen water accounts for about 69 percent of Earth's fresh water. Another 30 percent is groundwater. **Groundwater** is water stored in the cracks and spaces between particles of soil and underground rocks. Less than one percent is running water, such as rivers, and standing water, such as lakes. A tiny bit of Earth's water is found in the atmosphere as water vapor.

Fresh Water Sources

There are three main sources of usable fresh water.

Groundwater When water seeps into soil, it enters groundwater through aquifers. An **aquifer** is an underground layer of rock or soil that has connected pores, or holes, that water can pass through. As water flows through an aquifer, it eventually reaches a layer of rock that it cannot move through. Fresh water builds up on top of this rock. Groundwater is most useful when it is close to the surface. It can be reached by drilling or digging into the ground and pumping the water up through a well.

Running Water Many cities and towns are built next to sources of running water, such as streams or rivers. Thousands of freshwater rivers cross Earth's surface. Running water provides a source of fresh water for homes, farms, and businesses.

Standing Water Bodies of standing fresh water, such as lakes and reservoirs, are also sources of usable fresh water. A *reservoir* is an artificial lake that stores water. Reservoirs are usually made by building a dam on a river. Water is stored behind the dam and released when needed.

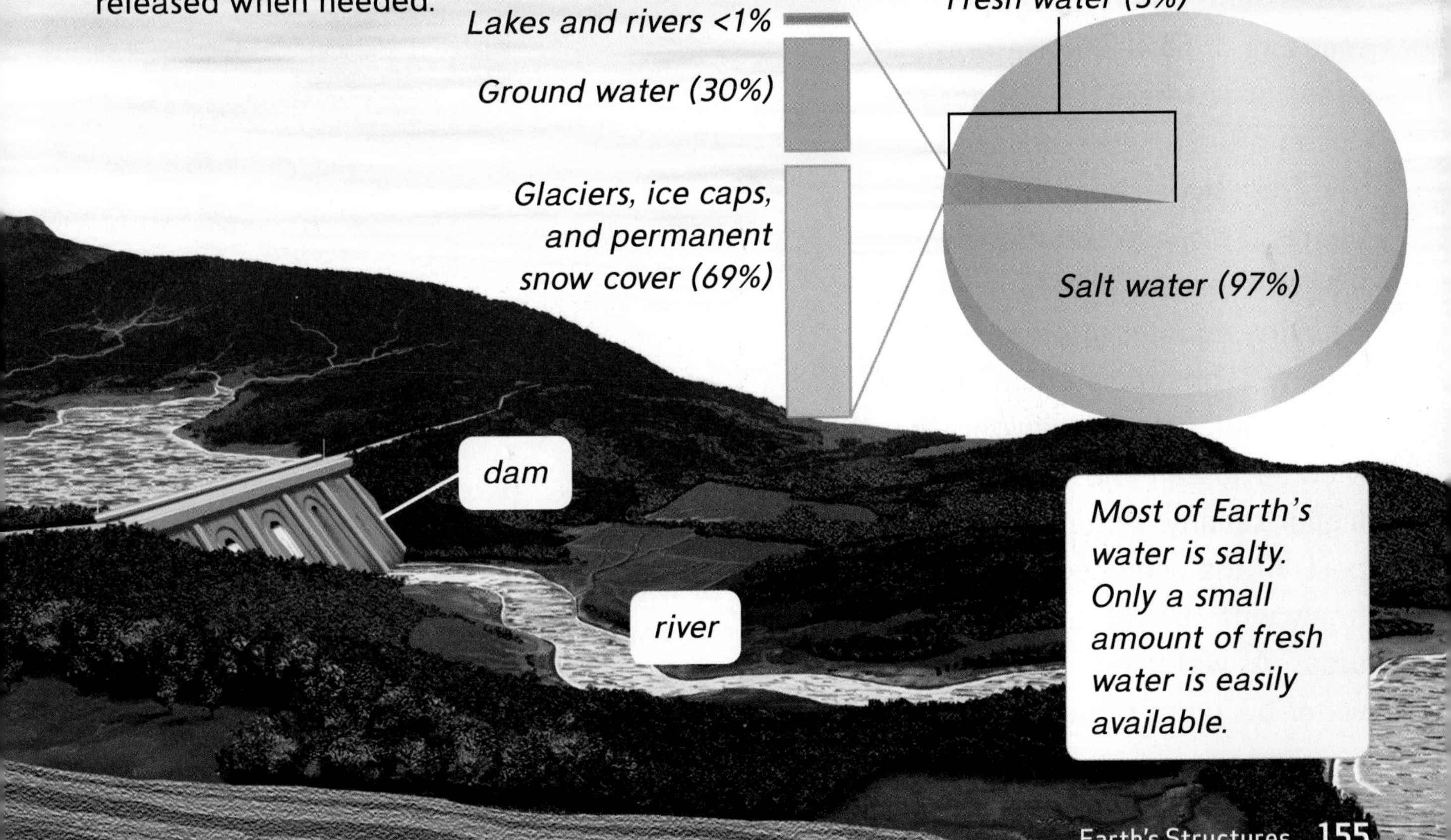

Most of Earth's water is salty. Only a small amount of fresh water is easily available.

Earth's Oceans

Covering 70 percent of Earth's surface, the oceans are the largest part of the hydrosphere. The water in the oceans is salty. This saltiness is because salt has been added to the water for billions of years. As rivers and streams flow toward the oceans, they dissolve minerals from the rocks and soil. The water carries these dissolved minerals to the ocean. At the same time, water is evaporating from the oceans, leaving the dissolved minerals behind. Sodium chloride, or salt, makes up the largest percentage of dissolved materials in ocean water.

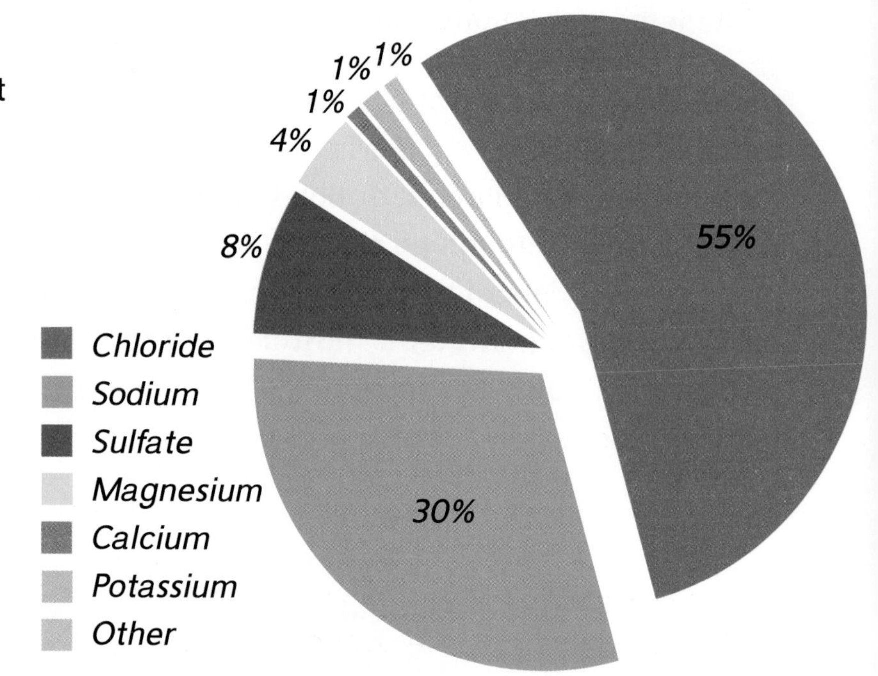

Chloride and sodium together make up the largest amount of the materials dissolved in the oceans. Together, these two materials make salt.

Some areas of the ocean are saltier than others. *Salinity* is a measure of how salty water is. Areas near where rivers enter the ocean often have lower salinity. The fresh water from the rivers dilutes the ocean water and makes it less salty. Ocean waters near the equator also have lower salinity. High precipitation in these areas dilutes the salty water. The Mediterranean and Red Seas have higher salinity. Water in these parts of the ocean does not mix freely with that of more open areas. As water evaporates, the water becomes saltier.

Salinity of water near the poles changes with the amount of sea ice. Freezing of water in this area leaves the dissolved salts behind, making the water saltier. When the ice melts, salinity decreases.

Lissa Harrison

Temperature of the Oceans

The temperature of ocean water decreases as the depth of water increases. Water in the oceans is classified into three layers according to temperature.

Surface Zone The surface zone is the warm, top layer of ocean water. It extends to about 300 meters (1,000 feet) below sea level. Sunlight heats the top third of this layer. Movement of ocean water mixes the heated water with cooler water below.

Thermocline The thermocline is a layer between about 300 m (1,000 ft) and 700 m (2,300 ft) deep. Here water temperature drops quickly.

Deep Ocean This layer of the ocean extends from the bottom of the thermocline to the ocean floor. Temperature in this zone averages 2° Celsius (36° Fahrenheit).

Ocean Currents

Ocean water moves in currents. Streamlike movements of water at or near the surface are called *surface currents.* Generally, warm water currents begin near the equator and carry warm water toward the poles. Cold water currents begin closer to the poles and carry cool water toward the equator. Surface currents affect the climate in many parts of the world. Some surface currents warm or cool coastal areas year round.

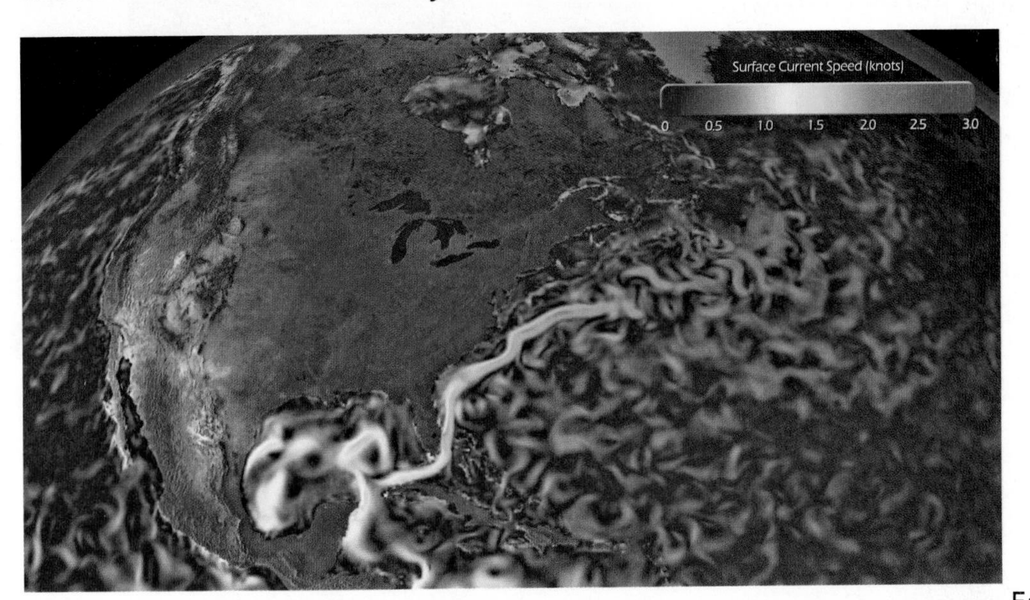

NOAA

The Gulf Stream is a surface current that carries warm water from near the equator to the North Atlantic Ocean. It warms the coastal areas of the British Isles.

Earth's Soil

There are dozens of different kinds of soils. Each has its own set of properties. Soil is made of small pieces of rocks, minerals, and humus. *Humus* is nonliving plant or animal matter. Water, air, and living things are also in soil.

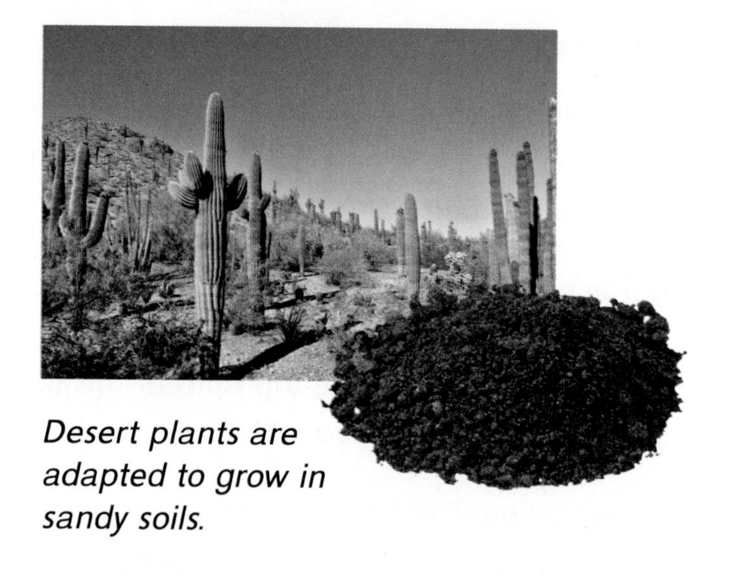
Desert plants are adapted to grow in sandy soils.

Soil Properties

Soils can be different colors. They can also have particles of different sizes.

The spaces between particles of soil are called *pore spaces.* The pore spaces in soil act like filters. They remove certain substances from water as it moves through the soil. This movement keeps the water clean. Materials with pore spaces are said to be *porous.* Porous soils hold air and water.

Medium-textured soils are good for growing many crops.

The sizes and numbers of pore spaces affect a soil's permeability. *Permeability* describes how fast water passes through a porous material. Sandy soil has high permeability. The large particles in sandy soil are packed loosely together and hold little water.

Remember, many plants need soil in which to grow. The different properties of soil result in different groups of plants being able to grow in each type of soil.

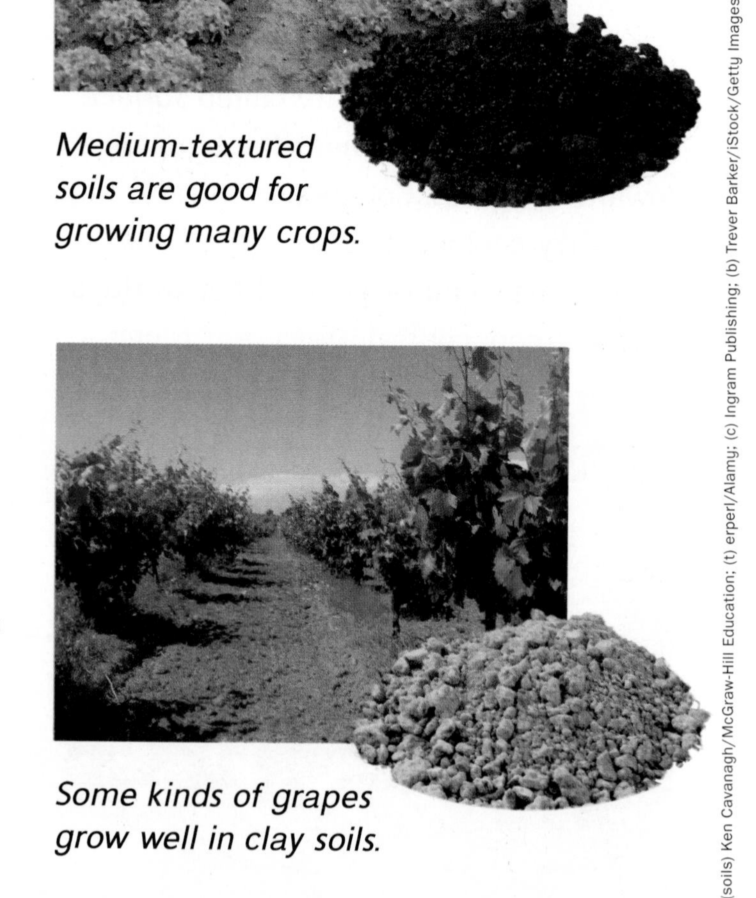
Some kinds of grapes grow well in clay soils.

How Soil Forms

Soil can take thousands of years to form. As rocks break down, they become sediment. Plants take root and break down more rock. Animals move and mix the sediment. When plants and animals die, bacteria and fungi decompose them, adding humus to the soil. Humus has new nutrients for plants to grow. In this way, living things renew the soil year after year.

Soil forms in layers called *horizons.* Each horizon has different amounts of rock and humus. A *soil profile* shows these horizons. The rock and humus that make up soil are not the same everywhere. That is why soil profiles differ from place to place.

The layer of soil at the surface is called the A horizon. It is rich in humus and minerals. Another name for the A horizon is *topsoil.* Topsoil is home to many living things.

The next layer down is the B horizon, or *subsoil.* It is often lighter in color and more compact than topsoil. It has bits of clay and minerals that trickle down from the topsoil. The roots of strong plants may grow down into this layer from the topsoil.

At the bottom of most profiles lies *bedrock,* or solid rock. The C horizon is above the bedrock and below the subsoil. It is made of broken pieces of bedrock.

Earth's Soil - Types of Soil

Each type of soil supports different plant and animal life. Most of the United States is covered by one of three types of soil: forest soil, desert soil, or grassland and prairie soil.

Forest Soil The soil in a forest has a thin layer of topsoil with little humus. Frequent rainfall carries minerals deep into the ground. Plants need long roots reach these minerals. Much of the forest soil in the United States is in the Northeast and Southeast region.

Desert Soil Desert soil is sandy and does not have much humus. However, desert soil is rich in minerals. Little rain falls to wash the minerals away. Animals can sometimes be raised in areas with desert soil. Crops can be grown only if water for the plants is piped to the area. Desert soil is found in the Southwestern region.

Grassland and Prairie Soil Grasslands and prairies are found between the Rocky Mountains and the eastern forests. Crops, such as corn, wheat, and rye, grow on land from Texas to North Dakota. The soil is rich in humus, which provides nutrients for crops, and holds water so minerals are not washed deep into the ground.

Skill Builder

Read a Map

Look at the colors in the key to find out how each soil is indicated. Then find each type of soil on the map.

Soil Types in the United States

- *desert soil*
- *grassland soil*
- *forest soil*
- *wetland soil*
- *weakly developed soil*

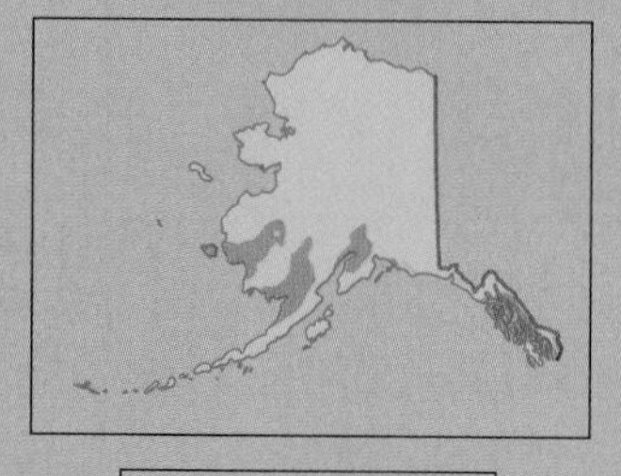

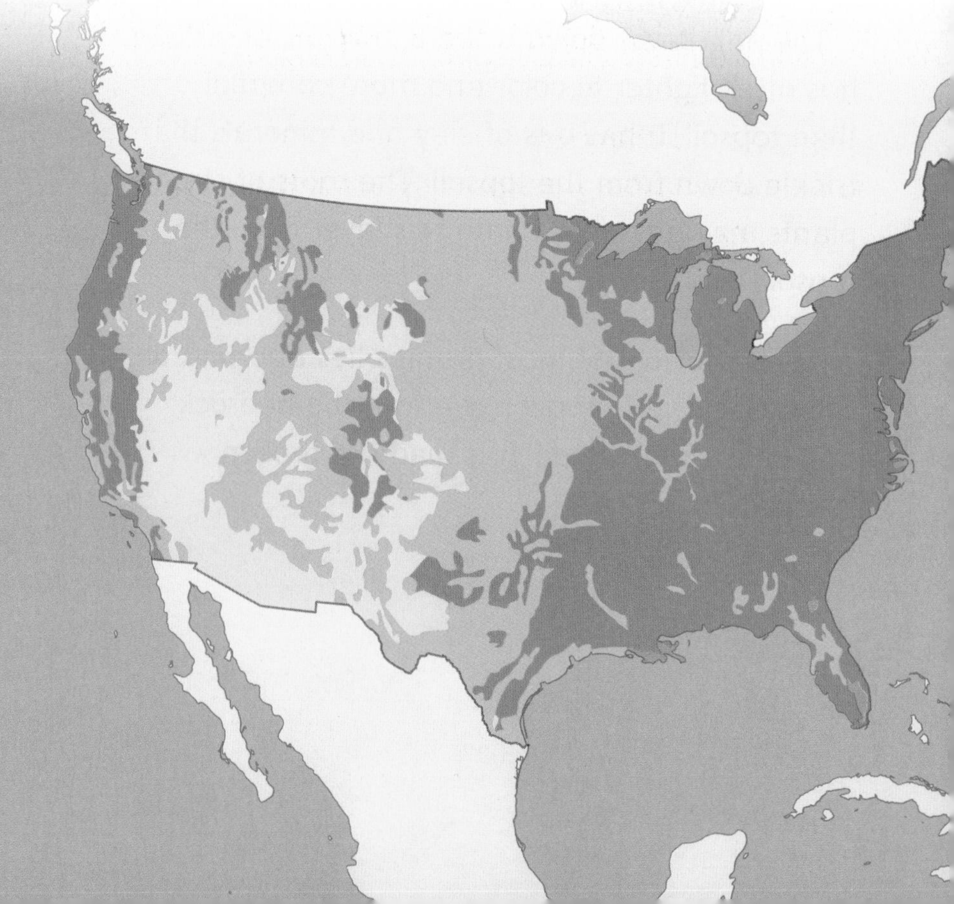

Making Soil More Fertile

Soil is a resource that can be used up, wasted, or spoiled. Soil can be washed away by flowing water or blown away by wind. Plant roots hold soil in place. If plants are removed, more soil may wash or blow away. Less soil may change the types of plants that can grow in an area. It can also make it difficult for any plants to grow.

The nutrients in soil are naturally removed by plants. The plants use the nutrients to grow. The nutrients are normally replaced when plants die and decompose. However, crops grown for food are not allowed to decompose on the land. Most parts of crop plants are removed when farmers harvest their fields. There are no plants left behind to decompose. Nutrients do not return to the soil. The land becomes less able to support the growth of new crops.

Farmers can return nutrients to the soil using chemicals called *fertilizers*. Farmers match the type of fertilizer to the crop they wish to grow. Matching the fertilizer ensures that the plants have the right nutrients to grow.

Growing the same crop year after year can also remove nutrients from the soil. To slow this process down, crops can be changed from year to year. This practice, called *crop rotation*, helps keep soil nutrient-rich.

What Could I Be?

Soil Scientist

Soil scientists do a lot more than dig in the dirt. They determine the properties of soil. They look for ways to make soil more fertile. They even engineer ways to keep soil from eroding. Learn more about becoming a soil scientist by turning to the Careers section.

Keeping soil nutrient-rich is important for growing crops.

Earth's Changing Surface

Forces within Earth build up the crust into mountains, plateaus, and other landforms. At the same time, processes such as weathering break down the crust. **Weathering** is the slow process that breaks down rocks into smaller pieces.

Physical Weathering

Rocks can change size and shape without changing their chemical properties. This process is *physical weathering.* Ice, wind, water, gravity, plants, and animals cause physical weathering.

Animal Actions Many animals, including moles, gophers, prairie dogs, worms, and ants, burrow in the ground. These animals can loosen and break apart rocks in the soil as they move.

Plant Growth Plant roots enter cracks in rocks. As roots grow, they force the cracks apart. Over time, the roots of even small plants can pry apart cracked rocks.

The actions of living things such as the burrowing of animals or the growth of plant roots can cause weathering.

Freezing and Thawing Water can seep into cracks in a rock during warm weather. When the water freezes during cold weather, the water expands and makes the crack bigger. This process, called *frost wedging,* also widens cracks in sidewalks and causes potholes in streets.

Abrasion In nature, the action of rocks and sediments grinding against each other and wearing away surfaces is called *abrasion.* Abrasion can happen in many ways. When rocks and pebbles roll along the bottom of a swiftly flowing river, they bump into and scrape against each other. This rubbing and scraping knocks the sharp edges off the rock. Often river rocks are smooth and round from being tumbled in moving water for many years.

Wind also causes abrasion. When wind blows sand against exposed rock, the sand wears away the rock's surface the way sandpaper scratches wood.

Abrasion also occurs when rocks fall and tumble against one another. During a rock slide, gravity pulls a mass of rocks down a hill. The rocks crash into one another as they fall. The force of the rocks hitting each other knocks pieces off the rock. Anytime one rock hits another, abrasion takes place.

The crashing of rocks together in a rock slide causes abrasion, a type of physical weathering.

Design Pics/Jack Goldfarb

Erosion and Deposition

Erosion is the process of weathered rock moving from one place to another. The process of eroded materials being dropped off in another place is **deposition.** Erosion and deposition work together to change the shape of the land.

Erosion and Deposition by Gravity

Gravity is a powerful agent, or cause, of erosion. Gravity causes *mass movement,* the movement of sediment downhill. Mass movements can be fast or slow. Mudslides, landslides, and rockslides are all fast movements of a large amount of material down a slope. All of these events may occur after an earthquake or a volcanic eruption causes the ground to move. This movement loosens the sediments. Gravity then pulls the materials downhill. A slow mass movement is *creep*. It occurs over many years as particles of soil slowly move downhill.

Erosion and Deposition by Running Water

As water runs downhill, it can wash away soil and erode rock. The water, soil, and rocks will eventually flow into a larger body of water, such as a river. Rivers with fast-moving water tend to follow straight paths and have deeper channels and steeper banks. Fast-moving water has more energy. It can wash away a larger amount of heavier sediments. Rivers with slow-moving water tend to follow curved paths and have shallow channels and low banks. Slow-moving water has less energy. It carries smaller particles of sediment.

The looping curves in this river are called meanders. Slow-moving water deposits sediments on the inside of a meander. Faster-moving water erodes sediments on the outside of meanders.

Alan Morgan

Rivers eventually flow into a larger body of water, such as a lake or an ocean. Since the water is no longer flowing downhill, it slows down. The sediment carried by the water is deposited on the bottom of the lake. Over time, this sediment builds up into a landform called a *delta*.

Rivers also deposit sediment when they flow out of a steep, narrow canyon. Here, the stream becomes wider and shallower. The water slows down as it spreads out. Sediments are deposited in an alluvial fan.

When water enters a still body of water, it slows and drops its sediment.

Erosion and Deposition by Wind

Wind can move sand from one place to another. When wind blows over the land, it picks up the small particles of sediment. The stronger the wind blows, the larger the particles it can pick up.

All the sediment eroded by wind eventually is deposited. Deposition occurs when the wind slows or when a clump of grass or a rock traps the sediment. A *sand dune* is a deposit of wind-blown sand. Sand dunes come in many shapes and sizes. Dunes move over time. Little by little, the sand shifts with the wind from one side of the dune to the other. Sometimes plants begin growing on a dune. Plant roots can help keep a dune from moving.

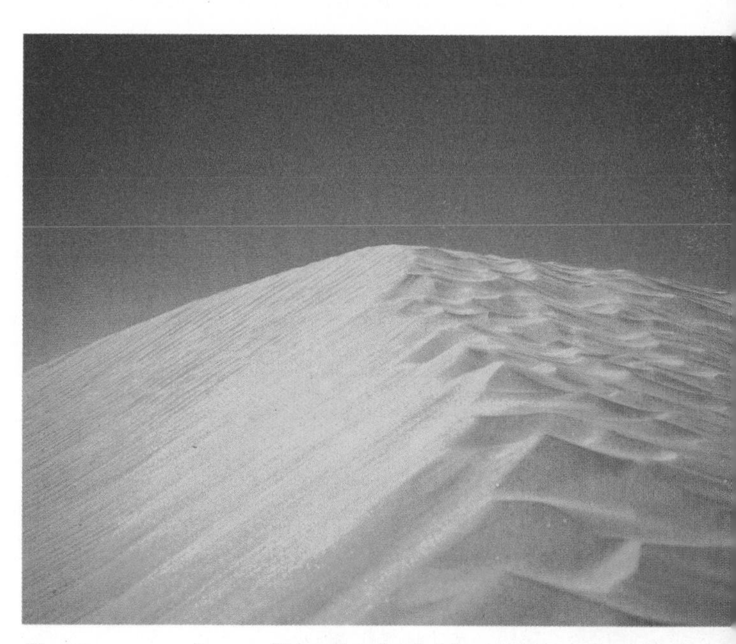

Dunes are deposits of wind-blown sand. As the wind continues to blow, sand moves up and over the top of the dune. This sand moves the dune forward.

Erosion and Deposition - Erosion by Glaciers

In very cold places, thick sheets of ice called *glaciers* creep over the land. Over one million years ago, glaciers began to cover much of Earth. Few places are cold enough for glaciers today.

Glaciers form where snow collects quickly and melts slowly. Year after year, the snow builds higher. The weight on top of the mound puts pressure on the snow below. The snow in the lower layers of the glacier slowly turns to ice. Near the ground, the pressure from the snow above causes some of the ice to melt.

As the weight of the ice increases, the glacier begins to flow. The bottom and sides of the glacier freeze onto rocks. As the glacier continues to move, it tears rock from the ground. It scratches, flattens, breaks, or carries away things in its path. A glacier can make a valley wider and steeper. It can even carve grooves into solid rock.

Skill Builder

Read a Diagram

The dark lines in the center of the glacier are eroded sediments. Trace these lines to the point where the sediment is deposited.

A Glacier Deposits Land

Deposition by Glaciers

Glaciers not only erode the land, they deposit the eroded rock. As glaciers melt, they leave behind the rocks they carried. The leftover rocks are called *glacial debris*.

Glacial debris can be made of large boulders or small particles. It can have bits of gravel, sand, and clay. This mix of material is called *glacial till*. The glacier drops most of this debris at its downhill end, or terminus.

Materials that glaciers pick up or push can be deposited along its edges and front. These long ridge-like mounds of broken rock are called moraines. Not all glaciers leave moraines behind. Sometimes the melted water from the glacier washes this material away. Today, glacial till is scattered across Canada and the northern part of the United States.

Glaciers can erode solid rock! These grooves were formed as a glacier moved across the land.

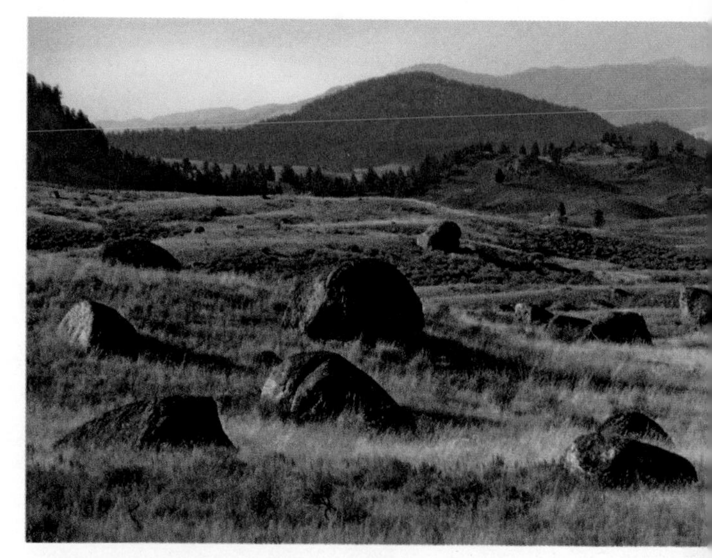

These large boulders were dropped by a glacier when it melted.

terminus

glacial till

Erosion and Deposition - Shorelines

Wind, water, and waves combine to erode and deposit sediments along shorelines. A *shoreline* is where land and a body of water meet.

A lot of energy is released when waves hit a beach. A large wave can break solid rock or throw broken rocks back against the shore. The rushing water in breaking waves can easily wash into cracks in rock, helping to break off large boulders or fine grains of sand and sediment. The loose sand picked up by the waves polishes and wears down coastal rocks. Waves can also move sand and rocks and deposit them in other locations, forming beaches.

A *beach* is any area of shoreline made of material deposited by waves. Some beach material is deposited by rivers. Other beach material is eroded from areas near the shoreline. The movement of sand along a beach depends on the angle at which the waves strike the shore. Waves often hit beaches at an angle or curve. As they strike the shore, the waves pick up sand and rocks and move these materials farther down the beach.

When waves reach a headland, or an area of land that has water on three sides, they curve around it and wash away at the sides of the headland. As the waves continue to erode the sides of the headland, an arch forms.

Sediments are also deposited along shorelines away from the beach. Waves moving at an angle to the shoreline push water along the shore. A *longshore current* is a movement of water near and parallel to the shoreline. Sometimes waves erode material from the shoreline, and a longshore current transports and deposits it offshore. A *sandbar* is an underwater or exposed ridge of sand or gravel.

Barrier islands are large sandbars that stretch for long distances along a shoreline. Barrier islands protect the coastline from erosion caused by large waves during storms. The waves hit the barrier islands first and erode them rather than the coastline beaches. After severe storms, barrier islands may be so completely eroded that they no longer appear above the water. Without the barrier islands, the coastline's erosion will be worse during the next storm.

Some coastal areas have one or more sets of dunes running along the shoreline directly inland from the beach. As in deserts, coastal dunes form when wind erodes sand and deposits it in another area. As the wind blows, it picks up sand from the dunes closest to the water and blows it farther inland. Dunes protect inland areas from the large waves that occur during storms. Dunes also shelter inland areas from the wind. If severe winds or waves occur, dunes may be completely eroded.

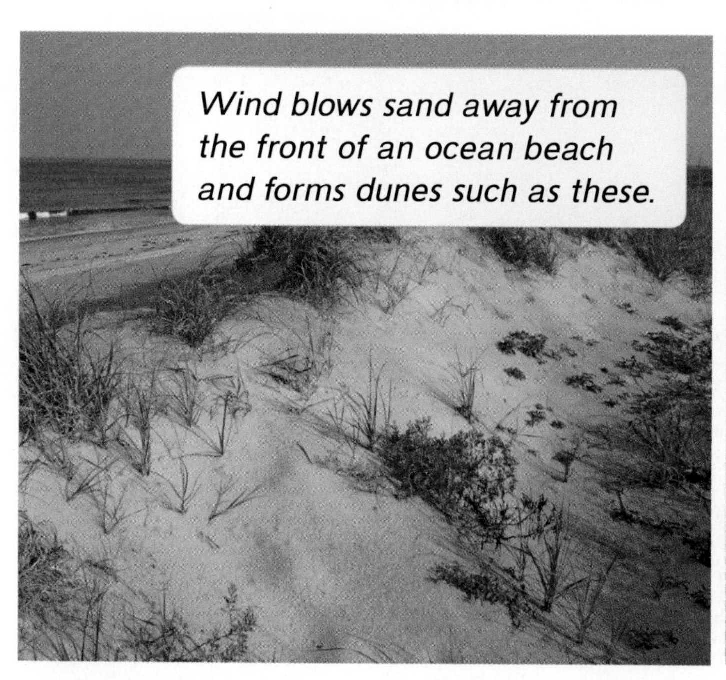
Wind blows sand away from the front of an ocean beach and forms dunes such as these.

Sandbars form where longshore currents flow parallel to the beach.

Plate Tectonics

Continental Drift

A *geologist* is a scientist who examines rocks to find out about Earth's history and structure. About 100 years ago, Alfred Wegener, a German geologist, noticed something curious about Earth's continents. Wegener noticed that the continents looked like pieces of a jigsaw puzzle.

Wegener thought that millions of years ago Earth's continents were joined together like a completed jigsaw puzzle. Then, as time passed, a force pulled the puzzle pieces apart. The continents slowly moved to the positions they are in today. Wegener's idea became known as the *theory of continental drift*.

Wegener concluded that all the continents had once been part of a "supercontinent." He called this landmass *Pangaea*. In Wegener's time, many people rejected the idea of continental drift. Scientists then did not know of any forces that could move continents. Over time, however, scientists found evidence that supported continental drift. One piece of evidence was that the mountains on the east coast of South America had the same types of rocks as the mountains on the west coast of Africa. These rocks were also the same age.

Theory of Continental Drift

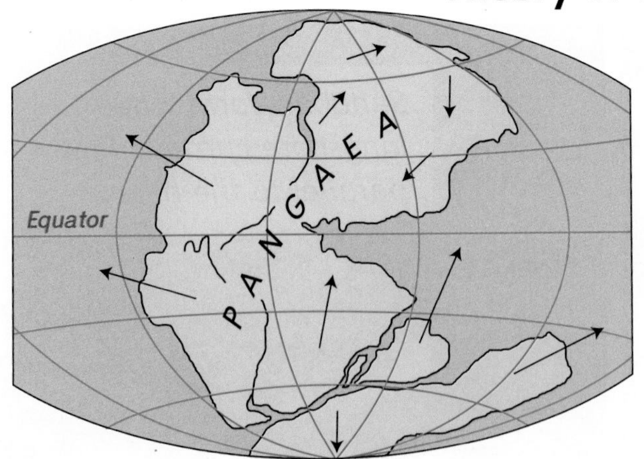

225 million years ago

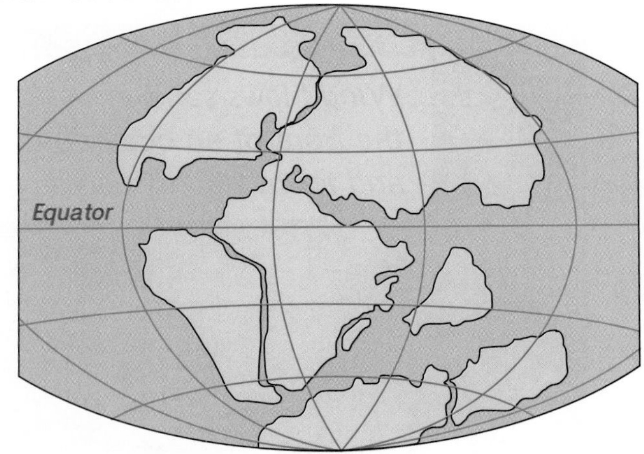

200 million years ago

Wegener was the first to suggest that the continents are constantly moving.

Additional evidence that the continents had moved came from plant and animal fossils. A fossil of a freshwater reptile called *Mesosaurus* has been found in very old rocks in southern South America and Africa. *Mesosaurus* could not have swum through the salt water in the Atlantic Ocean. However, if the continents had once been joined, this reptile could have traveled across the larger continent through freshwater rivers.

a *Mesosaurus* fossil

Evidence also came from plant fossils. *Glossoperis,* an ancient plant species, has been found in a band that stretches across the continents of South America and Africa.

New evidence supporting Wegener was discovered in the 1950s, when scientists mapping the floor of the Atlantic Ocean made an amazing discovery. They found that, in the middle of the Atlantic, there was an underwater mountain chain. By the 1960s, similar structures had been discovered in other oceans. On both sides of the mountain ranges, further study showed that the sea floor was spreading apart over time.

Using this and other evidence, scientists eventually concluded that South America and Africa *were* once joined. Over millions of years, they have drifted apart. The water that filled the space between them formed the Atlantic Ocean.

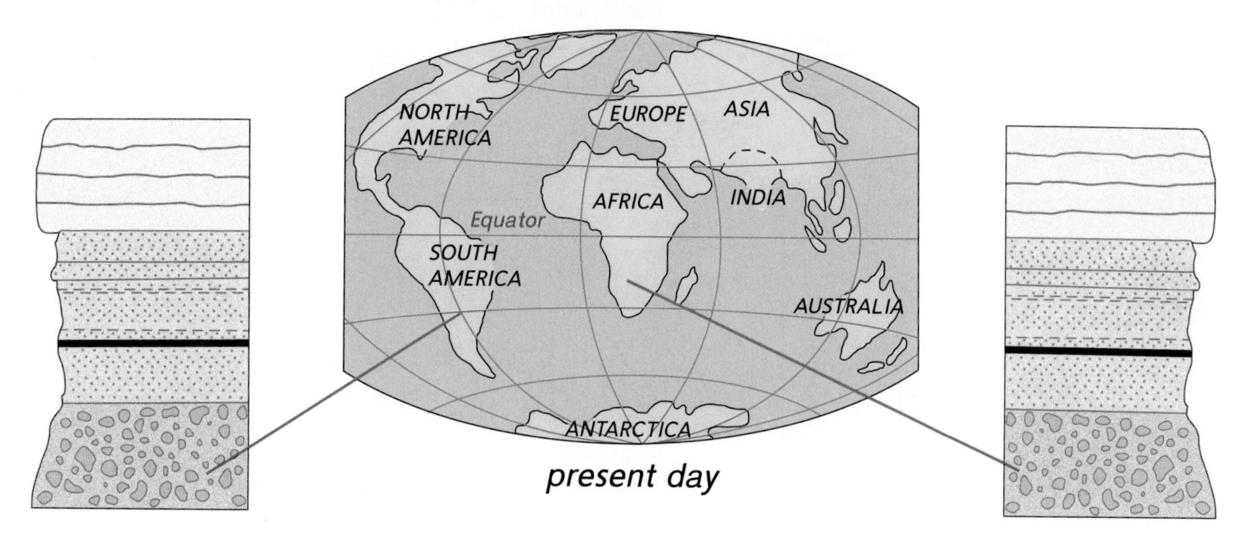

present day

Plate Tectonics - Moving Plates

The ocean floor between South American and Africa is spreading at a rate of about 4 centimeters (1.5 inches) every year. If this rate of movement has happened for the past 130 million years, South America and Africa should be 520,000,000 cm (5,200 km or 3,230 mi) apart. This distance is about the width of the southern Atlantic Ocean. The ocean has filled up the space between the two continents.

Scientists have developed a theory called **plate tectonics** to explain how forces deep within Earth can cause ocean floors to spread and continents to move. This theory describes the lithosphere as being made of huge plates of solid rock. Earth's continents rest on these plates. The almost melted rock in Earth's mantle acts as a slippery surface on which the plates can move.

Uneven heating in the mantle produces slow-moving currents of plastic-like, fluid rock. The cooler, rigid rock of the crust and upper mantle rests on top of this fluid rock. The slow movements in the fluid part of the mantle drag the plates sideways.

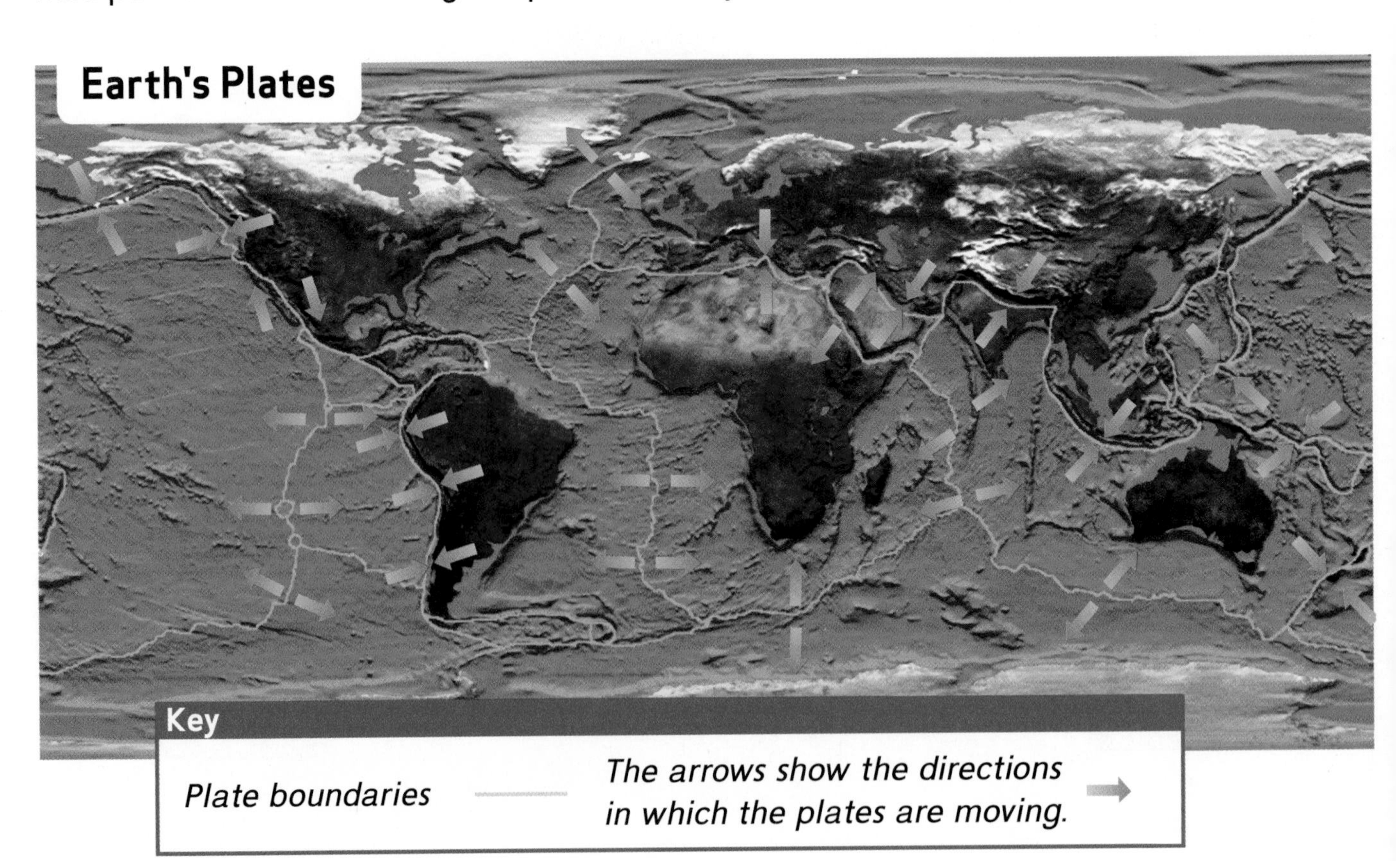

Seafloor Spreading

In the middle of the ocean where the plates are moving apart, magma rises from the mantle toward the surface. The upward movement of magma causes tension, or a stretch or a push, on the plates. This tension moves the ocean floor apart and separates the plates on either side of the mid-ocean ridge. Since the continents rest on these plates, they also move apart.

In Iceland, a divergent boundary is forming new land.

As the hot rock reaches the surface, it cools and builds up on both sides of the opening. The cooling rock forms the mid-ocean ridge and the rift valley along its top. The mid-ocean ridge runs roughly parallel to the separating continents. The ridge remains stationary while the ocean floor on both sides of the ridge grows wider. As the ocean floor grows wider, the continents move farther apart. This is called a *divergent boundary,* a location where plates move apart.

Divergent boundaries also can occur on land. Examples of this can be seen in the Rift Valley of Africa and in Iceland, which is located at the northern end of the mid-ocean ridge in the Atlantic Ocean.

Spread of the Ocean Floor

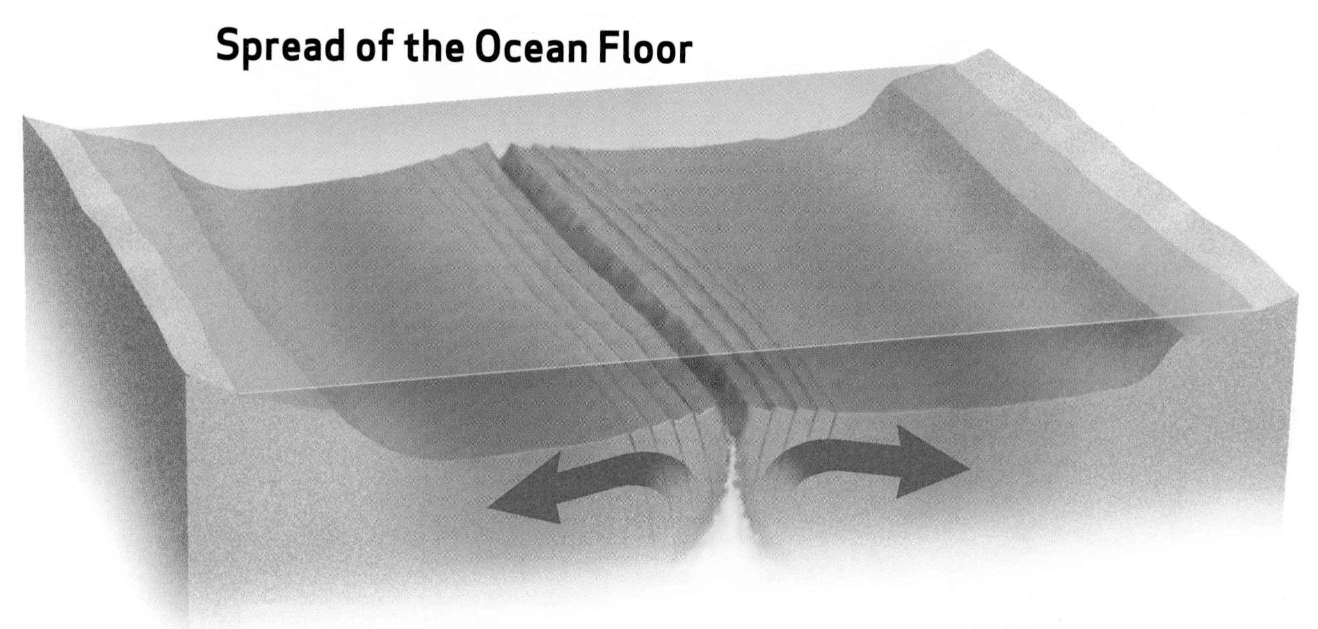

Plate Tectonics - Mountain Building

Tension, or forces that pull things apart, moves Earth's plates. Plates also can be moved by pushing forces. When plates are pushed together, the force that occurs is called compression. *Compression* is a squeezing or a pushing together of the crust. When a plate is compressed, the crust is forced upward, producing *folded mountains*.

The Himalayas are folded mountains. They began forming millions of years ago as India and Asia collided. As the plates continue to push into each other, the Himalayas grow about 5 millimeters (0.2 in.) taller every year.

When one plate rubs past another plate, this movement causes *shear*, or a force that twists, tears, or pushes one part of the crust past another. In some places, there are deep cracks in Earth's crust called **faults**. Shear can cause Earth's crust to break apart along a fault. When this happens, one side of the fault moves up and the other side moves down. This movement produces *fault-block mountains.* The Sierra Nevada in the western United States are fault-block mountains.

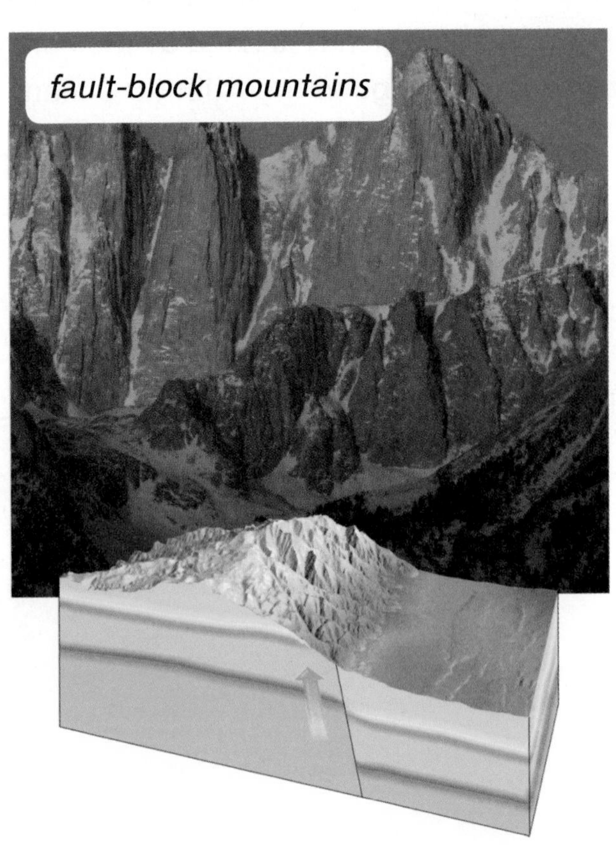

Island Formation

The Hawaiian Islands are an island chain, or a line of volcanic mountains, in the Pacific Ocean. The Hawaiian Islands rest on a slowly moving tectonic plate. As it moves, the plate passes over a stationary plume of magma called a hot spot.

Over millions of years, lava erupting from the hot spot formed a mountain. Eventually, the mountain grew taller than the ocean's surface and became a volcanic island. As the plate moved, that island moved away from the hot spot, and a new island began to form.

Skill Builder

Read a Diagram
Follow the arrows to determine which is the youngest and which is the oldest Hawaiian Island.

In areas where an ocean-floor plate is pushed under another ocean-floor plate, an island arc forms. As the plate is pushed down, it melts. Magma forms, rises upward, and erupts through the ocean floor. These eruptions form a series of volcanic islands along the plate boundary. The Aleutian Islands in Alaska form an island arc.

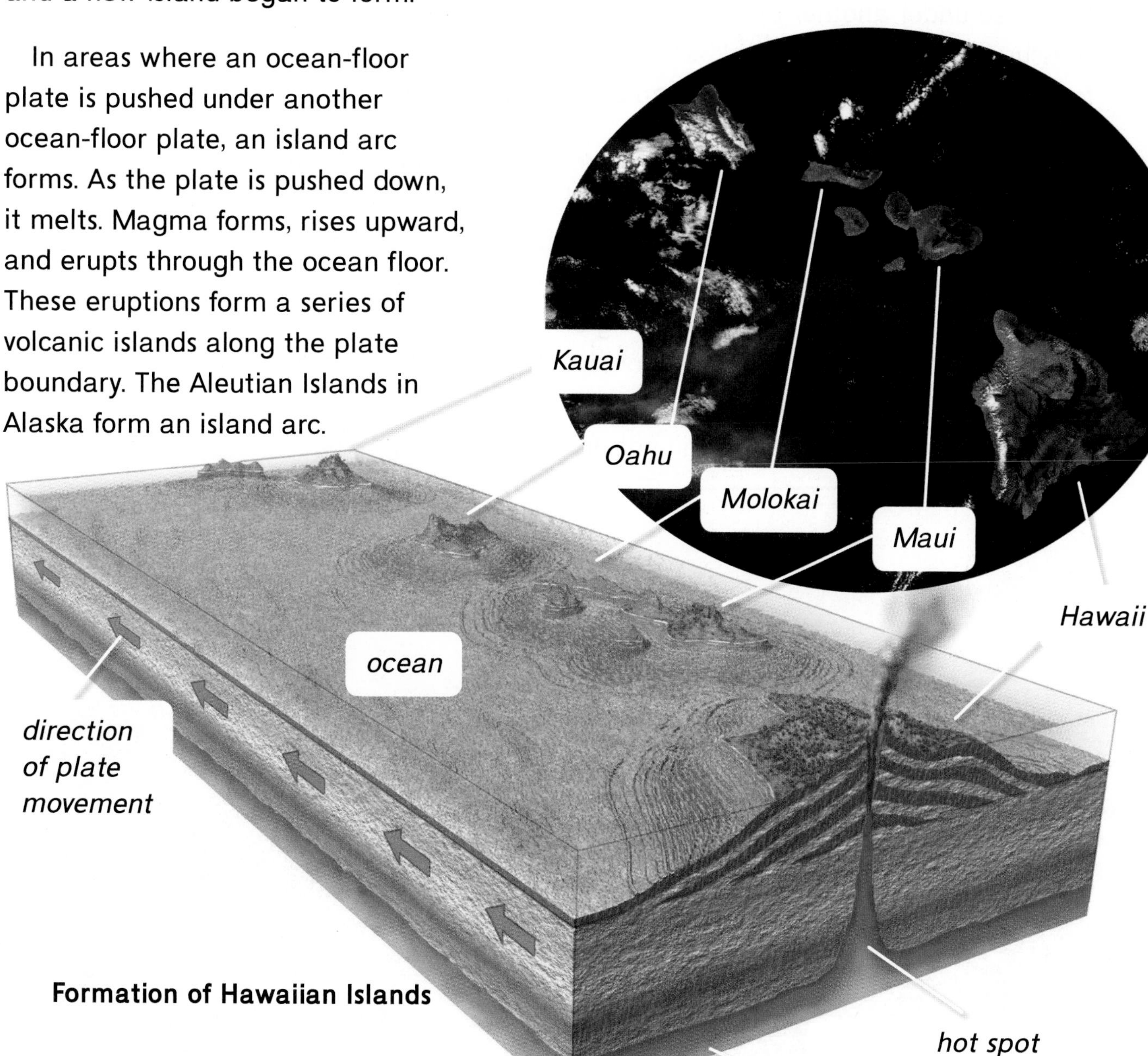

Formation of Hawaiian Islands

Volcanoes

Volcanoes form on land and on the ocean floor. A **volcano** is an opening in Earth's crust. Volcanoes are located only at certain places on Earth's surface. Earth's crust is broken up into a number of moving tectonic plates. Most volcanoes are located where these plates meet.

However, volcanoes do not erupt at all plate boundaries. After collecting data about the directions in which plates moved, scientists concluded that volcanoes tend to erupt where one plate is pushed under another plate. Plates melt under great heat and pressure as they are pushed down into the mantle. The melting forms magma, which pools in a chamber underneath the crust.

The magma may rest quietly for hundreds or thousands of years. Sometimes a crack forms above the chamber, or the pressure in the chamber grows too great to be held in by the rock above it. Then the magma rushes up toward Earth's surface. An *active volcano* is one that is currently erupting or has recently erupted. A volcano that has not erupted for some time, but that scientists think may erupt in the future, is called a *dormant volcano.* A volcano that scientists think will not erupt again is an *extinct volcano.*

an active volcano

J.D. Griggs/U.S. Geological Survey

How Volcanoes Build Land

All volcanoes have at least one vent, or opening. Once magma reaches the surface, it is called *lava*. When lava comes out of a vent, it is liquid. Lava forms solid rock as it hardens. Over thousand of years, new lava can increase the height of a volcano and form a volcanic mountain. This height is new land the volcano has built.

Active volcanoes differ in how they form and in the shapes of the mountains they build.

Shield Volcanoes Volcanoes built by thin, fluid liquid that spreads over a large area are called *shield volcanoes*. These mountains have a broad base and gently sloping sides. This type of volcano formed the Hawaiian Islands.

Cinder-cone Volcanoes Thick lava that is thrown high into the air and falls as chunks or cinders builds *cinder-cone volcanoes*. These mountains form as a cone shape with a narrow base and steep sides. They have a single bowl-shaped opening at their summit and usually grow to only about 300 m (1,000 ft).

Composite Volcanoes Layers of ash and cinders sandwiched between layers of hardened lava produce *composite volcanoes*. These volcanoes often have several vents and are generally cone shaped with steep sides. Composite volcanoes sometimes reach heights of 3,000 m (10,000 ft).

Types of Volcanoes

shield volcano

cinder-cone volcano

composite volcano

Earthquakes

An **earthquake** is a sudden movement of Earth's crust. It begins below ground and is caused by plates moving along a fault. Usually, rocks on both sides of a fault are stuck together. As plates grind against each other, energy builds up in the rock along the fault. Rocks may store this energy for many years. Then suddenly, this energy is released. The rocks move along the fault, and the crust shakes. Earthquakes are common in places with active faults, such as parts of California, Alaska, and Canada. Many smaller earthquakes, called aftershocks, can follow a major earthquake. Aftershocks are not as strong as the main earthquake. They can continue for days, weeks, or months after the first earthquake.

The sudden movement of an earthquake causes rock to vibrate. A vibration that travels through Earth and is produced by an earthquake is called a **seismic wave**. Seismic waves spread out in all directions from the *focus,* the point below ground where the earthquake began. The point on the surface directly above the focus is called the *epicenter*.

Earthquakes can also occur away from plate boundaries. Here, the condition of rocks and soil may cause movement and shifting that can produce earthquakes.

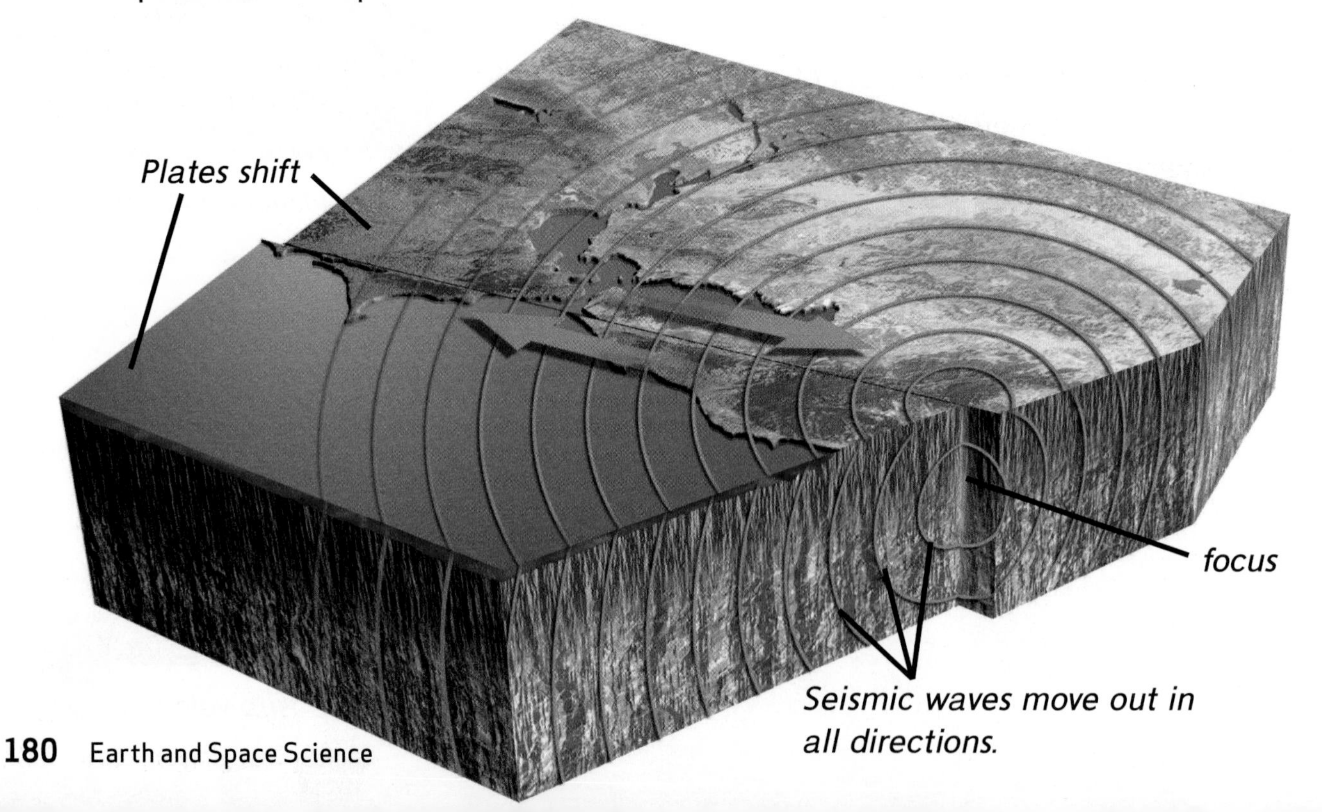

Earthquake Safety

Most earthquakes are too weak to notice. Others can cause extreme damage. During a major earthquake, buildings and roads may break apart, and bridges may collapse. You can stay safe during an earthquake by following a few simple rules. If you are indoors, duck under a table or doorway. Keep away from walls and windows. If you are outdoors, stay away from trees, power lines, and any large structures that might fall down.

People cannot stop an earthquake from occurring; however, they can take steps to reduce the damage it causes. Damage to buildings is reduced by placing layers of rubber and steel between a building and its foundation. These layers cushion up and down motion but still allow the building to move from side to side as the ground moves. Some of these motion dampeners, made of springs, ball bearings, and padded cylinders, act like the shock absorbers in a car.

Tsunamis

Sometimes earthquakes occur below the ocean. These earthquakes can generate a giant ocean wave called a **tsunami**. Scientists who detect the earthquake send out warnings. People warned of a possible tsunami should seek shelter on high ground away from shorelines.

What Could I Be?

Structural Engineer

Earthquakes can cause great damage. Structural engineers work to find ways to make buildings more earthquake-proof. Learn more about what structural engineers do by turning to the Careers section.

USGS photo by Walter D. Mooney

Where Earthquakes and Volcanoes Occur

Earthquakes happen along plate boundaries because pressure from the movement of the plates pushes on nearby faults. The map on this page shows the locations of earthquakes, volcanoes, and the plate boundaries.

Scientists have a good idea where most earthquakes are likely to happen. However, it is much more difficult for them to know exactly when an earthquake will happen. Scientists use instruments to continually monitor vibrations in Earth's crust. A change in the frequency or intensity of vibrations may signal that an earthquake will soon occur.

Scientists measure seismic waves with an instrument called a seismometer. A seismometer detects and records vibrations in Earth's crust. The device shows the waves as curvy lines on a diagram called a **seismograph**. The stronger the quake, the steeper the lines. Today, anyone can gather real-time data by using the Internet, which gives scientists and the public a continuous, global view of Earth's seismic activity.

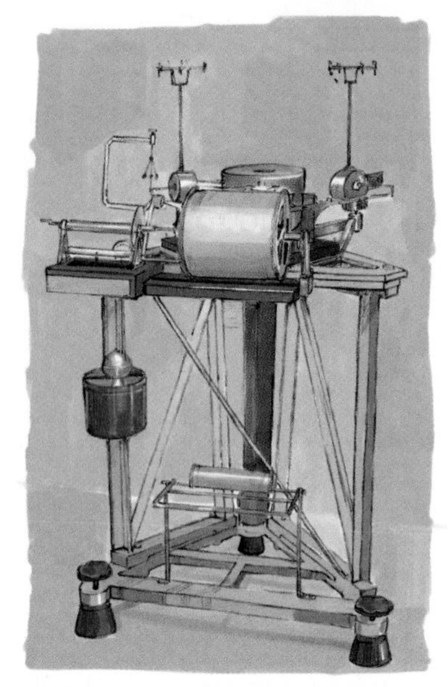

This early seismograph helped scientists track seismic activity more accurately.

Skill Builder

Read a Diagram

Examine the key to find out what each color and symbol on the map indicates.

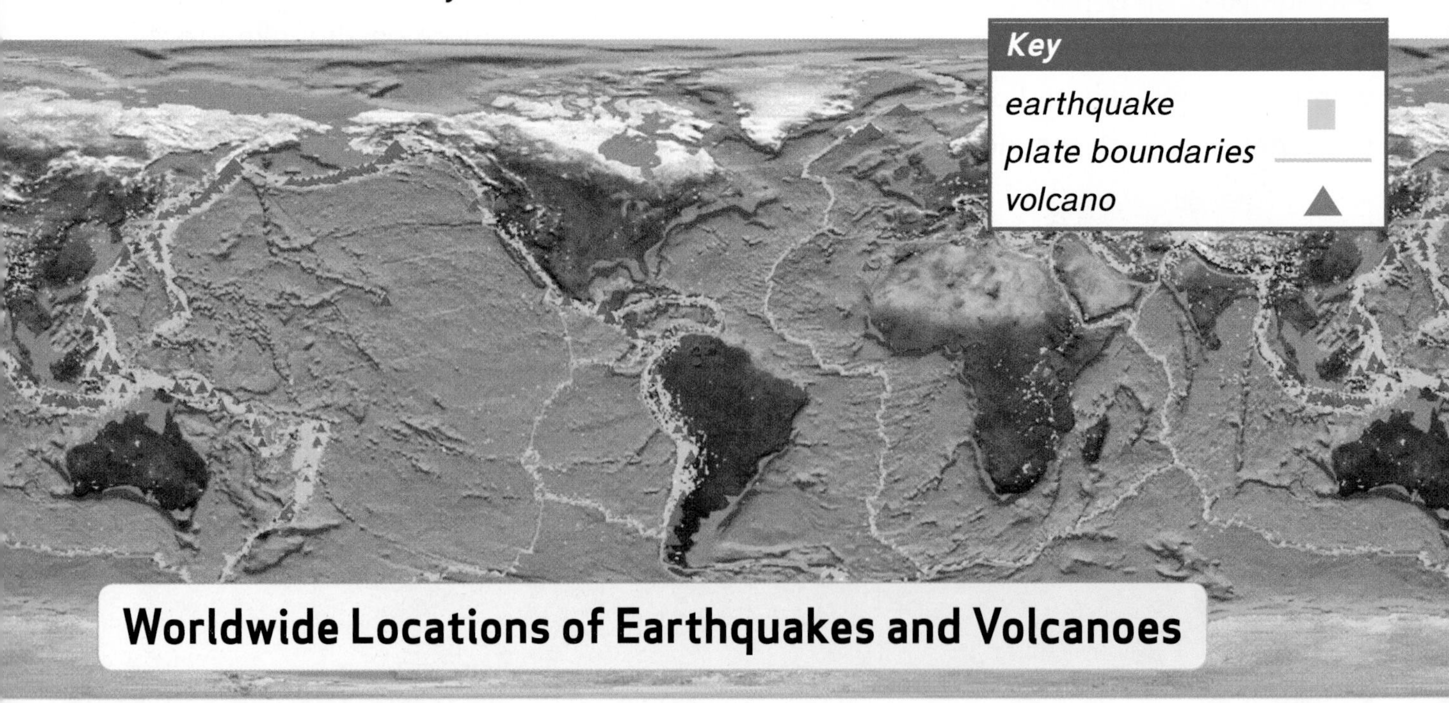

Worldwide Locations of Earthquakes and Volcanoes

The Ring of Fire

Volcanoes are more likely to erupt at plate boundaries than anywhere else on Earth. Volcanoes can occur where two plates pull apart. Here, plate movements cause the crust to fracture. The fractures in the crust allow magma to reach the surface.

Volcanoes can also occur where two plates push together. As the plates push together, one plate can sink beneath the other plate. This process is called *subduction.* The plate that is pushed down into the mantle is put under great heat and pressure, causing it to melt. Water that is brought down with the sinking plate eventually helps form magma, which rises to the surface.

About 90 percent of Earth's volcanic eruptions happen around the Ring of Fire. The *Ring of Fire* is a circle of volcanoes in the Pacific Ocean. The Ring of Fire follows the boundaries of plates that meet there. A circle of volcanoes around plate boundaries, such as the Ring of Fire, is called a *volcanic arc.* Volcanic arcs are associated with both earthquakes and volcanic eruptions.

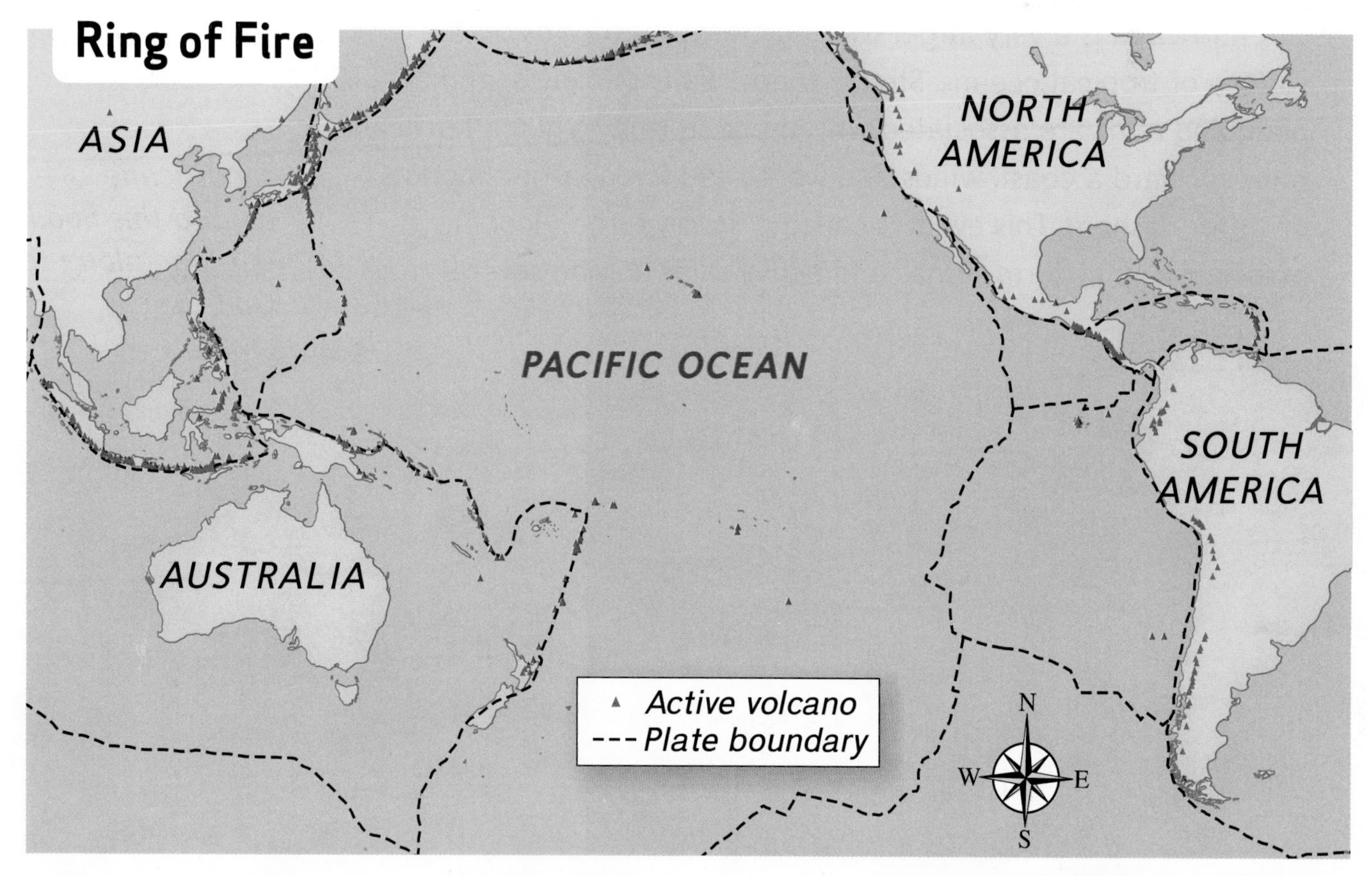

Floods

Water runs over the ground in streams and rivers. Sometimes, water enters a river faster than the river can carry it away. When water collects on land that is normally dry, it is called a **flood.** Floods occur when water from a body of water overflows banks or beaches. A flood may also occur during a heavy rainfall. Natural wetlands can soak up water and reduce the chances of a flood. Draining wetlands or cutting down plants along a riverbank may make floods more likely.

Floodwaters carry and deposit sediments over the land. A **floodplain** is a place that is easily flooded when river water rises. Floods can erode the shoreline of a body of water and change its shape or course. Floods also can cause damage by carrying mud into homes and streets. However, floods can also have a positive effect on natural systems. After a flood, new soil is deposited on the land. The nutrients in this soil help plants grow.

Hurricanes and Flooding

A hurricane is a very large, swirling storm that forms on the surface of tropical oceans. Strong winds, walls of clouds, and pounding rains are associated with these storms. When a hurricane moves toward a coast, winds and waves can force large amounts of water onshore. This event is called a *storm surge.* Flooding associated with storm surges and heavy rains can be severe.

A hurricane caused this flood damage along the Gulf Coast.

NOAA/Department of Commerce

Landslides

When rocks and soil on a slope are loosened, gravity pulls them downward. A **landslide** is the sudden movement of a large amount of rocks and soil down a slope. Sometimes earthquake vibrations cause a landslide. Volcanic eruptions and heavy rains can also cause landslides. Human activity, such as clearing the land of trees in hilly areas, can make landslides more likely
to occur.

When it rains, some water soaks into the ground. At some point, the ground can no longer absorb water. Then the remaining water mixes with the soil and forms mud. Eventually, the mud contains so much water that it becomes very heavy and cannot stay on the slope. If a lot of mud flows down the slope, it can knock down trees and destroy whatever is in its path. This event is a *mudslide,* and it can cause rapid erosion of rocks and soil.

A *lahar* is similar to a mudslide. In a lahar, a mix of mud, volcanic ash, and rock rush down the sloping side of a volcano. This kind of event is triggered by a volcanic eruption.

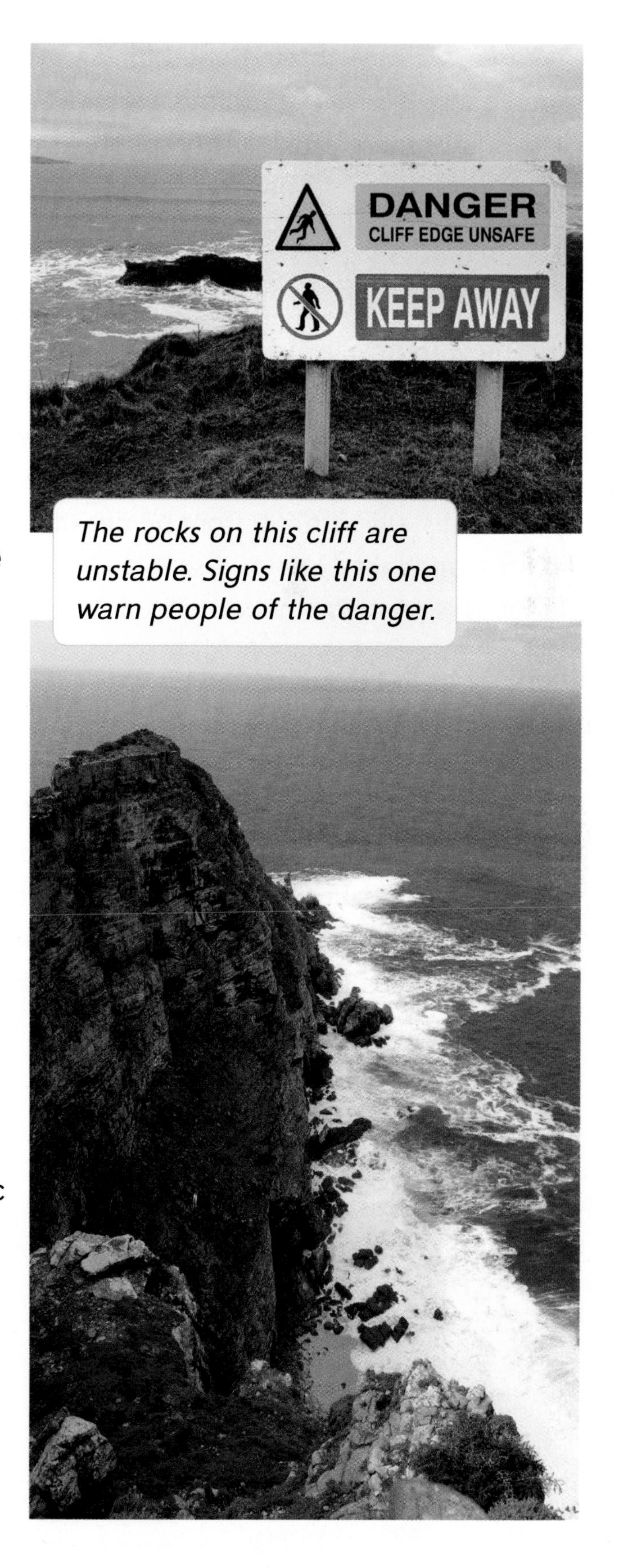

The rocks on this cliff are unstable. Signs like this one warn people of the danger.

(t) Difydave/iStock/Getty Images; (b) Chris Willig

The Water Cycle

Water is always moving and being recycled through the hydrosphere. The **water cycle** is the movement of water between Earth's geosphere and its atmosphere. Evaporation, condensation, and precipitation move water through the cycle.

In the Air

In the water cycle, water changes among liquid, gas, and solid states. The Sun is the energy source for this cycle. The Sun's energy causes water to evaporate. When water **evaporates,** it changes from a liquid to a gas in the form of water vapor. Water is constantly evaporating from bodies of water such as oceans, streams, and lakes. Water also evaporates from the leaves of plants. This process is called *transpiration.*

As it rises in the air, the water vapor cools and condenses. **Condensation** occurs when a gas changes to a liquid. Water vapor can condense onto dust particles in the air, forming clouds. Clouds are masses of condensed water vapor in the atmosphere. Inside a cloud, small water droplets may join together and form larger ones. If it is very cold, some droplets freeze into ice. During **precipitation,** water falls from clouds over land and water. Precipitation can fall as rain, sleet, snow, or hail.

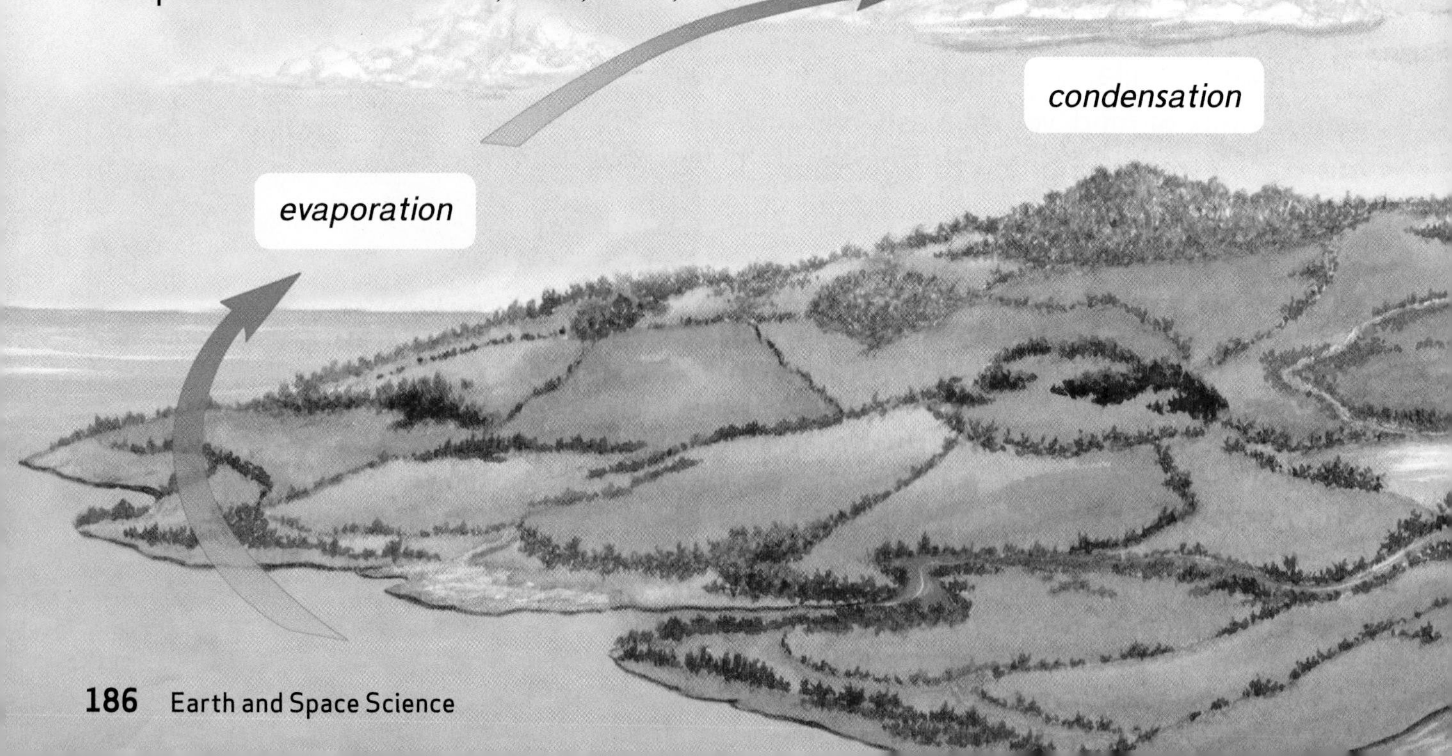

On and Below the Ground

When it rains, water flows over Earth's surface as **runoff**. Runoff gathers in lakes, oceans, and streams. Snow can become runoff when it melts. Runoff enters streams, which eventually become part of a river system. In a river system, many channels conduct the water into a main river. The region that contributes to a river or a river system is called a **watershed.** Landforms such as mountain ridges often form the boundaries of a watershed.

Water that soaks into the ground moves downward through small cracks and spaces. Some of this water becomes **groundwater**. Plants take up some water, and some evaporates. Groundwater moves until it is blocked by rocks so tightly packed that there are few places through which the water can flow. As the water backs up, it fills the spaces in the rocks and soil above. The top of this water-filled space is called the *water table.* This water feeds many wells, rivers, and lakes. Water is always moving from place to place, in one form or another.

Skill Builder

Read a Diagram

Follow the arrows to observe how water flows through the water cycle.

Weather and Climate

The Atmosphere

Even though air in Earth's atmosphere looks empty, it contains matter. The gases that make up the atmosphere are matter, including nitrogen, oxygen, carbon dioxide, and water vapor.

The air particles in the atmosphere have mass and weight. They press on Earth's surface and on objects they surround. The force put on a given area by the weight of the air above it is called *atmospheric pressure.* At sea level, the average air pressure is 1.04 kilograms per square centimeter (1.04 kg/cm^2), or 14.7 pounds per square inch (14.7 lb/in.^{2}). You do not feel this weight because atmospheric pressure pushes in all directions and is balanced.

Atmospheric pressure is higher at sea level than at higher elevations. *Elevation* is the height above or below sea level. Another term for elevation in the air is *altitude.* There are four layers of Earth's atmosphere: the troposphere, stratosphere, mesosphere, and thermosphere. Atmospheric pressure and temperature both change through the different layers.

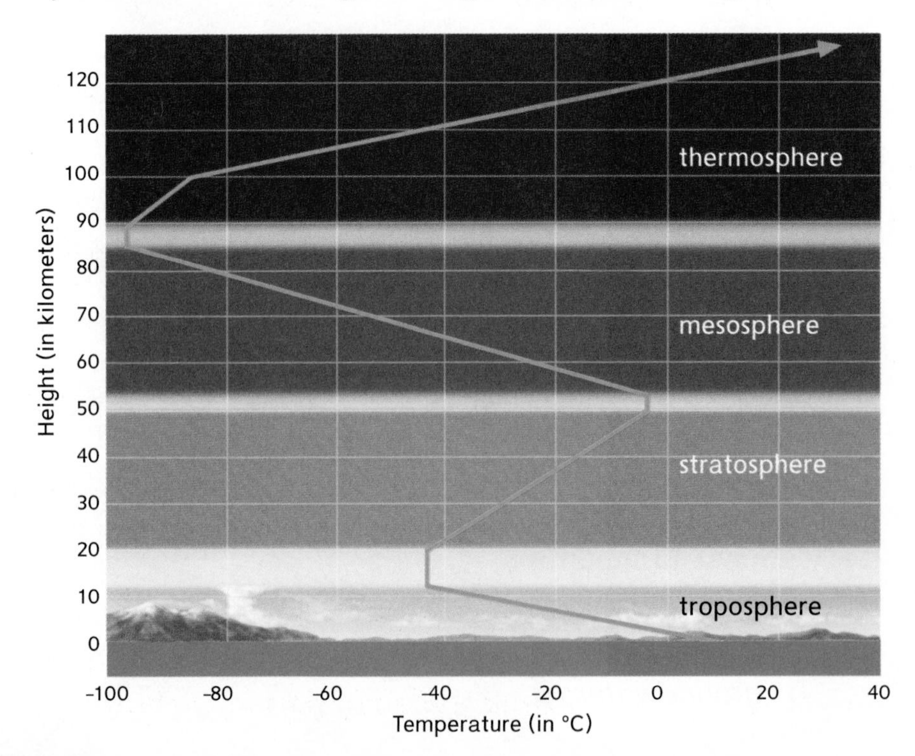

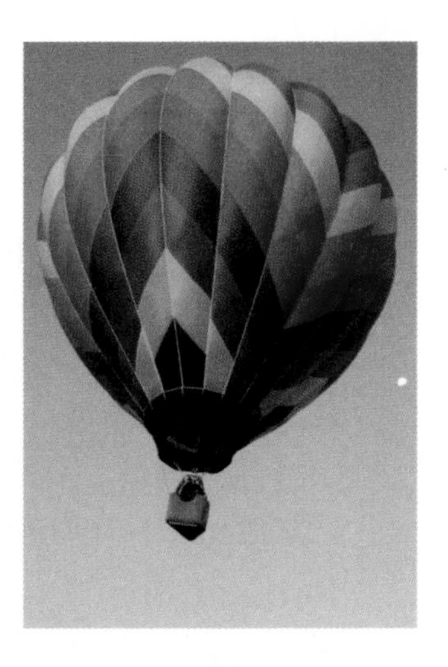

Air is matter. It takes up space in this hot-air balloon.

Skill Builder

Read a Graph

To identify the temperatures in Earth's atmosphere, follow the *x*-axis (Temperature in °C) and observe where the line is in different layers.

Key

The orange arrow in the graph shows how temperature changes throughout the layers of the atmosphere.

Layers of the Atmosphere

Troposphere The layer of Earth's atmosphere closest to Earth's surface is called the *troposphere*. The troposphere is 8–18 kilometers (5–11 miles) thick. The troposphere is thickest at the equator and thinnest at the poles.

Stratosphere Above the troposphere is the *stratosphere*. The stratosphere contains the ozone layer, which helps protect us from the Sun's harmful rays. This layer is also where jet airplanes typically fly.

Mesosphere Above the stratosphere is the *mesosphere*. Rock fragments from space often burn up in this layer. The coldest temperatures in Earth's atmosphere occur at the top of the mesosphere.

Thermosphere The outermost layer, or *thermosphere*, is where the International Space Station orbits. As altitude increases, the atmosphere becomes thinner until, eventually, conditions resemble the near-vacuum of space.

This rocket plane was designed to travel beyond the mesosphere. Rocket plane pilots that go beyond the mesosphere can officially be called astronauts.

When sunlight strikes Earth, about 50 percent of the Sun's energy is absorbed by Earth's surface. About five percent of the energy is reflected by Earth's surface. Clouds may absorb or reflect the remaining 45 percent.

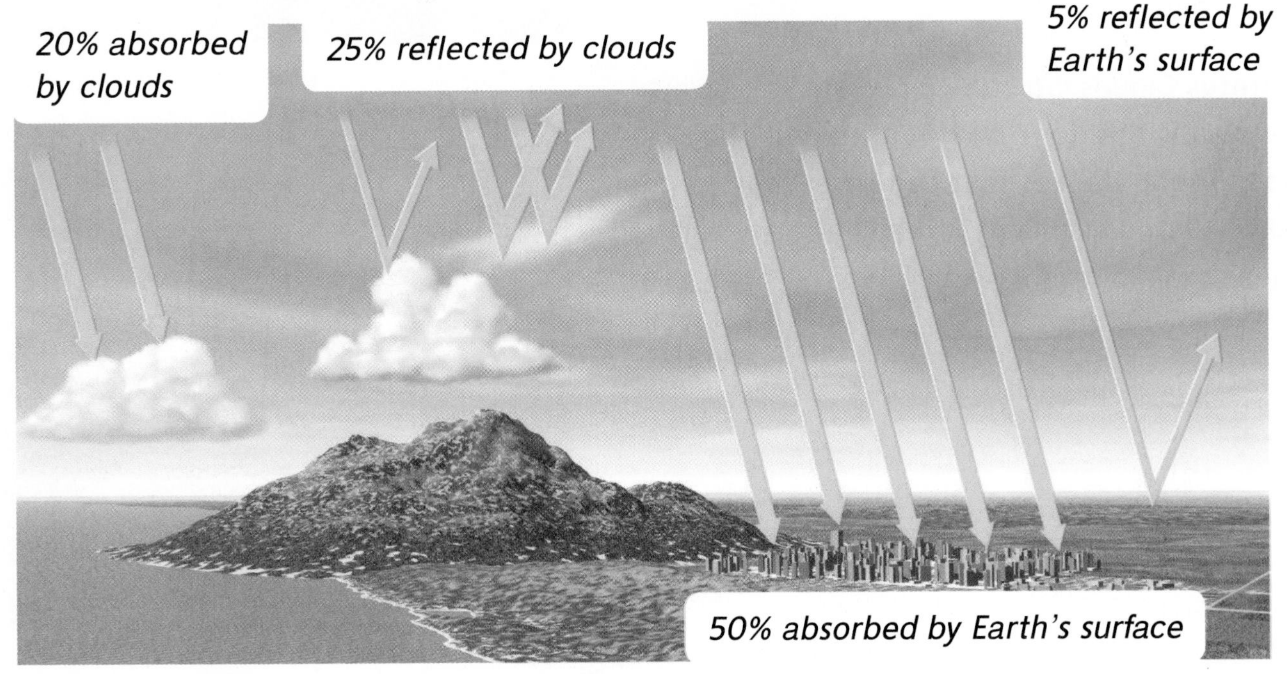

Clouds

Water vapor is one of the gases in the atmosphere. As water vapor particles are carried higher, they begin to cool. As they cool, they move more slowly and come closer and closer together. Eventually they condense around dust particles in the air. Clouds are made of these tiny drops of liquid water.

The appearance of a cloud depends on how high it forms in the atmosphere and what the temperature is there. Scientists classify clouds into three main types based on how and where they form.

Cirrus Clouds When water vapor condenses at high altitudes, it forms tiny bits of ice instead of water droplets. The thin, feathery clouds that form from these crystals are *cirrus* clouds. Rain does not fall from cirrus clouds.

Cumulus Clouds *Cumulus* clouds are puffy white clouds with flat bottoms. They are made from tiny bits of liquid water at middle altitudes.

Stratus Clouds *Stratus* clouds are flat layers of clouds that form at low altitudes. They look like a blanket that can cover most of the sky. What we call fog is really a stratus cloud near ground level.

cirrus clouds

cumulus clouds

stratus clouds

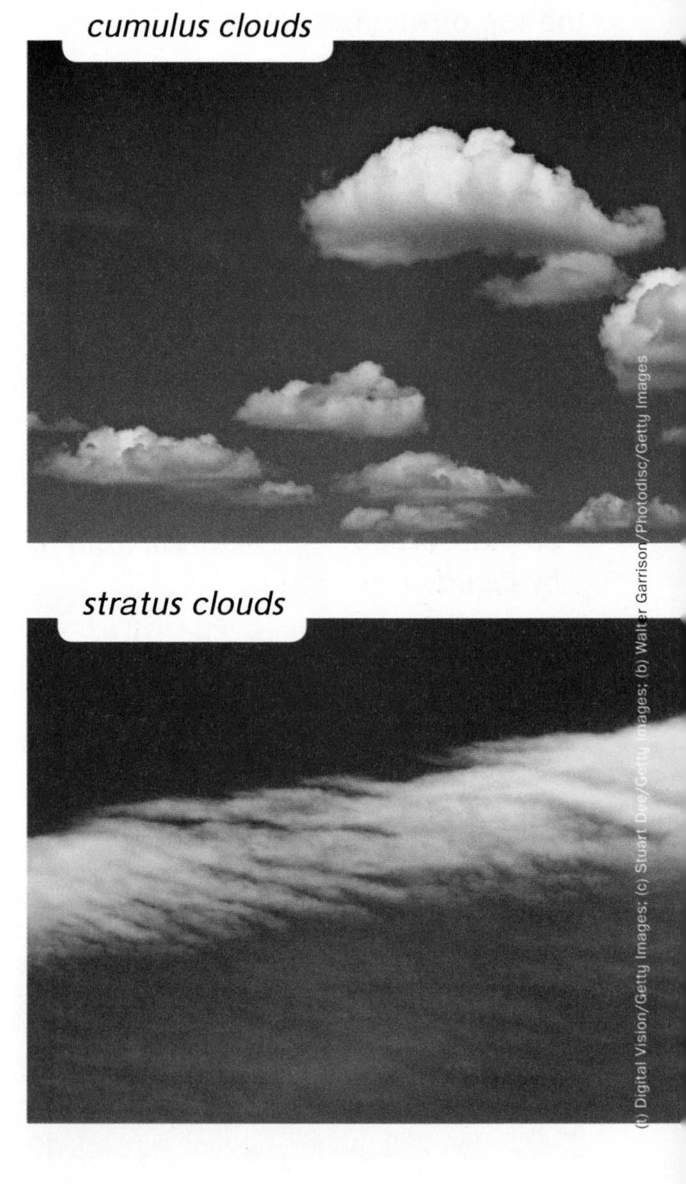

(t) Digital Vision/Getty Images; (c) Stuart Dee/Getty Images; (b) Walter Garrison/Photodisc/Getty Images

Clouds and Weather

Clouds can be used to predict the weather. Cirrus clouds appear during fair weather. Sometimes cirrus clouds form in long rows. These *cirrocumulus* clouds usually indicate fair, cold weather. Cirrus clouds may indicate approaching bad weather if they thicken and lower in altitude.

Cumulus clouds form as warm air rises. They generally indicate fair weather. However, when these clouds get larger they produce thunderstorms. A dark, thick cumulus cloud that produces thunderstorms is called a *cumulonimbus* cloud, or a thunderhead. Hail, thunder, and lightning often accompany the rain that falls from these clouds.

Stratus clouds often indicate light rain or drizzle in the near future. These clouds can sometimes bring days of light, steady rain. *Nimbostratus* clouds are dark sheets of clouds that block sunlight. Long periods of precipitation often follow the appearance of these clouds.

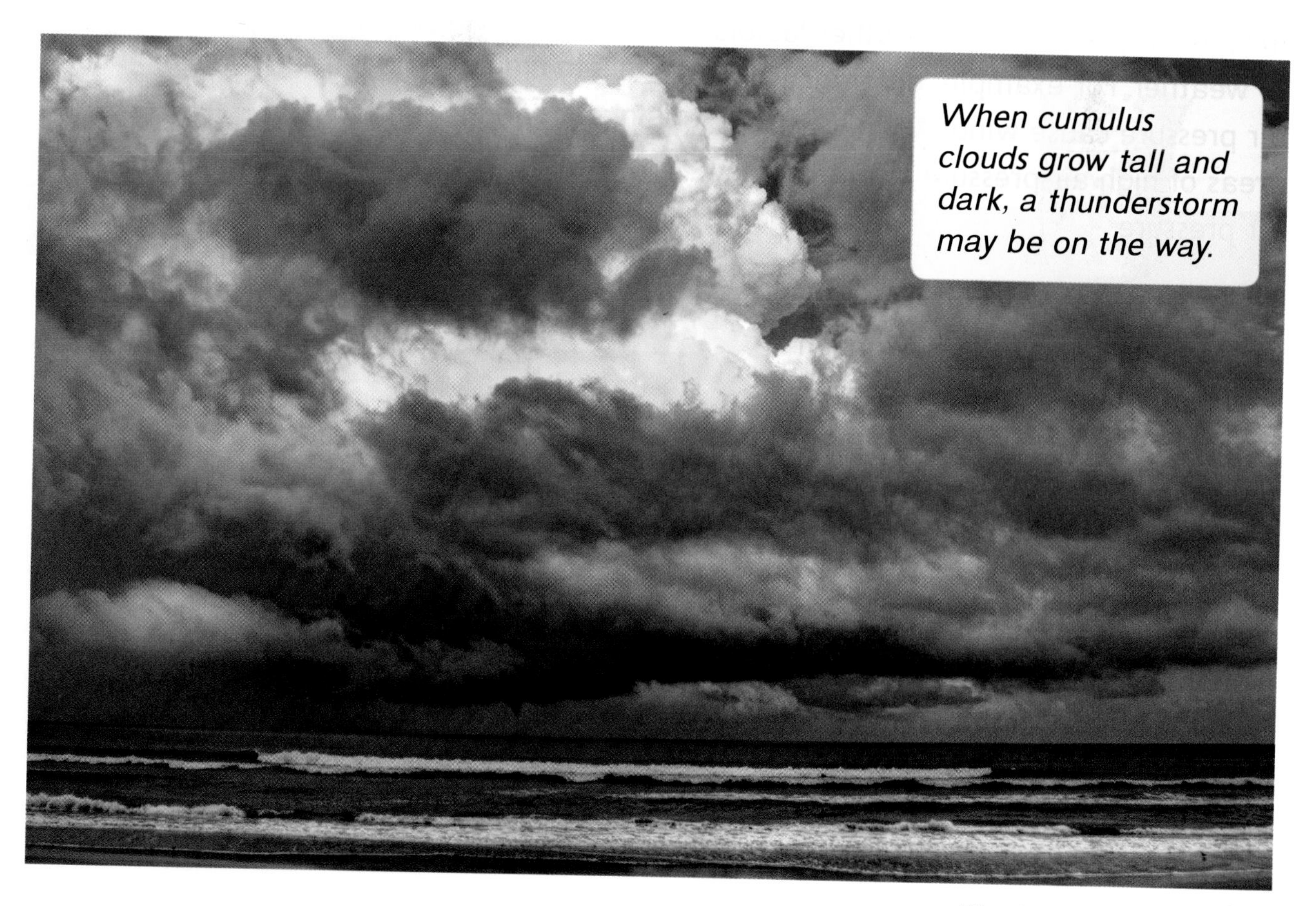
When cumulus clouds grow tall and dark, a thunderstorm may be on the way.

Weather - Air Masses and Fronts

The weather in an area is affected by the air mass that is passing over that area. An *air mass* is a large region of air that has a similar temperature and humidity. Air masses can cover thousands of square kilometers of land and water.

Depending on where they form, air masses can be cold or warm and dry or humid. An air mass that forms above a warm area of water will be warm and humid. An air mass that forms over a cold area of land will be cold and dry. The weather in one part of an air mass is similar to the weather throughout the rest of the air mass.

Air masses form all the time, usually near the poles or equator. The area from which an air mass gets its characteristics is called its *source region.* The map shows some common paths air masses follow.

Air Masses in North America

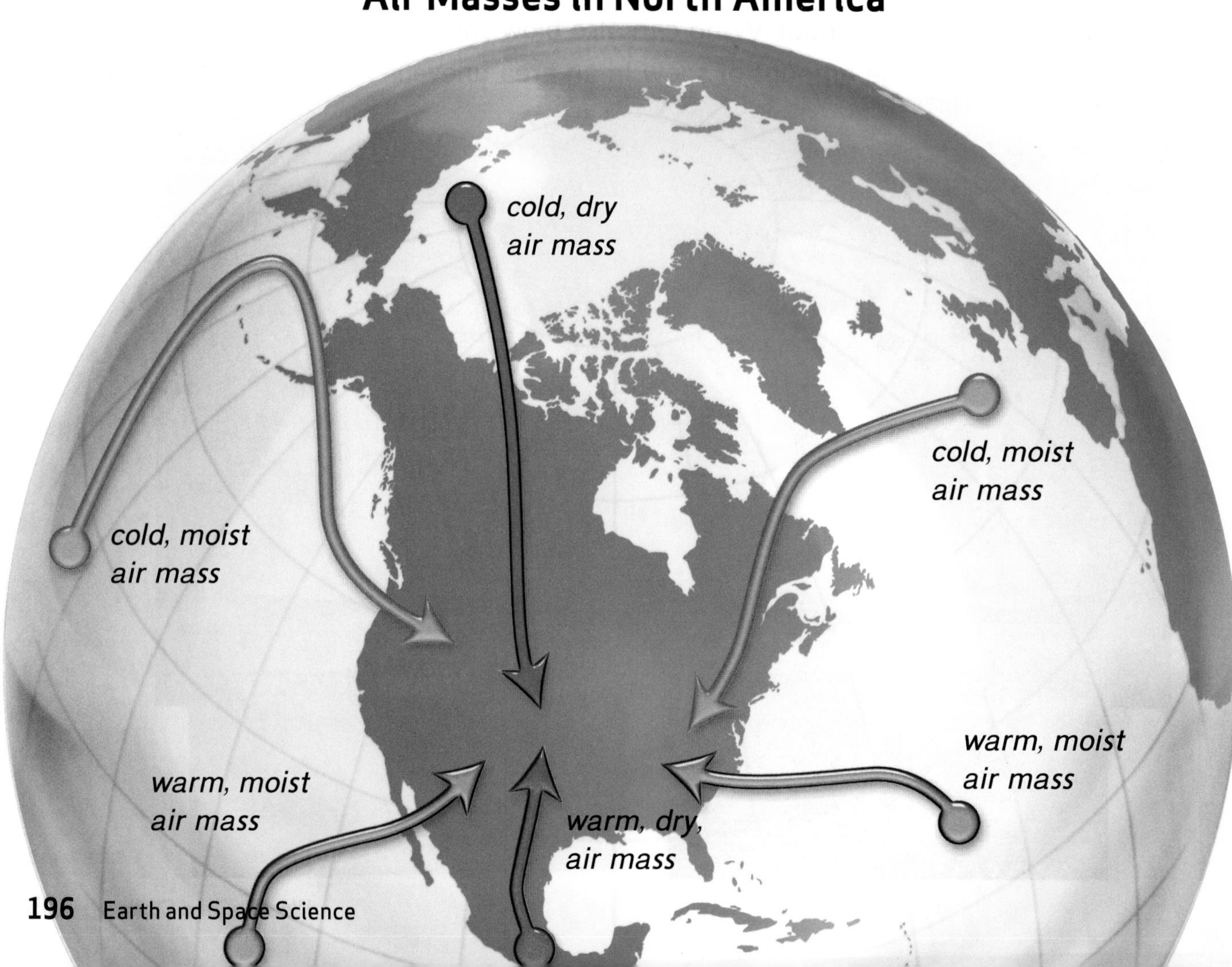

When one air mass meets a different air mass, the meeting place is called a **front**. A front is the boundary between two air masses that have different temperatures. Along fronts, weather can change rapidly. There are three basic types of fronts.

Thick clouds that can bring heavy rain announce the approach of a cold front.

Warm Fronts When a warm air mass pushes into a cold air mass, a *warm front* forms. The warm air mass slides up and over the cold air mass. Layers of clouds form, and the cold air retreats. A warm front often brings light, steady rain. After the front passes, the air temperature rises.

Cold Fronts A *cold front* forms when a cold air mass pushes under a warm air mass. The cold air mass forces the warm air mass upward quickly. Thick clouds form as the warm air rises and cools. Cold fronts often bring stormy weather.

Stationary Fronts Sometimes rainy weather lasts for days. This kind of weather can be caused by a stationary front. A *stationary front* is a boundary between air masses that are not moving.

> **Skill Builder**
>
> ### Read a Diagram
>
> The arrows in the diagram indicate temperature as well as movement. Red arrows indicate warm air movements. Blue arrows indicate cold air movements.

Different Fronts

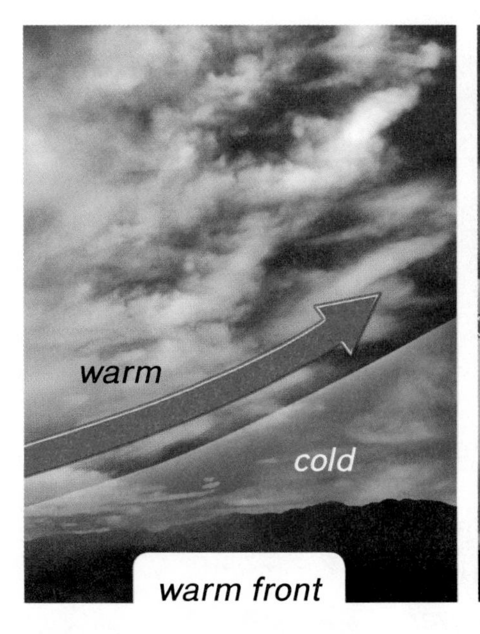

warm front

cold front

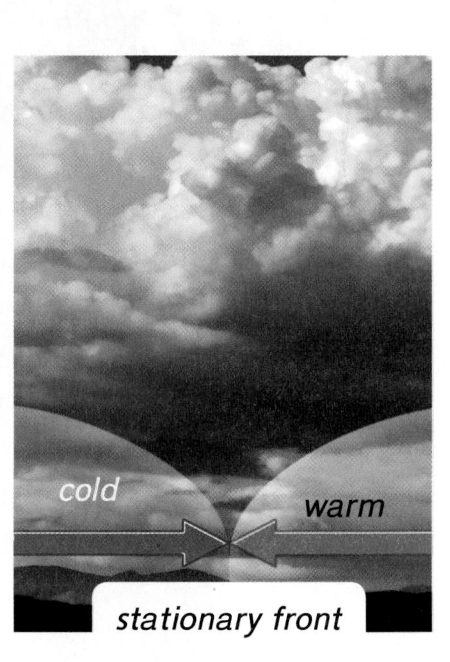

stationary front

Weather - Gathering Weather Data

The information in weather reports comes from a variety of sources. On the ground, weather instruments at weather stations record data such as air temperature, wind direction, wind speed, and humidity.

Weather Balloons Scientists also use weather balloons to take measurements from high up in Earth's atmosphere. Weather balloons rise to 35 km (22 mi) above the ground, recording data up into the stratosphere. Eventually, the balloons burst. The data recorders then return to the ground by parachute.

Satellites Using weather balloons is expensive, because the balloons must be replaced constantly. In some cases, satellites can perform similar functions. The satellites orbit Earth, taking photographs and sending them to computers. Satellite images show large-scale weather patterns, such as the movement of air masses and where they meet to form fronts. Satellites are useful in tracking hurricanes. They collect information about cloud patterns. They record temperatures at the tops of clouds and at the sea surface. Satellites also measure the direction and speed of winds. This information helps scientists track the size, path, and intensity of a storm.

A wind vane points into the wind to indicate direction.

The cups on an anemometer spin to indicate wind speed.

Weather balloons carry instrument packs that collect weather data.

Weather satellites provide data about temperature, winds, moisture, and cloud cover.

Radar Another important tool for weather scientists is radar. Radar, which stands for *r*adio *d*etection *a*nd *r*anging, uses radio signals to detect precipitation. The equipment sends pulses of energy at the area under study. Radar measures the time it takes for echoes of the signal to return and records any changes to the signal. The signals provide data about precipitation in the atmosphere. Doppler radar is another tool that meteorologists use. **Doppler radar** sends out radio waves from an antenna. Objects in the air, such as raindrops, reflect the waves back to the antenna. Doppler radar can measure the direction and speed of storms that are moving.

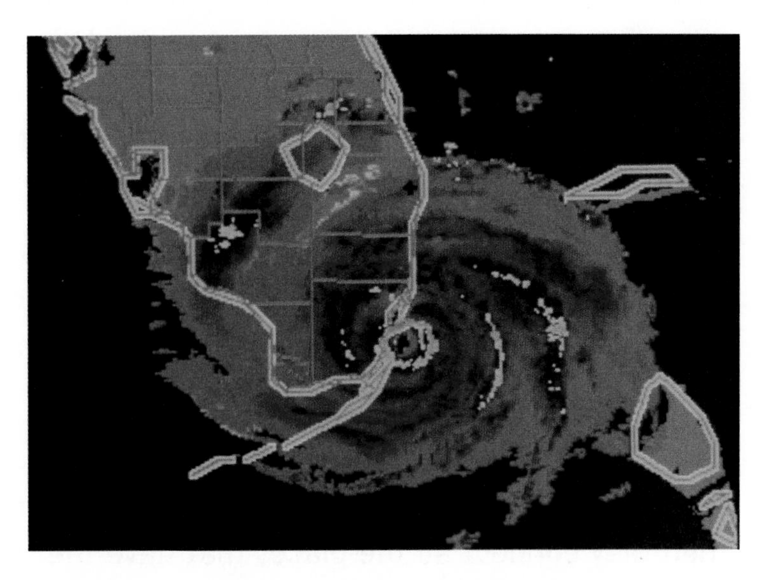

Scientists use computer images made from Doppler radar to track thunderstorms, hurricanes, and other storms.

Buoys Weather buoys scattered across Earth's oceans measure conditions such as surface wind, waves, temperature, and fog. Scientists keep track of this and all other data by entering the information into supercomputers. The data helps them produce models of weather that they can adapt as conditions change.

By combining information from ground measurements, weather balloons, satellites, and radar, scientists can form detailed pictures of weather conditions.

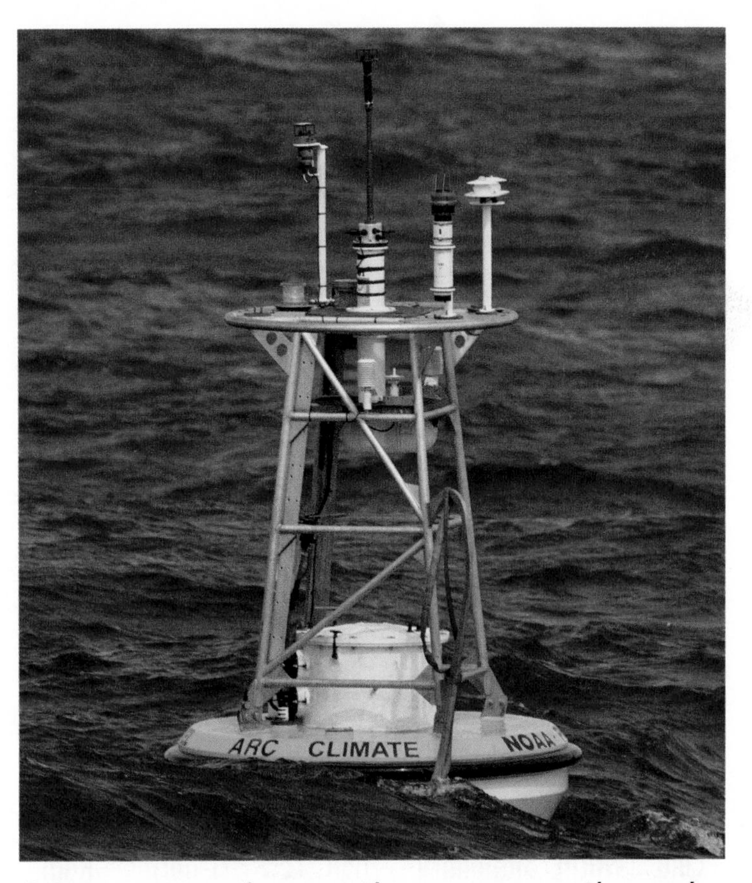

Buoys scattered across the oceans continuously record and transmit weather data.

Weather - Weather Maps and Forecasting

Each day, scientists make and share weather maps. Weather maps show weather conditions at a certain place and time. They tell about air temperature, pressure, precipitation, and winds. Weather maps may also show the locations of fronts. The fronts appear as lines of triangles or half circles.

Weather maps also show low- and high-pressure systems. A *low-pressure system* is a large air mass with low air pressure in the center. A *high-pressure system* is a large air mass with the highest air pressure in the center.

To figure out where high and low systems are at a particular time, scientists plot the air pressure of different areas on a map. Then they connect all the places that have the same air pressure with a line called an *isobar*.

Skill Builder

Read a Map

A weather map includes several legends. Colors represent temperature. Lines of triangles and half circles represent fronts. *H* and *L* stand for high pressure and low pressure. Other symbols indicate types of precipitation.

U.S. Weather Map

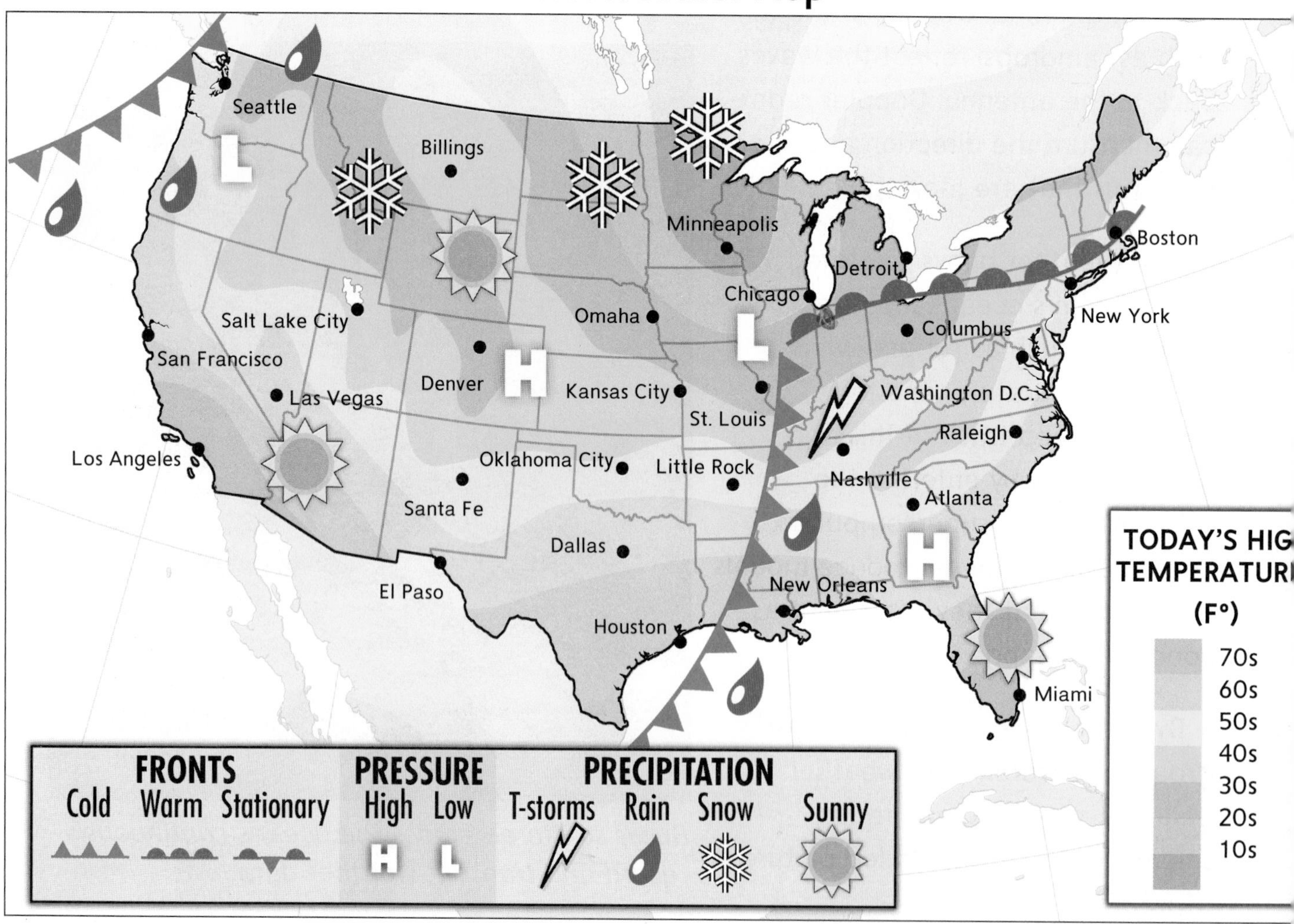

Forecasting Scientists use weather maps to make forecasts. To *forecast* is to predict weather conditions. Locate the cold front running from St. Louis to Houston on the weather map. The triangles point toward the east. Like most fronts in the United States, this one is moving from west to east. A forecast based on this map may predict a chance of rainy weather for New Orleans.

Scientists use many technologies in forecasting. Satellite data and computers help scientists analyze weather data and make better, more sophisticated weather maps.

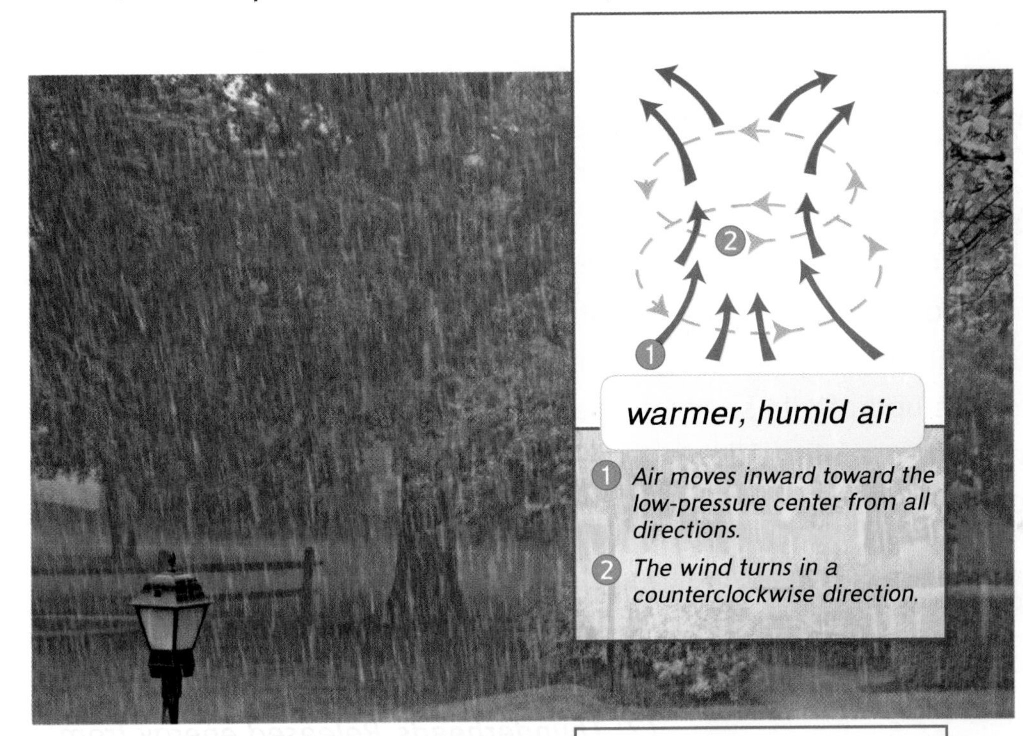

Low-pressure systems usually bring warm, stormy weather. Moisture in a low-pressure air mass condenses as it rises and cools, which brings clouds and precipitation.

High-pressure systems usually bring dry, clear weather. Any moisture carried in a high-pressure system tends to evaporate, clearing the sky of clouds and precipitation.

Weather Events - Tornadoes

The strongest thunderstorms can produce tornadoes. A *tornado* is a rotating, funnel-shaped cloud with wind speeds up to 500 kilometers per hour (300 miles per hour).

Tornadoes form in some severe thunderstorm clouds. Warm, moist air lifted upward produces areas of low pressure. When high-speed winds blow across the top of the cloud, it may cause a rising column of air to spin. Some of these spinning air columns may extend downward from the cloud. If they reach the ground, they are called tornadoes.

From the ground, the shape of the cloud looks like a funnel. Warm air rises up the center of the spinning funnel cloud. When the tip of the funnel cloud touches the ground, it becomes a tornado. Heavy downpours of rain often fall from tornado producing thunderheads.

Since only a relatively small section of a tornado actually touches the ground, tornadoes have been known to destroy houses on one side of a street while leaving houses on the other side untouched.

How a Tornado Forms

A tornado forms when air rushes upward into a low-pressure area. Then a funnel cloud forms.

(1) Warm air moves upward in a thunderhead.

(2) A funnel cloud forms when the air starts rotating.

Tornadoes can move either quickly or slowly. They can also change direction abruptly, moving first in one direction and then in another. Tornadoes can cause terrible damage, breaking up buildings, uprooting trees, and lifting cars into the air.

Tornadoes can occur anywhere, but some areas are more prone to experience these storms than others. Globally, areas between about 30° and 50° north or south of the equator have the best conditions for tornadoes to form. In the United States, two regions experience the most tornadoes: Florida and "Tornado Alley." Tornado Alley is a nickname given to an area in the south-central United States. Tornadoes in this region most often occur in the late spring and early fall.

Derechos

A *derecho* is a widespread, long-lasting windstorm associated with some thunderstorms. These storms can cause damage similar to that of tornadoes. However, the damage occurs in one direction along a straight path. As a result, the term "straight-line wind damage" is often used to describe damage from these events. To be classified a derecho, the wind damage path must be longer than about 400 km (240 mi), and the storm must have wind gusts of at least 93 km/h (58 mph) or greater along most of its length.

(3) The funnel cloud becomes a tornado when it touches the ground.

Weather Events - Tropical Storms

A *tropical storm* has rotating winds with low pressure at its center. Tropical storms form near the equator where the ocean is warm. Water vapor evaporates from the warm water. As the warm, moist air rises, cooler air flows toward the space where the warm air had been. Water continues to evaporate, lowering the air pressure even more. Surrounding, high-pressure air moves into the area of low pressure and causes rotating winds.

This hurricane is headed toward the Gulf Coast of the United States.

A tropical storm turns into a **hurricane** when wind speeds reach more than 119 km/h (74 mph). From space, a hurricane looks like a spiral of clouds with a hole in the center. The hole is the center of low pressure, called the "eye." A hurricane's fastest winds and heaviest rains occur next to the eye. Winds near the eye of a hurricane can reach speeds of almost 300 km/h (190 mph). The area inside the eye is calm, with no wind or precipitation. Hurricanes can also push ocean water onto the shore in a *storm surge*. Storm surges can cause extensive coastal flooding.

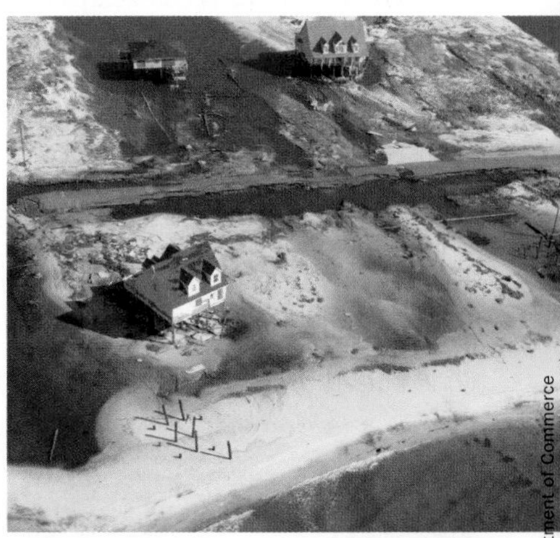

Storm surges associated with hurricanes can damage shorelines as well as buildings near the water.

Hurricanes are dangerous storms, and their effects are felt long after the storm is over. If possible, leave areas threatened by a hurricane. If you must stay, board up windows and stay away from both windows and doors. Bring in all outdoor furniture, decorations, garbage cans, and anything else that is not tied down. Store food, bottled water, flashlights, and battery-operated radios. Wait for floodwaters to subside before leaving a safe area.

Cyclone is a general name given to any large area of low air pressure with a circular wind pattern. These low pressure systems may produce thunderstorms, tornadoes, and hurricanes. Not all low pressure cyclones cause such storms.

Winter Storms

Winter storms occur when a cold, dry air mass meets a warm, humid air mass. In one type of winter storm, the main types of precipitation, such as snow or sleet, occur with cold air temperatures. Another type is a rainstorm in which ground temperatures are cold enough that ice forms on outside surfaces.

To stay safe during winter storms, make sure to stay indoors. Keep warm clothes and blankets available in case the power is lost. Assemble an emergency supply kit that includes flashlights, extra batteries, food, and bottled water.

Blizzards People often refer to any severe snowstorm as a blizzard. In the United States, a *blizzard* is a snowstorm with 56 km/h (35 mph) winds and enough snowfall that you can see up to only one quarter of a mile. Blizzards often drop several feet of snow on an area over several days. Winds can move the snow into drifts that can be significantly deeper than the total amount of snow that has fallen.

Ice Storms If rain falls through cold air, it can freeze into sleet or freezing rain. If the ground is also cold, the sleet or freezing rain will coat everything it lands on with a thin layer of ice. An *ice storm* is a storm in which freezing rain forms a layer of ice on outside surfaces. Ice can make power lines and tree branches break. The ice can also make streets slippery and dangerous for driving or walking.

Winter Storms

snow storm

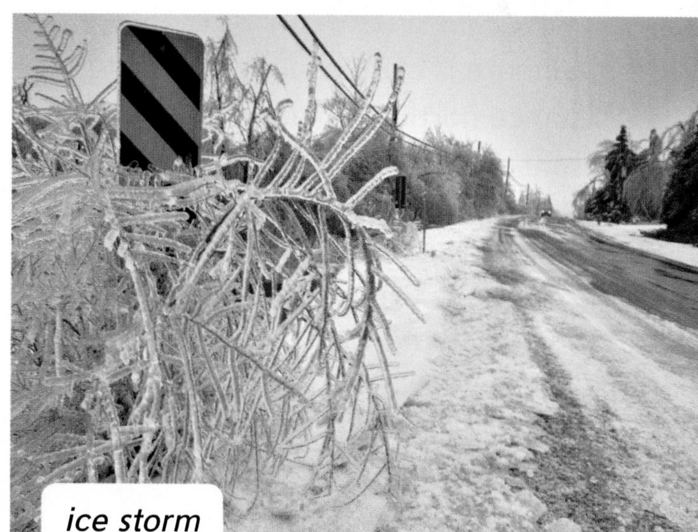

ice storm

Climate

Weather changes every day. However, the weather in any area tends to follow a pattern. **Climate** is the average weather pattern of a region over time. Climate varies from place to place. Climate differs from weather. Even if the climate in a region includes hot summers, the weather there on a summer day could be cool.

Average temperature and average rainfall, in addition to other weather information, are important variables in determining climate. Climate can be tracked for different areas, periods of time, and locations. There are several factors that affect climate.

Latitude

Climate is closely related to latitude, which is a location's distance north or south of the equator. Because Earth is shaped like a sphere, areas closer to the equator receive more direct energy from the Sun than do areas farther from the equator. On a global scale, the United States is in the temperate zone. The United States can also be divided into more specific local climate zones.

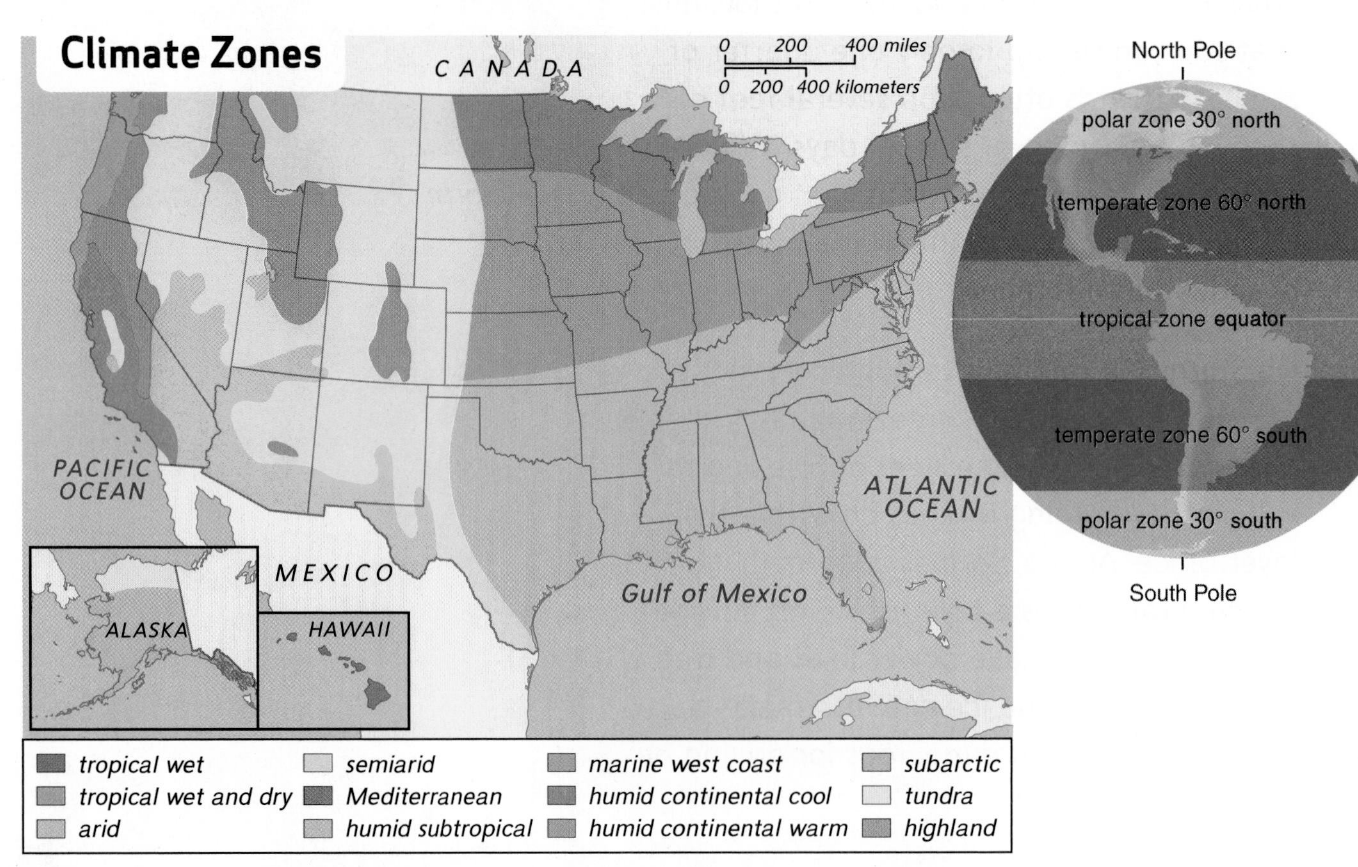

Water

Water makes up Earth's hydrosphere. It affects the atmosphere, including the temperature and precipitation of an area.

Distance from Water Most of Earth's surface is covered by water. However, some places can be located far from a body of water. The temperature of an inland city is generally warmer in summer and cooler in winter than the temperature of a city near the ocean. Precipitation is often higher near large bodies of water, where evaporation adds moisture to the air.

Ocean Currents A *current* is the constant movement of ocean water. The Gulf Stream, a current that moves along the east coast of the United States and across the Atlantic Ocean, carries warm water from near the equator toward the poles. Other currents, such as the California Current, carry cold water from the poles toward the equator. The temperature of a current affects the climate of the land nearby. For example, the warm water in the Gulf Stream causes mild temperatures in the British Isles. Areas near cool currents often have cool temperatures. Ocean currents moderate worldwide temperatures throughout the year.

Skill Builder

Read a Diagram

Use the key to identify the temperatures of the different ocean currents.

Ocean Currents of the World

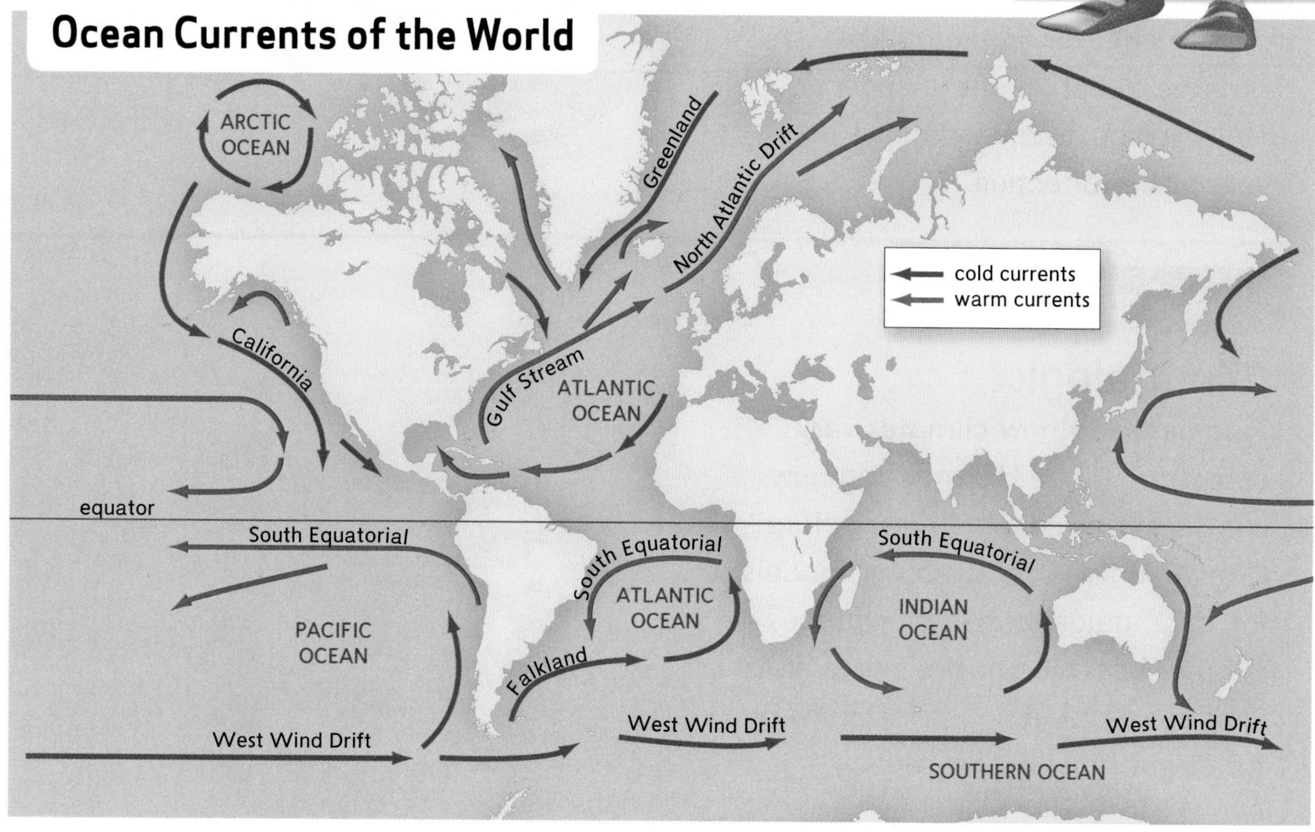

Climate - Global Winds

Air moves in a system of winds called global winds. A *global wind* blows steadily over long distances in a predictable direction. Winds blow from areas of higher pressure to areas of lower pressure. However, Earth's rotation pushes the wind to either the right or the left. This shift is called the *Coriolis effect*. As Earth rotates, places near the poles travel a shorter distance than places near the equator. This causes winds away from the equator to curve. In the Northern Hemisphere, winds curve to the right, or clockwise. In the Southern Hemisphere, winds curve to the left, or counterclockwise.

Air temperature also affects the movement of global winds. As air is heated at the equator, it rises up within the troposphere. As the air goes higher, it begins to cool. Because heated air continues to rise, the cooled air above it is pushed to the north. The air warms and cools in three different bands as it moves north from the equator. In the middle band, the direction of the cooling air is reversed, and wind blows in the opposite direction.

Global Winds

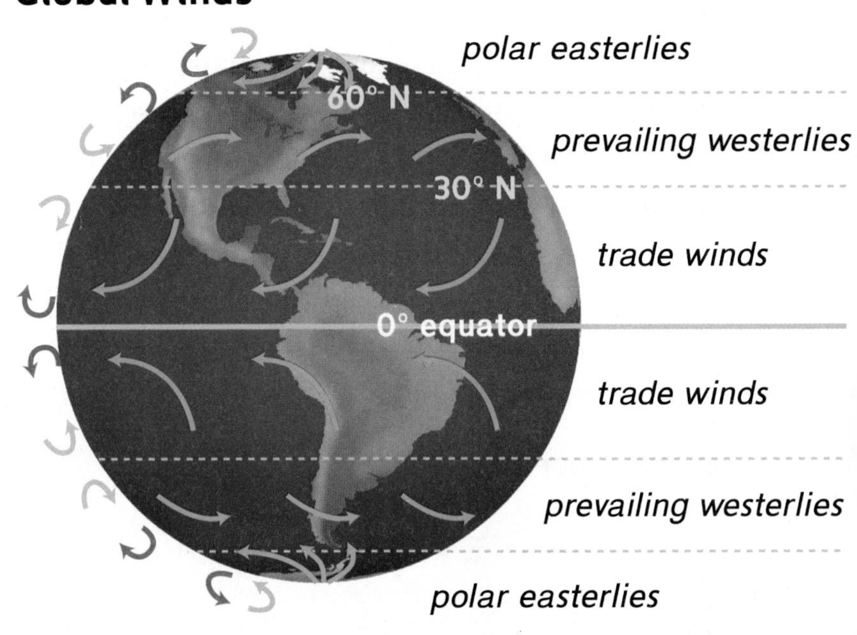

What Could I Be?

Climatologist

Fascinated by how climates vary across the globe? Wondering how climates may change in the future? Climatologists use computer models to forecast major, long-term changes in climate based on past climate history. Find out more about climatology by turning to the Careers section.

Johner Images/Alamy

Volcanoes Affect Climate

Erupting volcanoes send dust, ash, and gases into the atmosphere. Usually, the effects of an eruption are local. Condensation of water vapor on the particles added to the air can cause heavy rainfall. Fog often forms after eruptions, as well.

During large, explosive eruptions, however, the materials may reach the stratosphere. These eruptions can affect climate on a world-wide basis. Scientists are still studying these effects, which seem to be related to the size of the particles ejected from the volcano. Some volcanoes eject relatively large particles. These particles allow sunlight to strike Earth's surface, but do not allow heat to be re-radiated to space. This additional heat causes the climate to become warmer.

More commonly, these eruptions push sulfur dioxide gas into the stratosphere. The gas reacts with water vapor to form sulfuric acid droplets. Billions of these droplets can form after a large eruption. These droplets absorb radiation from the Sun, resulting in a cooling climate. These lower temperatures last only two to three years, because the droplets eventually sink into the troposphere where rain and wind scatter them.

Materials ejected from an erupting volcano can have short-term and long-term effects on climate.

Climate - Mountain Ranges

Mountain ranges affect both temperature and precipitation patterns in their region.

Altitude The higher a place is above sea level, the cooler its climate. Along the base of a mountain near the equator, you may find tropical plants growing. However, at the peak of this mountain, you may find permanent ice and snow.

Rain Shadows Mountain ranges affect precipitation patterns. As warm, moist air moves up a mountain, it gets colder. Water vapor condenses, and precipitation falls on the windward side of the mountain. The air that moves down the leeward side of the mountain is dry and hot. The dry area on the leeward side of the mountain is called a *rain shadow*.

Skill Builder

Read a Diagram

Examine the arrows to understand how air temperature changes as it moves up and over a mountain.

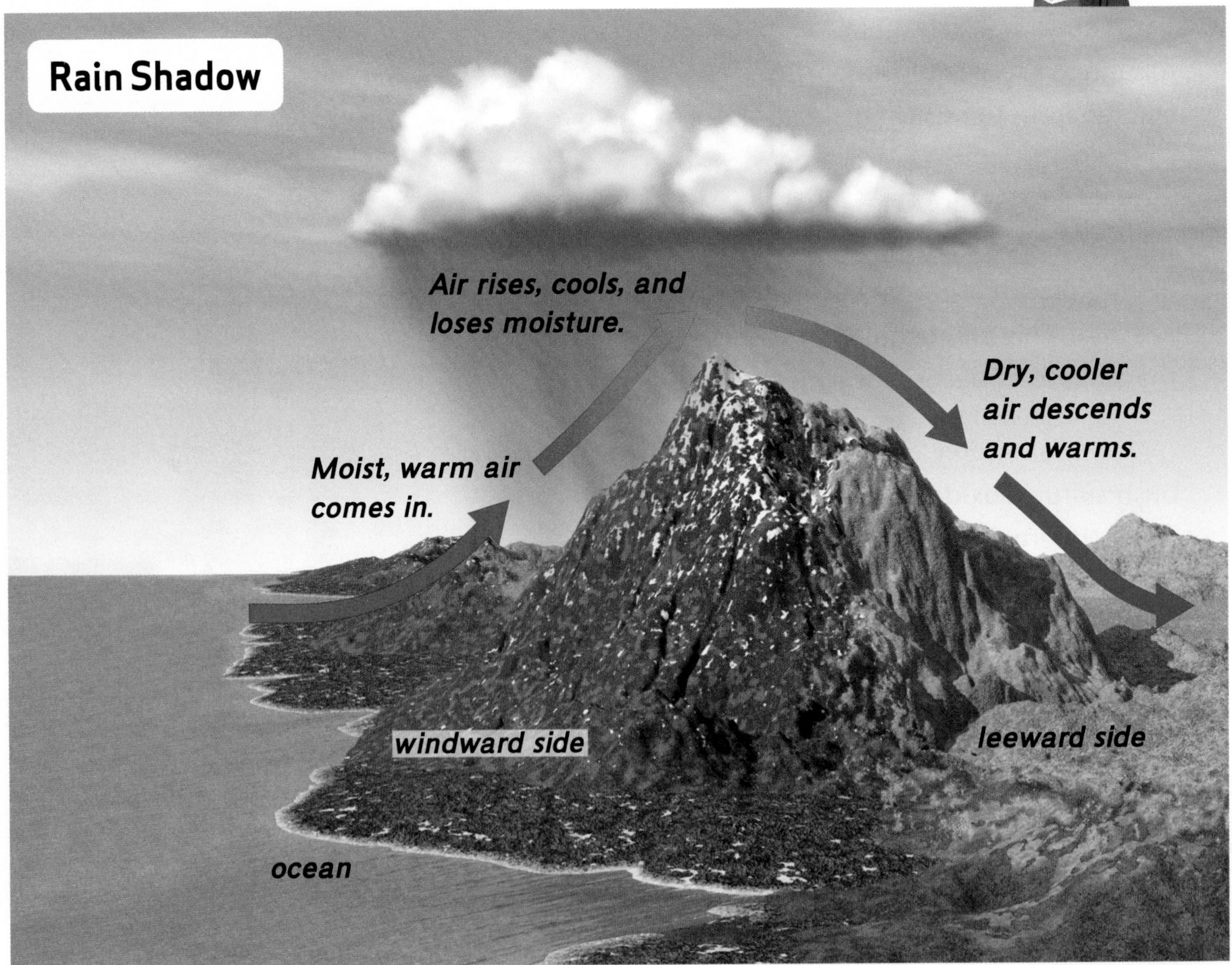

Climate and Plants

Another way to describe an area's climate is by the plants that grow there. Plant types require different levels of precipitation, sunlight, and temperature. For example, temperate forests have different weather during different seasons. Temperate trees are those that respond by losing leaves before the cold, dry winters.

The Sun's energy that gets absorbed by Earth's surface radiates back into the atmosphere as the ground cools at night. Much of this radiated heat is absorbed by a layer of gases. These gases include water vapor, carbon dioxide, and ozone. Some of the energy absorbed by the gases radiates back and warms Earth.

When fossil fuels are burned, they release greenhouse gases, including carbon dioxide. Burning trees also increases the amount of carbon dioxide in the atmosphere. As the amount of greenhouse gases in the atmosphere increases, more energy is captured in the atmosphere and radiated back to Earth.

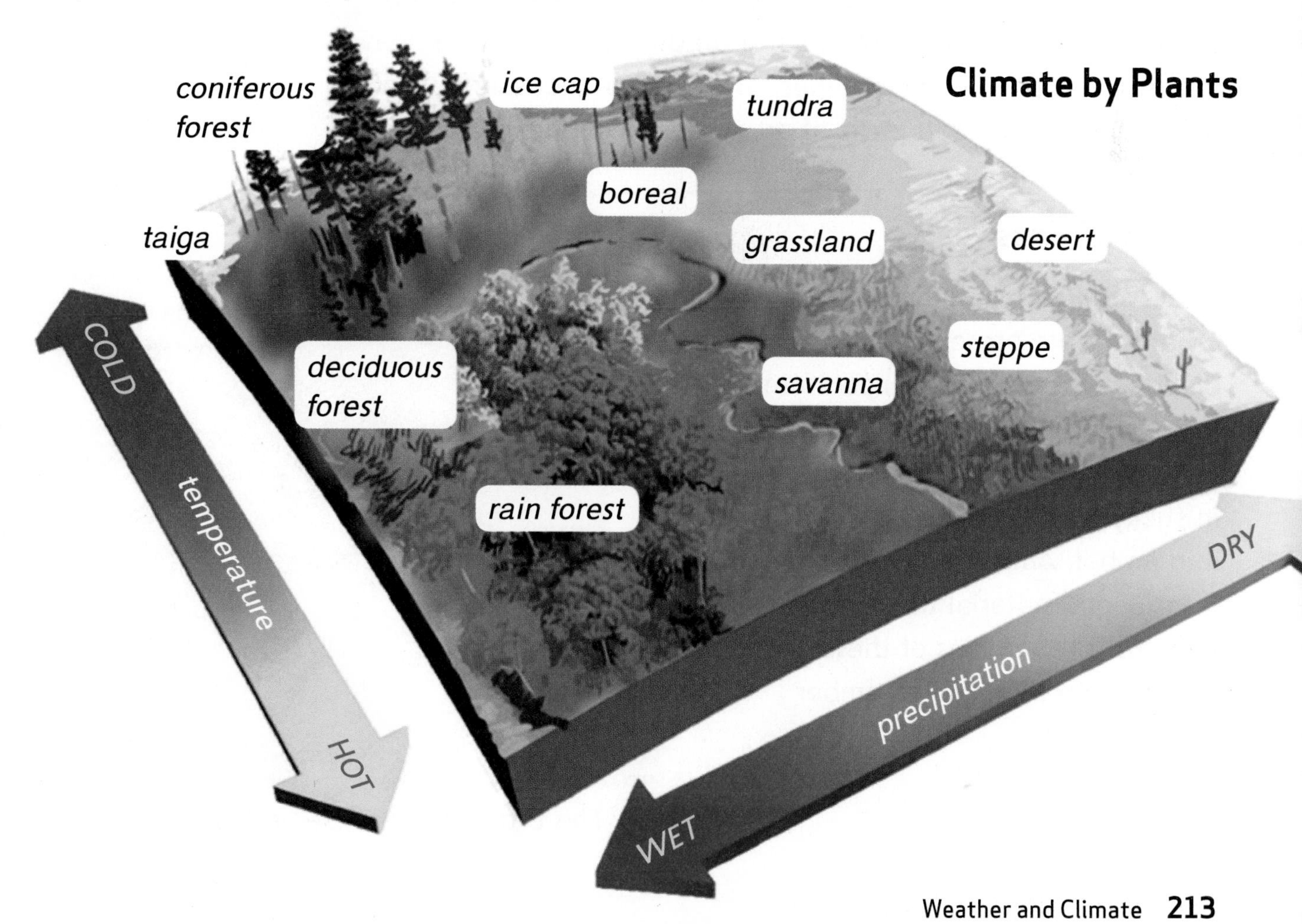

Earth's History

Earth dates back to over 4 billion years ago. About 3.5 billion years ago, the first unicellular organisms appeared. Over millions of years, other organisms developed. We know about these organisms through the traces they left in rocks.

Fossils

Scientists use fossils to learn about the past. When some ancient organisms died, their remains were covered with soil, sand, or some other sediment. Over many centuries, these sediments hardened around the organisms' remains. Fossils are the remains or traces of ancient organisms preserved in soil or rock. Almost all fossils are found in sedimentary rock.

When a plant or animal dies, the soft parts quickly decay or are eaten. That is why hard parts of an organism generally leave fossils. These parts include bones, shells, teeth, seeds, and woody stems. It is rare for soft parts of an organism to become a fossil.

Types of Fossils

Preserved Remains Sometimes entire organisms are preserved as fossils. Large elephant-like animals called mammoths have been found frozen in tundra ice. Insects and spiders can become trapped in sticky tree sap. The sap can harden into amber. *Amber* is a hard, smooth, translucent material that preserves these animals. The remains of these organisms can be seen through the amber.

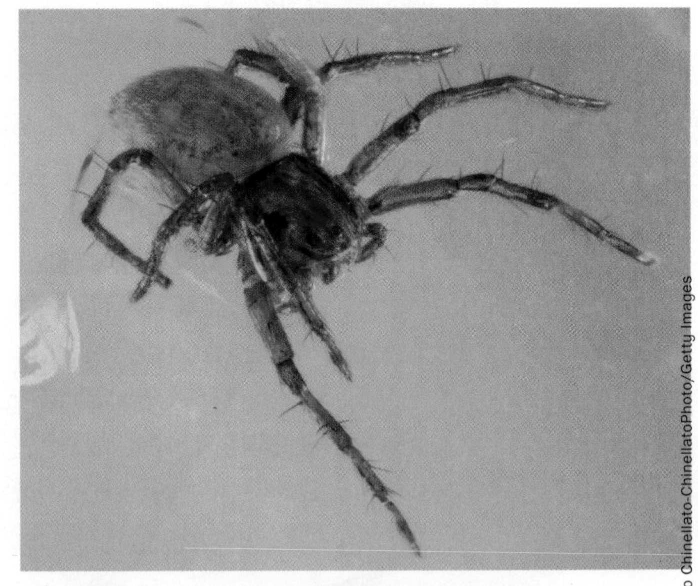

This organism is preserved in amber.

Petrified Fossils When minerals replace all or part of an organism, a petrified fossil forms. Dinosaur bones are this type of fossil. Wood can also be petrified. Water carrying minerals seeps into spaces in the organism's cells. Over time, the water evaporates, leaving the minerals behind.

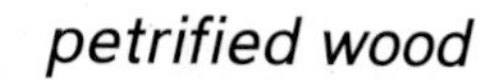

petrified wood

Carbon Films Leaves and insects are sometimes preserved as carbon films. When sediment buries an organism, some materials escape, leaving carbon behind. Over time, only a thin film of carbon remains.

carbon film

Trace Fossils Sometimes living things leave traces of their activity. An *imprint* is a mark made by an object pressing into materials such as mud. An imprint can become a fossil if it is not weathered away. Trails and burrows can also become trace fossils.

Molds and Casts The most common fossils are molds and casts. Shells often leave behind fossils known as molds and casts. A *mold* is a hollow area in sediment in the shape of an organism or part of an organism. A mold forms when an organism buried in sediment breaks down, leaving an empty space. Later, water may deposit minerals and sediment into a mold, forming a cast. A *cast* is a solid copy of the shape of an organism.

foot imprint

Word Study

The word *petrified* mea
"turned into stone."

Fossils - What Fossils Tell Us

Fossils give scientists information about environments of the past. Tropical forests once grew in what is now grasslands in Arizona. Tall fern plants once covered what is now icy Antarctica. Much of the United States was once covered by a shallow ocean.

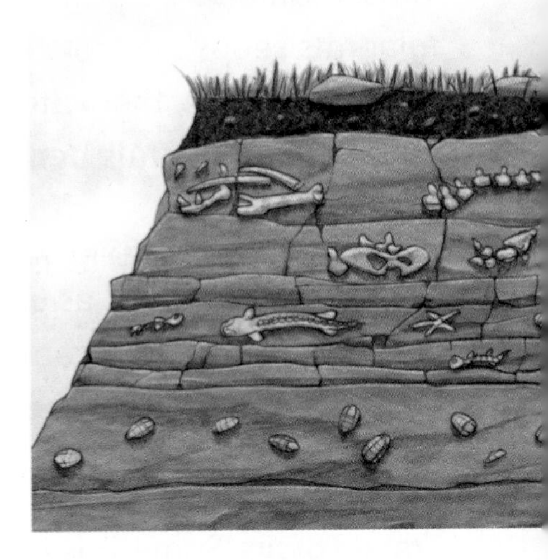

Fossils can help scientists estimate the age of rock layers.

When scientists study a fossil site, they investigate more than just fossils. They investigate what the environment was like long ago. Ammonites once lived in Earth's oceans. Ammonite fossils are found in rock that is now on dry land. Ammonite fossils indicate that the land was once covered by water.

Since fossils are found in rock layers, scientists can determine the relative ages of the fossils based on the layers in which they are found. Some fossils also provide clues to a rock layer's relative age. *Index fossils* are remains of living things that were widespread, but lived during a relatively short part of Earth's history. For example, trilobites were sea animals that lived on Earth for millions of years. They suddenly died out, or became *extinct*, all over Earth at about the same time. The presence of fossil trilobites of the same kind in various rock layers indicates that those layers all formed at about the same time. This fact is true even if the layers are in different areas.

Ammonites lived in water. These fossil ammonites were found on land.

How deep an organism is buried gives clues about when the organism lived. Fossils found in sediment layers closest to the surface are usually youngest. Fossils found in deeper layers are usually older. Ammonites are usually found in relatively deep rock layers. Trilobites are other organisms found in older, deeper rock. These organisms looked much like horseshoe crabs, an animal species that is still alive today. Unlike horseshoe crabs, trilobites have died out and become extinct. *Extinction* is the disappearance of a species when the last of its kind dies. Trilobites became extinct about 250 million years ago. Scientists have evidence that the horseshoe crab is the closest living relative of the trilobite.

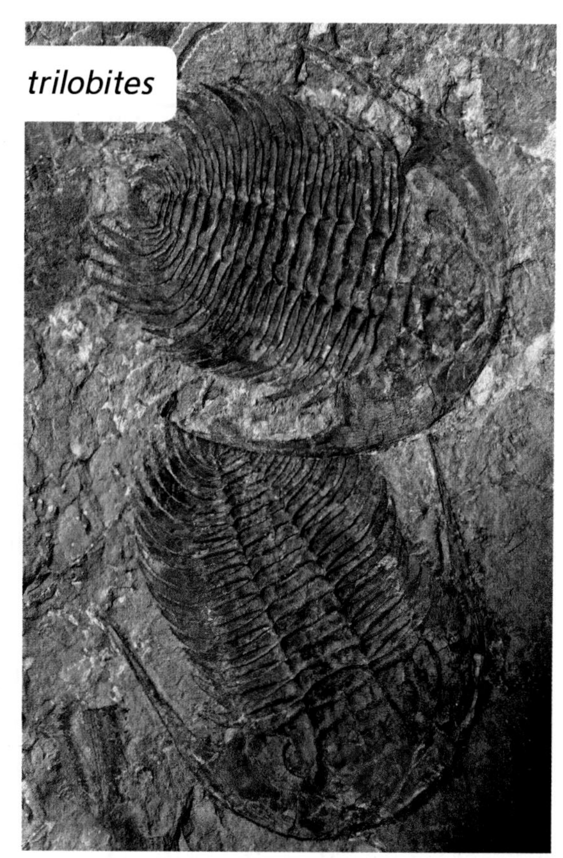
trilobites

Fossilized bones can be used to compare extinct species with modern animals. In the 1880s the fossilized bones of an organism named *Smilodon* were discovered. *Smilodon* is a type of large cat, often called a saber-toothed cat because its large fangs resemble the blade of a saber, or sword. The fossilized fangs are larger than the fangs of modern big cats. Scientists generally agree that, like modern cats, *Smilodon* used its fangs for hunting.

Compare the teeth of the modern tiger (left) to the Smilodon *skull (center) and* Smilodon *model (right).*

Space

The universe is gigantic. It contains all the matter and space that exists. There are billions of galaxies, each of which has billions of stars. A **star** is a sphere of hot gases that produces its own energy, including heat and light. The Sun is the closest star to Earth and, although it may look different from other stars, it is rather ordinary. Our Sun is a medium-sized star with an average surface temperature. Without the Sun, however, life on Earth could not exist.

The Solar System

The Sun and all the objects that orbit around it make up our *solar system*. The Sun has many satellites. A **satellite** is any object that orbits another larger body. An **orbit** is the nearly circular path that an object travels around a larger object. The Moon is a satellite that orbits Earth. On a clear evening you may see a planet or two in the sky, as well. A *planet* is a large, round object in space that orbits a star. Planets in our solar system, including Earth, orbit the Sun.

Skill Builder

Read a Diagram

Diagrams that show objects in the solar system are not to scale. Instead, these diagrams are meant to give you a general understanding of the size and positions of various space objects as compared to each other.

The Solar System

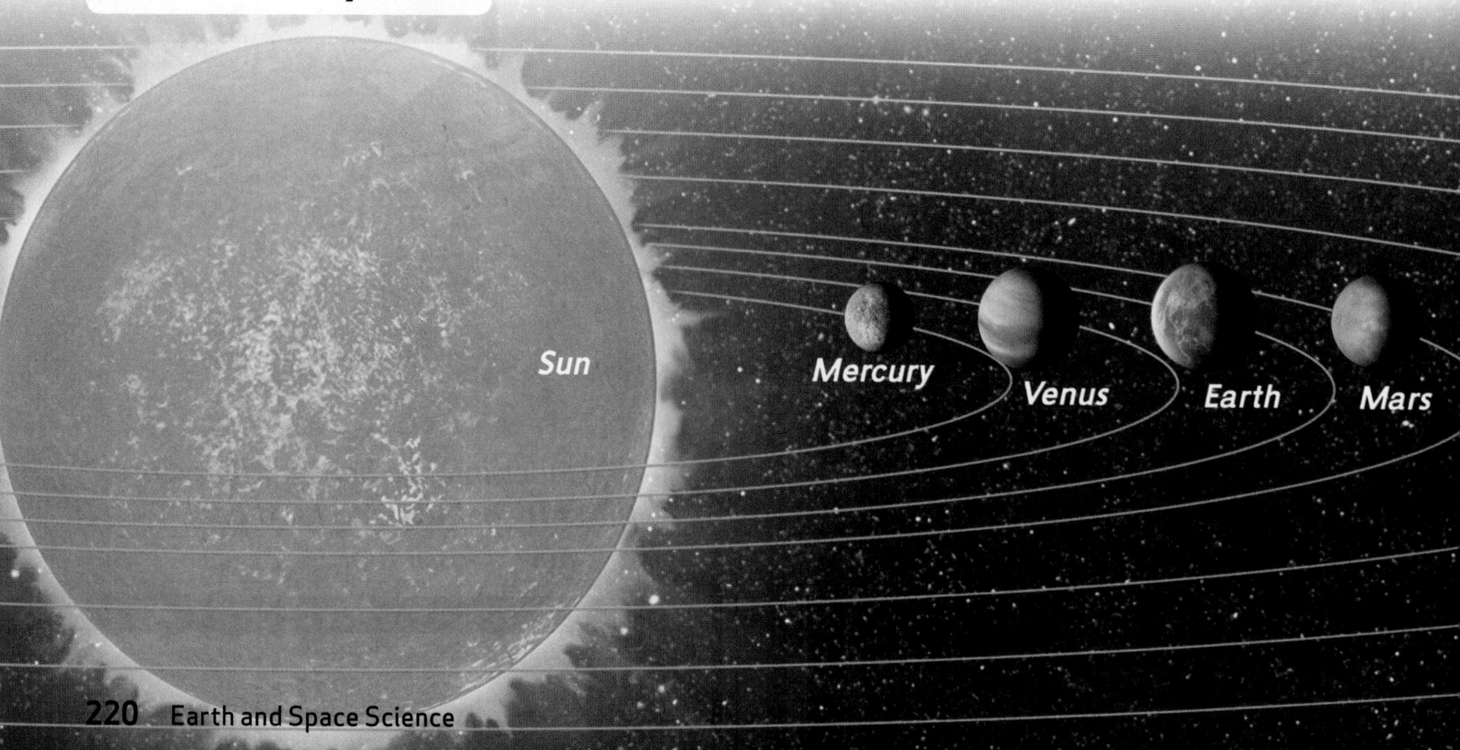

Planets of the Solar System

From nearest to farthest from the Sun, the planets are Mercury, Venus, Earth, Mars, Jupiter, Saturn, Uranus, and Neptune. The planets travel in elliptical, or nearly circular, orbits around the Sun.

Several planets are visible in the night sky from time to time, even without the use of a telescope. A **telescope** is an instrument that makes distant objects seem larger. Visible planets include Mercury, Venus, Mars, Jupiter, and Saturn. Planets do not make their own light. They reflect the light of the Sun. These lights move slowly in the night sky as the planets orbit the Sun.

Dwarf Planets Pluto was once known as the ninth planet. However, Pluto's elongated orbit and small size are different from those of the eight planets. These differences, among others, caused scientists to debate whether Pluto should be called a planet. In August 2006, the International Astronomical Union officially reclassified Pluto as a dwarf planet. *Dwarf planets* are made of rock and ice, and their orbits cross the paths of other space objects.

Scientists have discovered many dwarf planets in the solar system, but Pluto is the best known. Other dwarf planets include Ceres and Eris. Eris is larger than Pluto and even farther from the Sun.

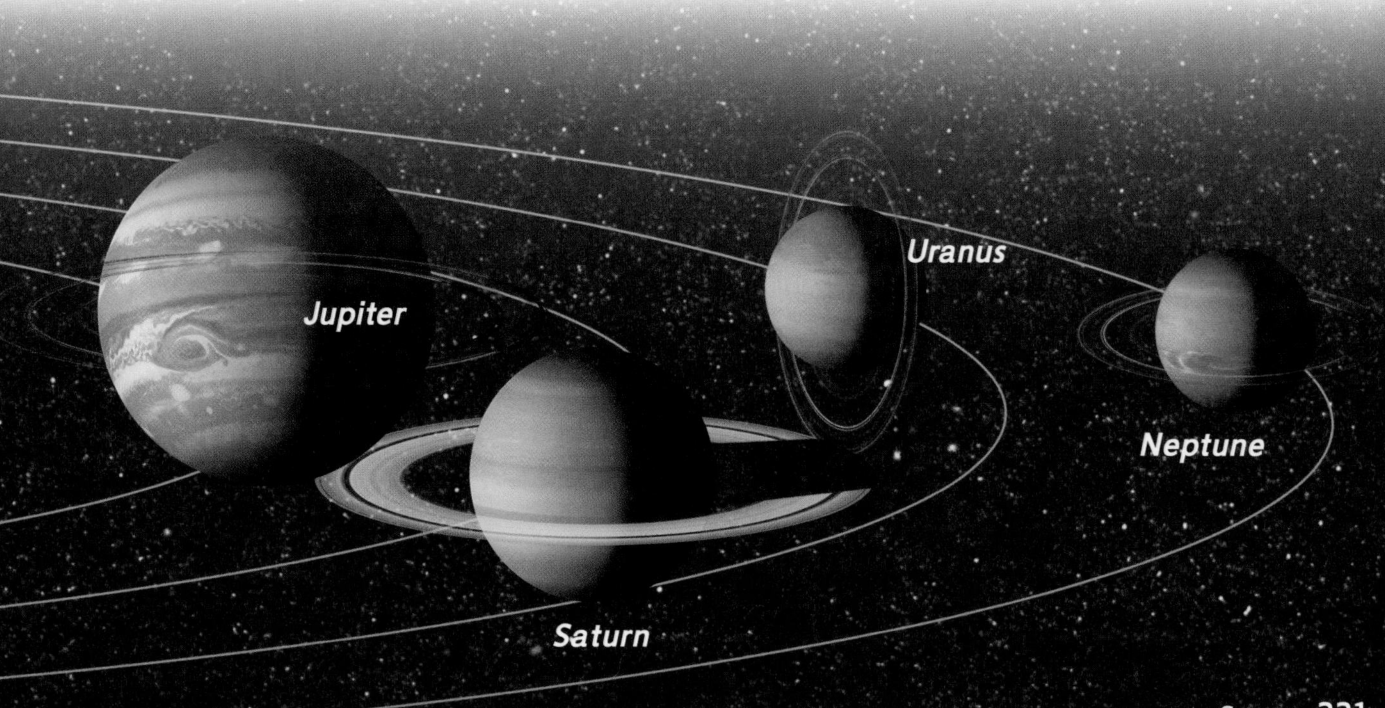

Gravity in the Solar System

Gravity is a force of attraction, or pull, between any two objects. The strength of gravity is affected by the total mass of the two objects and the distance between them. The pull of gravity decreases when the total mass of the two objects decreases and when the two objects are farther apart.

At the center of the solar system is the Sun, the most massive object in the solar system. It would take 109 Earths to fit across the Sun. The more massive an object is, the stronger its gravitational attraction. The Sun's gravitational attraction keeps Earth in orbit. In fact, gravitational attraction holds the entire solar system together.

The planets are held in their orbits around the Sun by the force of gravity between the Sun and each planet. Yet if gravity were the only force acting on a planet, the planet would be pulled into the Sun. This event does not happen because of another property, called inertia. **Inertia** is the tendency of an object to keep moving in a straight line. As Earth orbits the Sun, it is pulled toward the Sun because of gravity. At the same time, Earth's inertia makes it move away from the Sun. Together, these factors produce Earth's nearly circular, elliptical orbit.

Gravity and Inertia

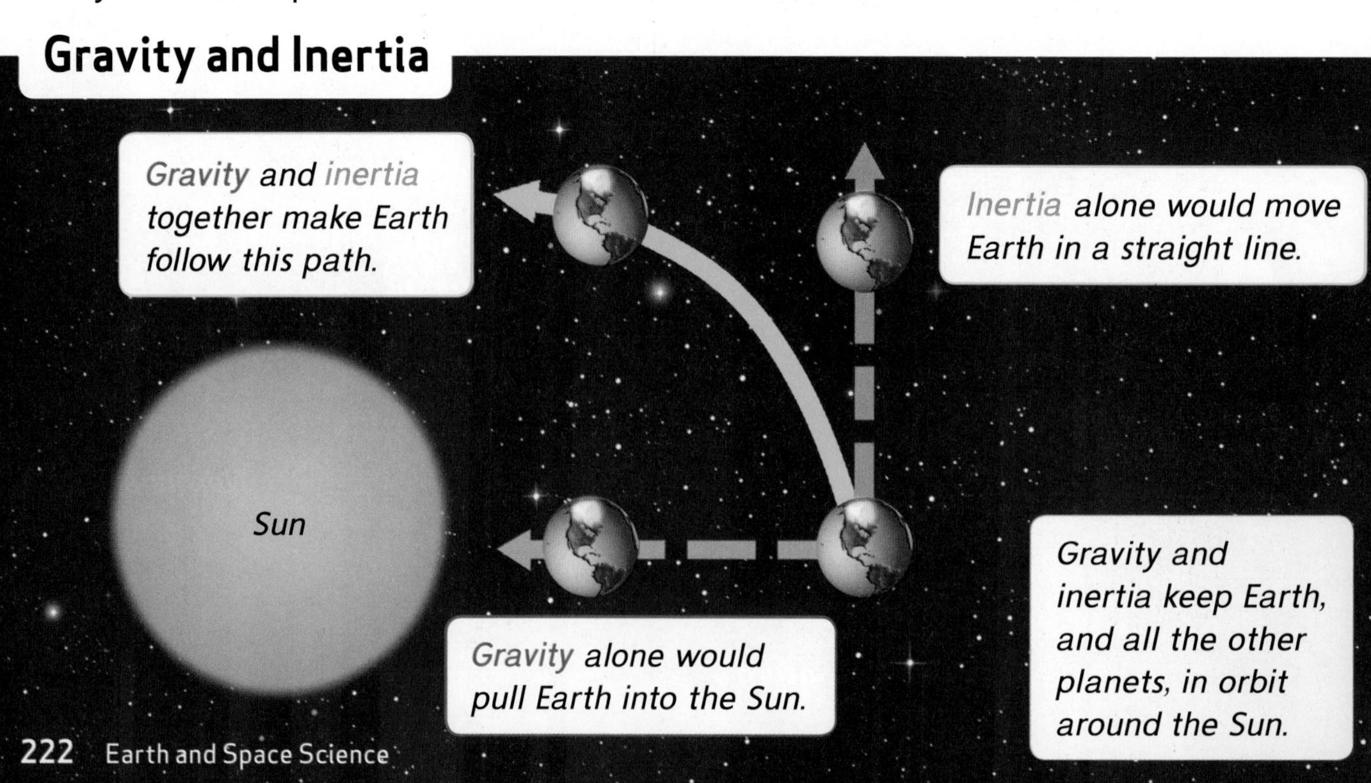

Earth's Gravitational Attraction

The Moon is Earth's only natural satellite. It is also Earth's closest neighbor in space. Some planets, such as Mercury and Venus, do not have moons. Jupiter has the most moons of any planet in the solar system. Scientists have observed 63 moons of Jupiter. Moons orbit planets for the same reason that planets orbit the Sun—because of gravitational attraction and inertia.

The Moon is less massive than Earth, so it has a weaker pull of gravity than Earth. In fact, the Moon's gravity is about one sixth of Earth's gravity. Think how high you can throw a ball. On the Moon, you could throw it about six times higher because the force of gravity is not as strong.

Earth's gravity affects the Moon, and the Moon's gravity also affects Earth. The Moon's gravitational force causes Earth's tides, the regular rise and fall of water levels along the shore. Earth's water bulges on the Moon-facing side of Earth. A bulge also forms on the side facing away from the Moon. The water level rises where the bulge is and falls where it is not. This bulge causes changing tides as the Moon orbits Earth.

This photo shows what Earth looks like from the Moon.

In this photo, you can see the height of astronaut John Young's jump on the Moon. He can jump higher on the Moon than on Earth because the Moon's gravity is about one-sixth of Earth's gravity.

Inner Planets of the Solar System

The four planets closest to the Sun are called the *inner planets*. They are also called the terrestrial, or rocky, planets because they are made mostly of rock. Evidence suggests that each has a core of iron. They have relatively similar sizes and closely spaced orbits. They have few, if any, moons. All the inner planets rotate relatively slowly and none of them have rings. Despite these similarities, each planet has its own unique features.

Mercury Mercury is the closest planet to the Sun. That closeness makes it very hot. It has almost no water and very little air. The surface has many craters, like Earth's Moon. A *crater* is a hollow area or pit in the ground. Craters form when large space rocks crash into other space objects. Mercury is the smallest inner planet. At its equator, it is less than half the size of Earth.

Venus Venus is the second closest planet to the Sun. It has a thick atmosphere that is made mostly of carbon dioxide, with atmospheric pressure 90 times greater than that of Earth. The atmosphere does not allow heat to easily escape. This atmosphere makes Venus the hottest planet. There are many volcanoes on Venus, and its surface is covered in lava flows.

Mercury

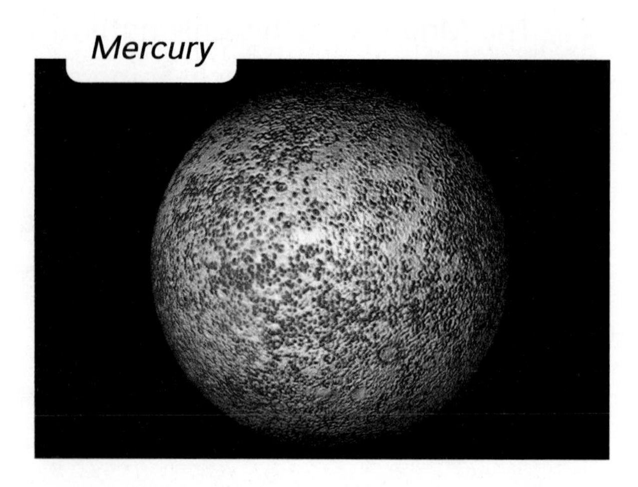

Distance to the Sun: 58 million km
Diameter: 4,880 km
Fast Fact: Mercury's surface is covered with craters.

Venus

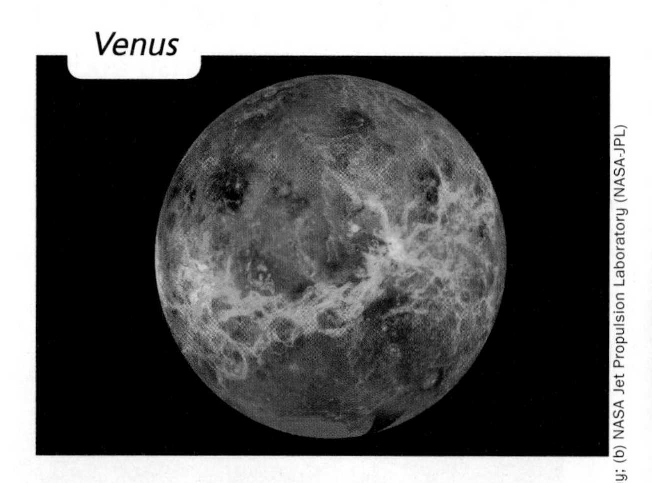

Distance to the Sun: 108 million km
Diameter: 12,100 km
Fast Fact: Temperatures on Venus can reach 500° Celsius (932° Fahrenheit).

Did You Know?

Mercury's Sun-facing side is hot enough to melt zinc. However, on the night side of the planet, temperatures can drop to -170°C (-274°F).

Earth Earth is unique in our solar system. It has oxygen and liquid water. Earth is the only planet known to support life. Earth's atmosphere keeps temperatures from getting too hot or too cold to sustain life as we know it. It is the largest of the inner planets.

Mars Of all the planets, Mars is the most like Earth. It has two small moons and a thin atmosphere. Mars has volcanoes, but they are no longer active. The surface has many features that show evidence of erosion by floods and rivers. Today, Mars is much colder than Earth. Its water is frozen in ice caps near both poles. NASA has sent probes to Mars and hopes to send astronauts to the red planet one day.

Beyond the Inner Planets

Beyond the orbit of Mars is a belt of space rocks called **asteroids**. These are rocky or metallic objects that orbit the Sun. Scientists have accumulated a great deal of information about asteroids in recent years. Space probes have sent back information that provides pictures of these orbiting objects.

Ida is a heavily cratered, irregularly shaped asteroid.

Earth

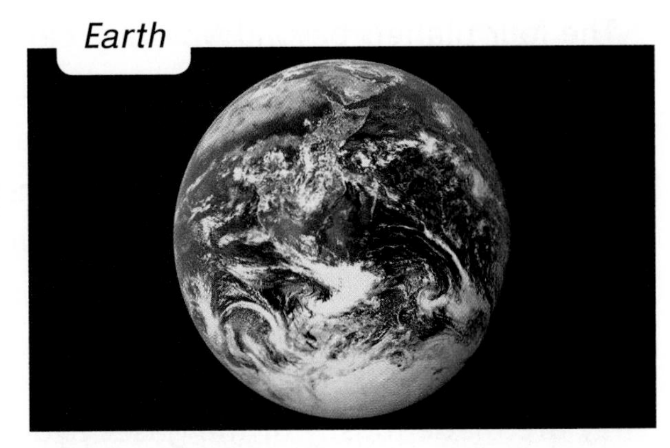

Distance to the Sun: 150 million km
Diameter: 12,756 km
Fast Fact: Earth's atmosphere makes it suitable for life.

Mars

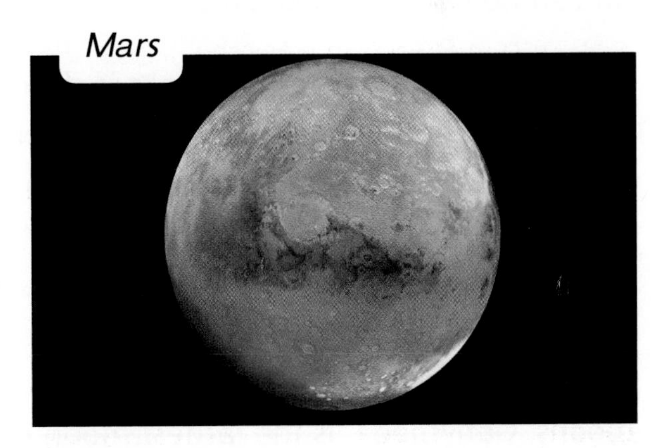

Distance to the Sun: 228 million km
Diameter: 6,794 km
Fast Fact: Iron oxide, or rust, gives Mars its reddish color.

Did You Know?

The largest asteroid in the asteroid belt is about one fourth the diameter of Earth's moon.

Outer Planets of Earth's Solar System

The four planets beyond Mars are called the *outer planets*. They are also called *gas giants* because they are huge compared with the inner planets, and because they consist mostly of gases. The largest gas giant, Jupiter, is five times farther from the Sun than Earth is.

The gas giants do not have solid surfaces. They are made mostly of the gases hydrogen and helium. Scientists have evidence that they may have some rock and ice at their cores. Each has a ring system, although most are difficult to see. They also have many moons and some have atmospheres.

Jupiter This planet's atmosphere is divided into bands of strong winds. The winds in each band blow in directions opposite the bands on either side of it. One band has a large red spot the size of Earth. The red spot is a storm that has been blowing for over 400 years. The storm is known as The Great Red Spot. One of Jupiter's moons, Ganymede, is the largest moon in the solar system. Another, Europa, may have an ocean of water beneath its icy crust. The moon named Io has active volcanoes.

Saturn Saturn is the second largest planet. It is famous for its system of rings. The rings are made of pieces of ice and rock. Most of these pieces are less than a couple meters in diameter. Saturn has at least 34 moons. The largest is named Titan.

Jupiter

Distance to the Sun: 778 million km
Diameter: 143,000 km
Fast Fact: Jupiter's four largest moons were first observed by Galileo in 1610.

Saturn

Distance to the Sun: 1 billion, 429 million km
Diameter: 120,536 km
Fast Fact: Winds on Saturn can blow at 500 meters per second.

(t) NASA-JPL; (b) Reta Beebe (New Mexico State University), D. Gilmore, L. Bergeron (STScI), and NASA

Uranus Uranus is sometimes known as the "sideways" planet. The axis is tilted so much that it rotates on its side. An *axis* is an imaginary line down an object's center that it appears to rotate around. Uranus's tilted axis means that one pole faces the Sun during parts of Uranus's orbit. The unusual blue-green color of this planet is due to gases in its upper atmosphere, including methane. Uranus has at least 27 moons. One of its moons, Miranda, looks as though it broke apart and the pieces clumped back together several times as it formed.

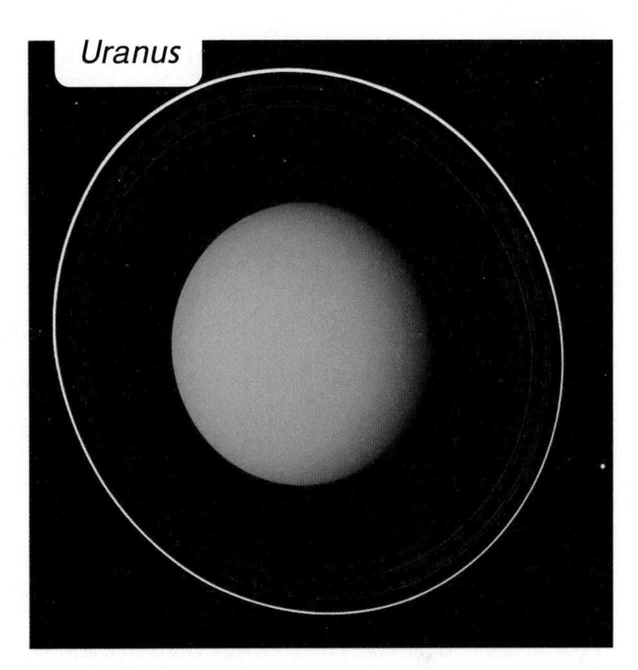

Distance to the Sun: 2 billion, 871 million km
Diameter: 51,118 km
Fast Fact: The axis of Uranus is tilted toward the Sun.

Neptune Neptune is the farthest gas giant from the Sun. Winds on Neptune can blow at speeds of 2,000 kilometers (1,200 miles) per hour. Its atmosphere, like that of Uranus, is mostly hydrogen, helium, and methane. There may be an ocean underneath Neptune's clouds. Scientists have observed 13 moons orbiting Neptune. Triton is the largest moon. Triton is known to have "ice volcanoes" that shoot material up to 8 km (5 mi) high.

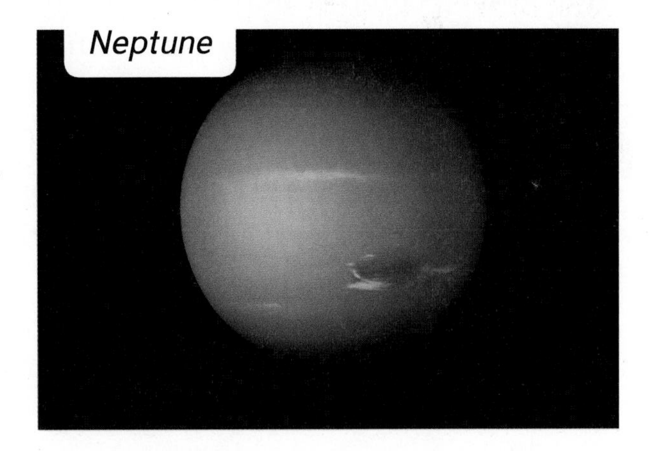

Distance to the Sun: 4 billion, 504 million km
Diameter: 49,528 km
Fast Fact: Neptune takes 165 Earth years to orbit the Sun.

Beyond the Outer Planets

Beyond Neptune's orbit lie the Kuiper Belt and the Oort Cloud. Both of these regions are composed of small, icy bodies and are the origin of comets. Scientists hypothesize that the bodies that make up these regions are the remnants from the formation of the solar system.

Other Objects in the Solar System

Other than the Sun, planets and moons are the largest objects in the solar system. The next largest objects are dwarf planets. However, many smaller objects are also found in our solar system.

Meteoroids A *meteoroid* is a small, rocky or metallic object that orbits the Sun in both the inner and outer regions of the solar system. The craters on the Moon were formed by meteoroids colliding with its surface.

This is a meteorite.

Meteors A *meteor* is a meteoroid that enters Earth's atmosphere. It appears as a bright streak in the sky. If a meteor does not break apart and burn up in the atmosphere, it can hit Earth's surface. You may have heard meteors called shooting stars. Some evenings you can observe many meteors in the night sky. These events are called meteor showers.

Meteorites A meteor that strikes Earth's surface is called a *meteorite*. Many places on Earth show evidence of meteorite impacts. One such place is Meteor Crater in Arizona. About 50,000 years ago, a large meteorite crashed there, forming a crater that is as wide as 11 football fields.

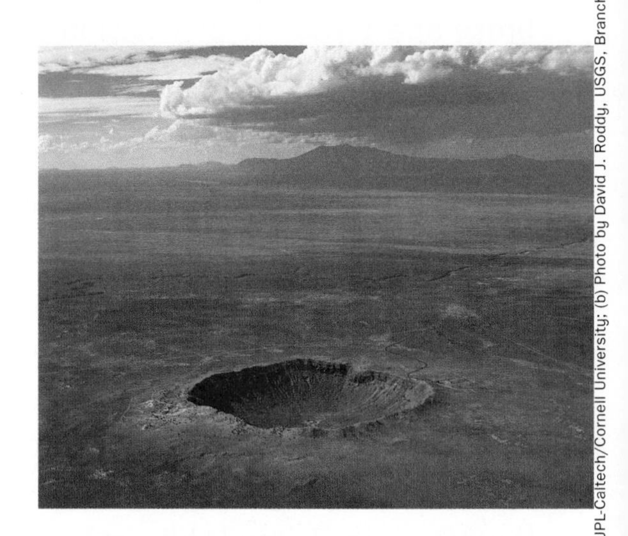

A meteorite caused this crater when it collided with Earth's surface.

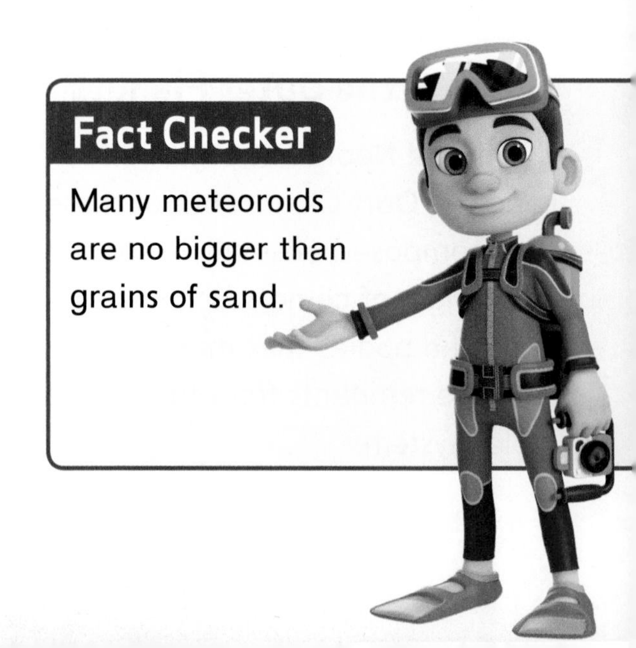

Fact Checker

Many meteoroids are no bigger than grains of sand.

Comets

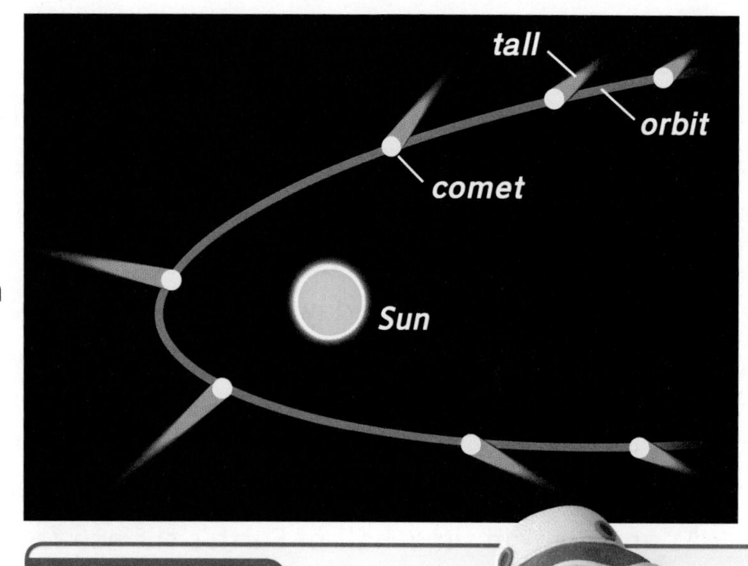

A *comet* is a mixture of frozen gases, ice, dust, and rock that moves in an elliptical orbit around the Sun. Comets are thought to be bits of material left over from the formation of the solar system about 4.6 billion years ago.

When a comet is farther away from the Sun, the gases and ice in the comet are frozen. As the comet moves toward the Sun, the core, or nucleus, of the comet warms up. Some of the ice and dust in the nucleus form a cloud or *coma* around the nucleus. Together, the coma and the nucleus make up the head of the comet.

Skill Builder

Read a Diagram

Notice the relationship between the comet's tail and the Sun.

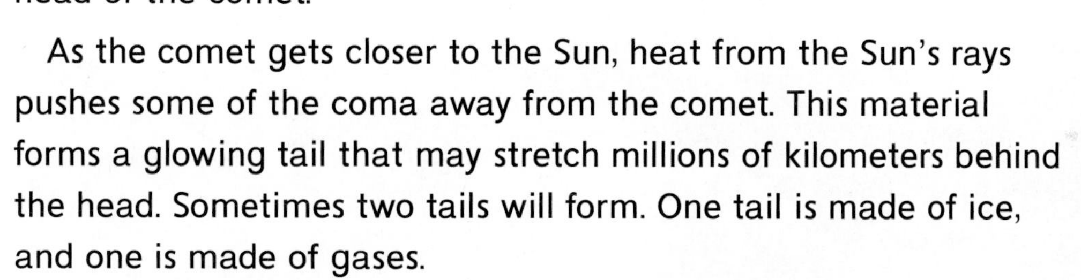

As the comet gets closer to the Sun, heat from the Sun's rays pushes some of the coma away from the comet. This material forms a glowing tail that may stretch millions of kilometers behind the head. Sometimes two tails will form. One tail is made of ice, and one is made of gases.

Heat moves out from the Sun in every direction. As a comet moves around the Sun, the head stays closest to the Sun and the tail trails out behind it. No matter where the comet is in its path around the Sun, the comet's tail always points away from the Sun.

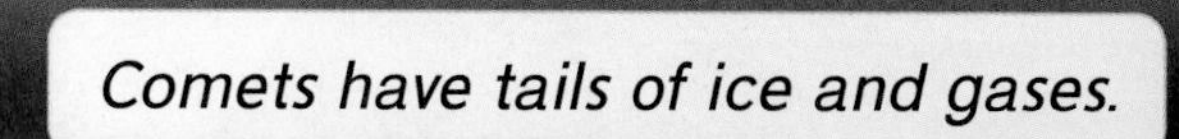

Comets have tails of ice and gases.

StockTrek/Getty Images

Earth in Space

Earth is moving at 30 km/s (19 mi/s) as it orbits the Sun. A **revolution** is a complete pass around the Sun, taking 365¼ days or one year. Earth is also spinning on its axis at about 1,600 km/h (1,000 mph). One **rotation** is a complete spin on the axis. Earth makes one rotation every day or 24 hours. Living things do not feel these movements because they are moving with Earth.

Earth's Rotation

At any point in time, half of Earth's surface faces the Sun and is in daylight. The other half of Earth's surface faces away from the Sun and is in darkness.

The tilt of Earth's axis affects the length of the day. If the axis were not tilted, day and night would each be 12 hours long. Instead, there are more hours of daylight and fewer hours of darkness during the summer. In winter, the amount of daylight is shorter.

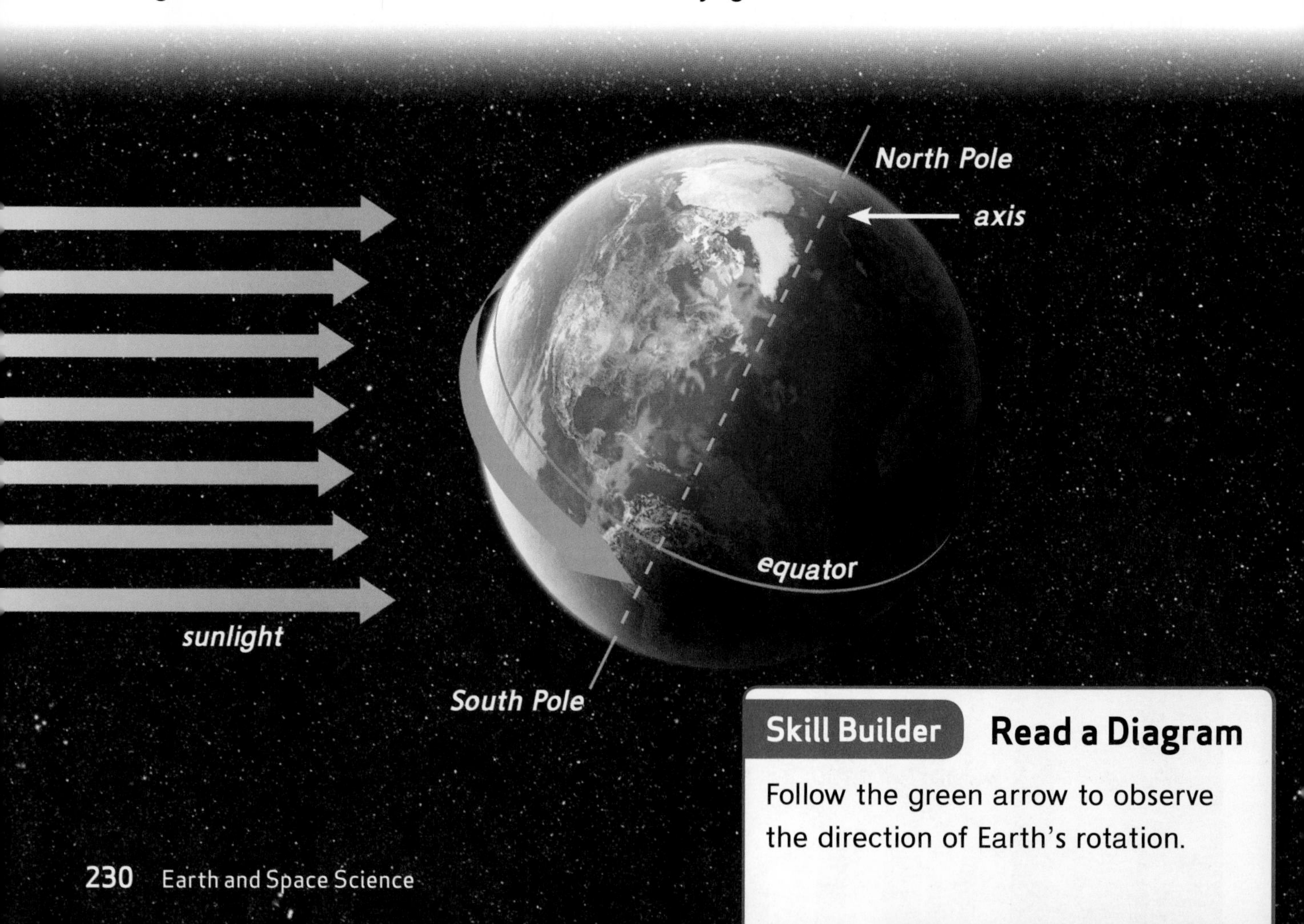

Skill Builder **Read a Diagram**

Follow the green arrow to observe the direction of Earth's rotation.

Apparent Motion

As Earth rotates, you see different parts of space. During the day, the side of Earth you live on faces the Sun. As that part turns away from the Sun, it becomes night. The rotation of Earth changes day into night and night into day again.

If you watch objects in the sky, such as the Sun, they appear to rise in the east, move across the sky, and set in the west. What you are seeing is the apparent motion of these objects, not their real motion. *Apparent motion* is the way something appears, or seems, to move. As Earth rotates from west to east, objects in the sky appear to move from the east to the west. Earth's rotation causes the apparent motion of many objects in space. Stars only seem to move. The Moon and planets do not always move in the same direction as their apparent motion.

Shadows

A shadow forms when light is blocked. The light strikes an object but cannot pass through it. You cast a shadow when your body blocks sunlight. Your shadow always points away from the Sun. As the position of the Sun in the sky changes, your shadow changes, too. Early in the morning, your shadow is long. It shrinks until midday. Then it grows longer again until sunset.

When the Sun is high in the sky, an object has a shorter shadow.

When the Sun is low in the sky, an object has a longer shadow.

Earth's Revolution

Earth revolves around the Sun. To revolve means to move around another object. The path a revolving object follows is its **orbit**. Earth's orbit is shaped like an ellipse, or a slightly flattened circle. Earth's orbit around the Sun takes 365¼ days, or one year.

Recall that Earth's axis, the imaginary line about which it rotates, is tilted. It is tilted at an angle of 23.5°. The tilt causes sunlight to strike different parts of Earth at different angles. At any given time, each hemisphere, or half, of Earth gets more or less sunlight than the other. The seasons result from both Earth's tilted axis and its revolution around the Sun.

Did You Know?

Every four years is a "leap" year—a year in which a 366th day is added to the calendar. This day is necessary because Earth takes slightly less than 365 ¼ days to orbit the Sun. Adding an extra day every four years keeps the calendar in line with the seasons.

How Seasons Change in the Northern Hemisphere During a Year

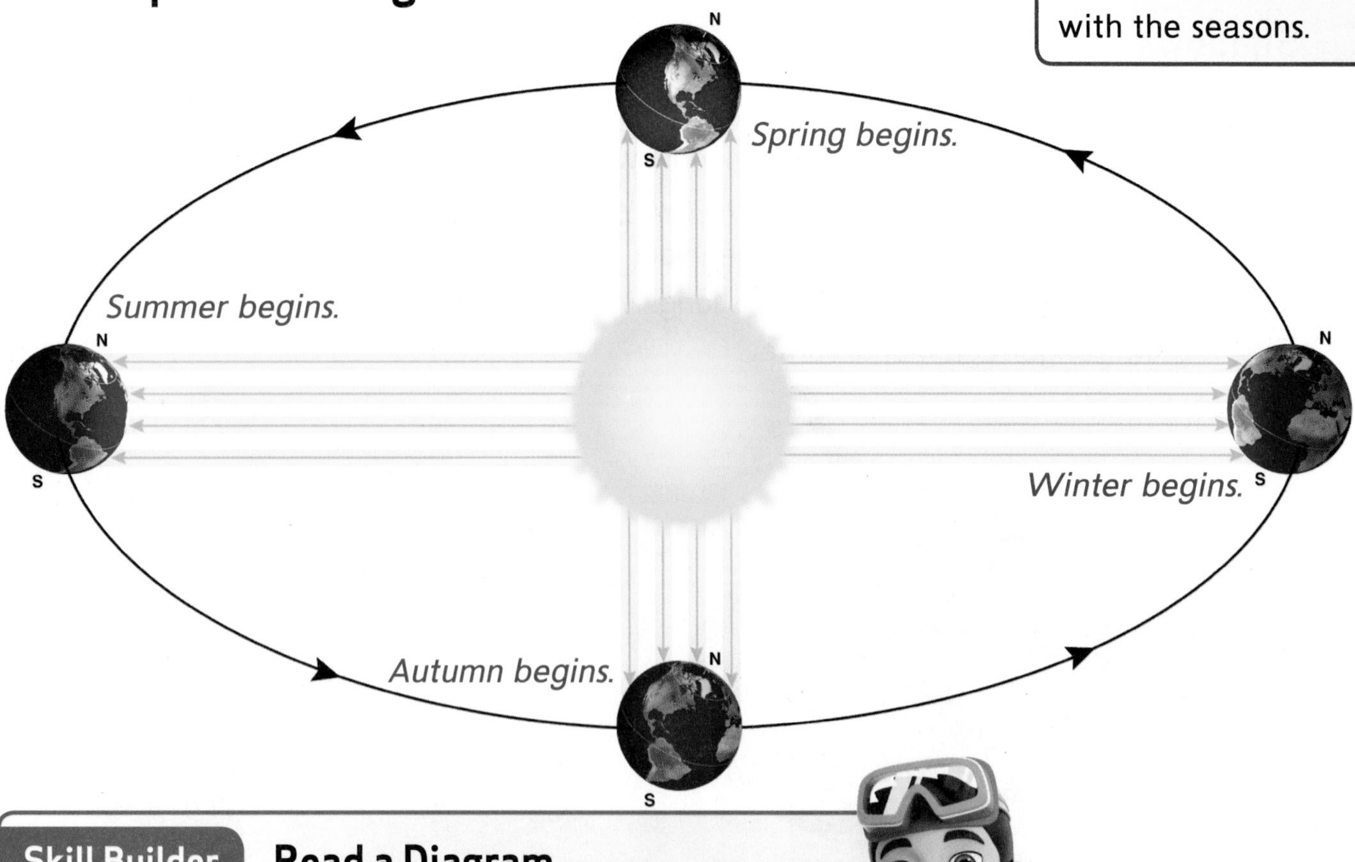

Skill Builder **Read a Diagram**

Identify a point on the Northern Hemisphere. Follow this point throughout Earth's orbit to see how the angle of light that falls on that area changes.

Seasons

As Earth revolves around the Sun, the tilted axis always points in the same direction. When the Northern Hemisphere is tilted away from the Sun, Earth's surface does not receive as much heat energy, and temperatures are lower. In the Northern Hemisphere, this is winter.

At the same time, it is summer in the Southern Hemisphere. The Southern Hemisphere is angled toward the Sun, so the heat energy of sunlight is more concentrated. The surface receives more heat energy, and temperatures are warmer.

Because the tilt of Earth's axis always points in the same direction, the seasons in the Northern Hemisphere and the Southern Hemisphere are always opposite. In spring and autumn, both hemispheres receive equal warmth from the Sun, making temperatures mild.

Fact Checker

Seasons do not have anything to do with how close Earth is to the Sun at any point in time. In fact, Earth is closer to the Sun in January than in July. Seasons occur because of Earth's tilted axis.

The Four Seasons in the Northern Hemisphere

winter
December 21 - March 20

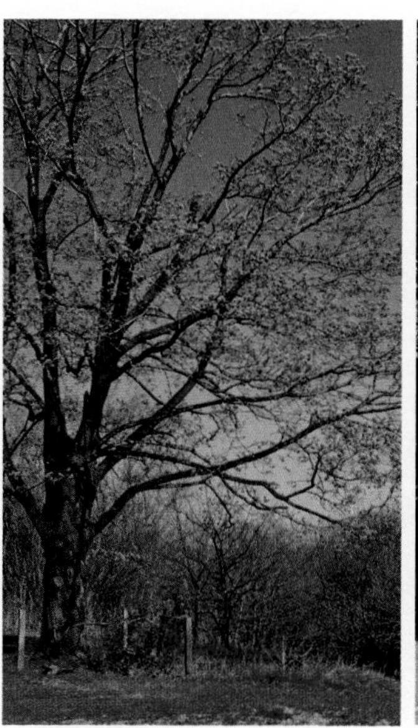

spring
March 20 - June 21

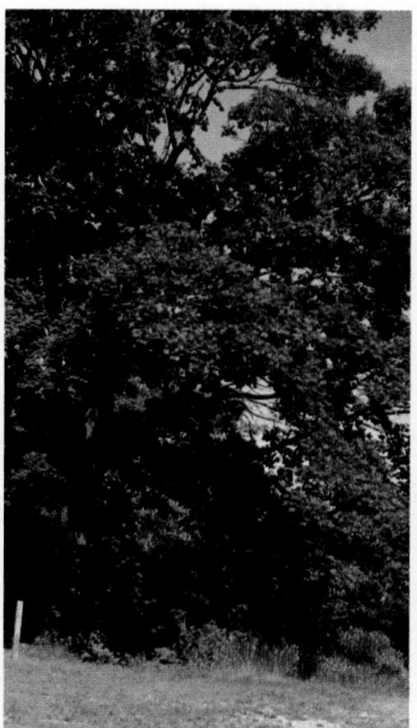

summer
June 21 - September 22

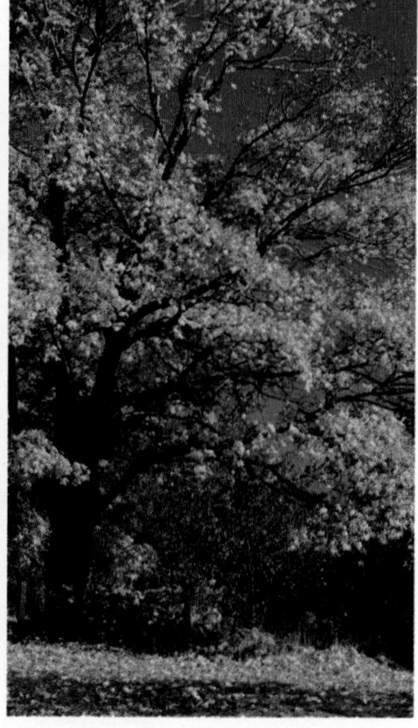

autumn
September 22 - December 21

Earth's Revolution - Seasons and the Sun

The Sun's apparent path changes from season to season. The diagram shows the Sun's apparent path across the sky during the day. Each yellow circle represents the Sun's position at midday. Notice that the Sun rises much higher in the sky during a summer day. The day on which the Sun appears highest in the sky is known as the summer solstice. In the Northern Hemisphere, the summer solstice occurs around June 21 each year. During this time of year, the Northern Hemisphere is tilted more toward the Sun.

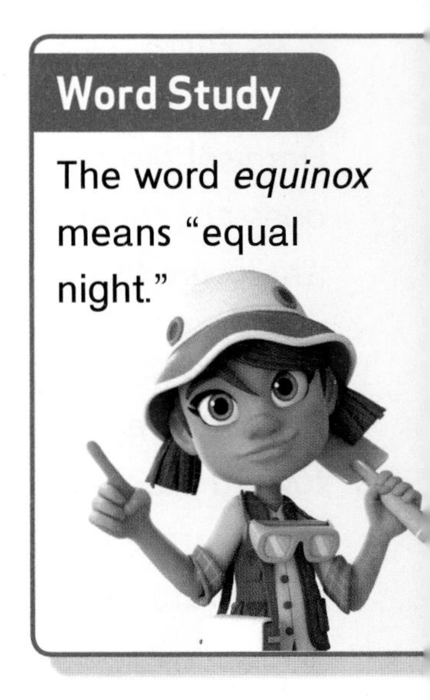

Word Study

The word *equinox* means "equal night."

In winter, the Sun appears much lower in the sky. In the Northern Hemisphere, the winter solstice occurs around December 21. This is the day on which the Sun appears lowest in the sky. At this time, the Northern Hemisphere is tilted away from the Sun.

Halfway between the solstices, neither hemisphere is tilted toward the Sun. The noon Sun is almost directly overhead. Each of these days is known as an equinox. During an equinox, day and night are each about 12 hours long. In the Northern Hemisphere, the spring equinox occurs around March 21. The fall, or autumnal, equinox occurs around September 22.

Apparent Path of the Sun

summer

spring and fall

winter

Fotosearch Premium/Getty Images

Note that the diagram showing the Sun's apparent path does not apply to all parts of the world. At the equator, the Sun's apparent path changes much less during the year. The farther away you travel from the equator, the greater the change in the Sun's height throughout the year. For example, near the poles in summer, the hours of daylight are great. During winter, the Sun hardly appears above the horizon. Examine the three graphs, which show the number of daylight hours throughout the year for three cities in the Northern Hemisphere. The height of the bars indicates the number of daylight hours on the 15th of each month throughout the year.

Hours of Daylight

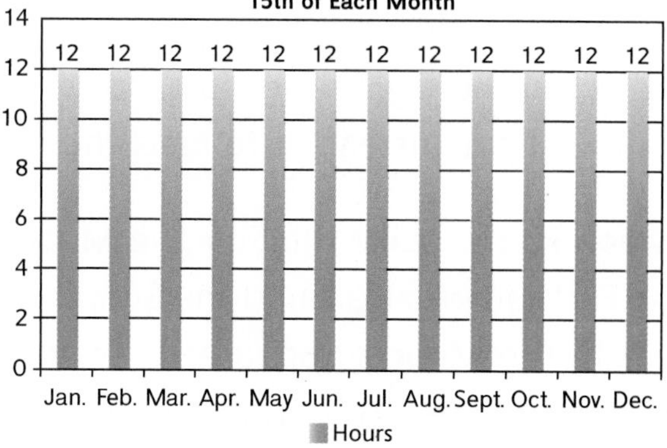

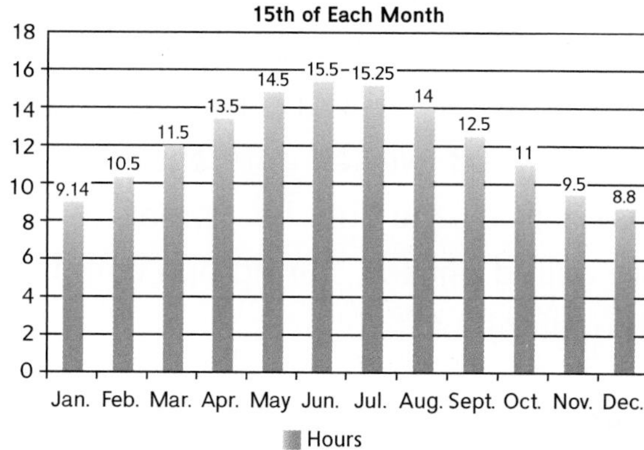

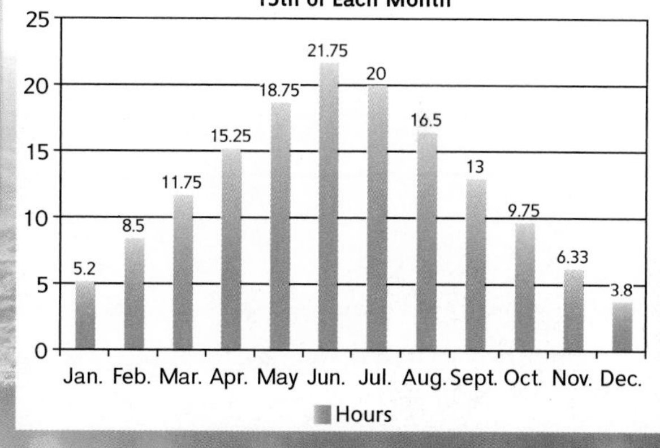

Skill Builder **Read a Bar Graph**

Macapá, Brazil, is just north of the equator. Minneapolis is 45° north of the equator. Fairbanks is near the North Pole. Note how distance from the equator changes the number of daylight hours.

This time-lapse image was taken in the polar region of the Northern Hemisphere during winter. Note how low the Sun is in the sky, even at its highest point.

alohaspirit/E+/Getty Images

Earth's Moon

On many nights, the Moon appears to be the largest, brightest object in the sky. Unlike stars, however, the Moon does not make its own light. Instead, it reflects the light of the Sun.

Moon Phases Like the Sun, the Moon appears to rise and set. As Earth revolves around the Sun, the Moon revolves around Earth. The Moon's appearance changes as it revolves. The Moon completes one orbit around Earth in just over 29 days. This amount of time is almost as long as an average month.

As the Moon orbits Earth, the Sun is shining. The Sun lights one half of the Moon at a time. The other half is dark. During the Moon's orbit, we see different amounts of the half of the Moon that is lit by the Sun. The apparent shapes of the Moon in the sky are called **phases**. During one complete orbit, the Moon cycles through all of its phases.

As the Moon appears to get larger, it is *waxing*. As it appears to get smaller, it is *waning*. A *crescent moon* appears to be a sliver, while a *gibbous moon* is almost full. During the *new moon* phase, the Moon cannot be observed at all.

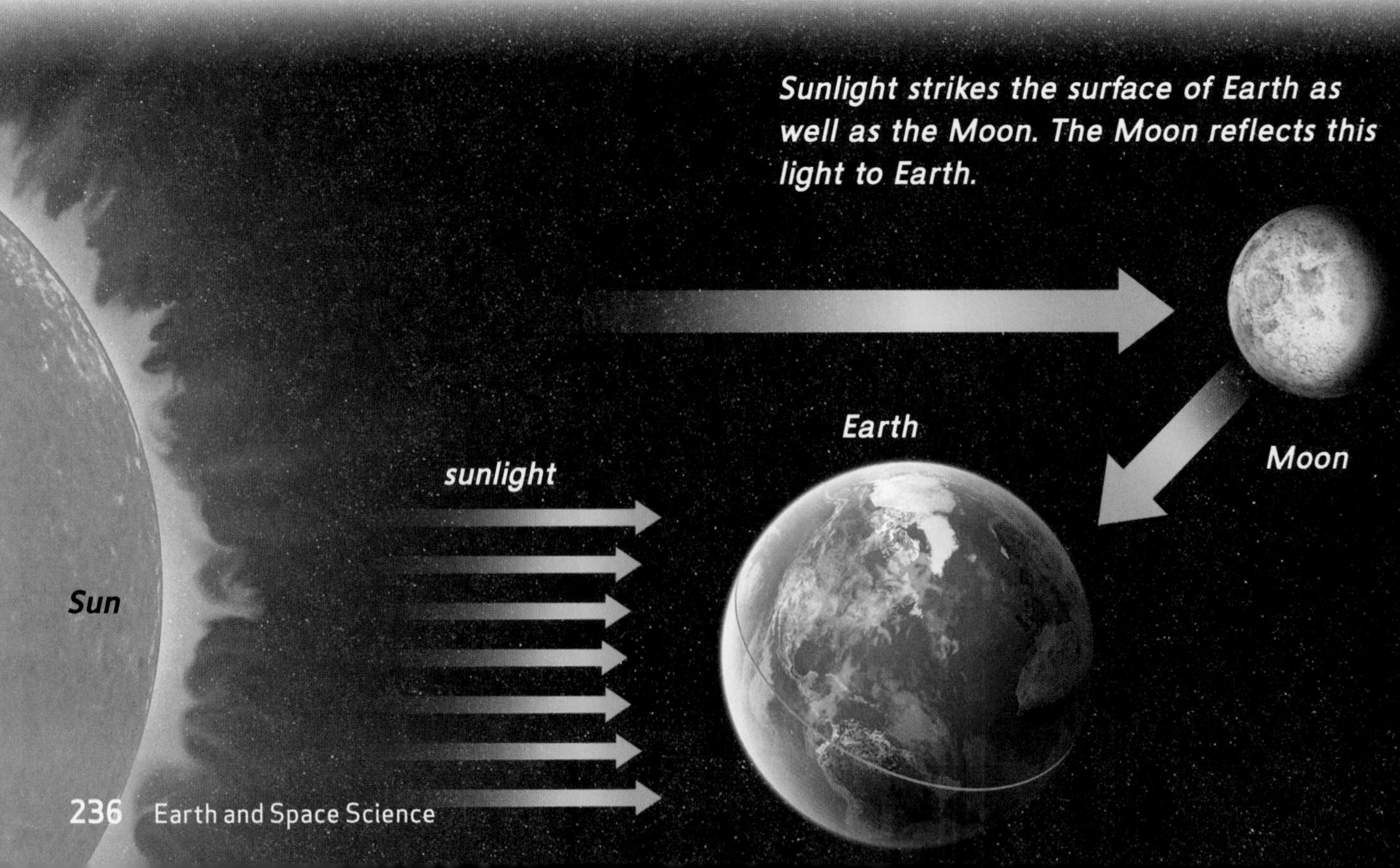

Sunlight strikes the surface of Earth as well as the Moon. The Moon reflects this light to Earth.

Phases of the Moon

third quarter moon
The Moon is three quarters of the way around Earth.

waning gibbous moon
Slightly less of the lit side can be seen.

full moon
The entire lit side can be seen.

waxing gibbous moon
The Moon is almost full.

first quarter moon
The Moon is a quarter of the way around Earth.

waxing crescent moon
Some of the lit side can be seen.

new moon
The lit side cannot be seen from Earth.

waning crescent moon
The left sliver of the Moon is the only part you can see.

Skill Builder Read a Diagram

Observe the small moons along the blue circle. Use these "mini-moons" to infer the direction from which the Sun's rays are shining.

Stars

A star is a sphere of hot gases that gives off light and heat. The only star you can observe during the daytime is the Sun. The Sun is the closest star to Earth. Other stars are much farther away. Throughout the universe, stars are found in large groups called *galaxies*. Our Sun is near the edge of a galaxy with billions of other stars. You may know this galaxy as the Milky Way. Our galaxy's nearest neighbor is the Andromeda Galaxy. The universe may have many more galaxies that have yet to be discovered, each with billions of stars.

Star Colors and Temperature

Stars are different colors. These colors occur because of the surface temperature of each star. Think about the flames of a bonfire. Different parts of the fire are different temperatures. Cooler areas are red. The hottest areas are orange-yellow. This same relationship between color and temperature applies to stars. The Sun's temperature makes it look yellow. Cooler stars are red or orange. Warmer stars are white or blue.

A star glows for a very long time. Our Sun is about 5 billion years old. Evidence suggests that it will glow for another 5 billion years or so.

Like the Milky Way, the Andromeda Galaxy is shaped like a spiral. It is wider than our own Milky Way Galaxy.

Star Distances

The Sun is about 150 million km (93 million mi) from Earth. It takes about eight minutes for its light to reach Earth. Most stars are much farther away. After the Sun, the next closest star to Earth is Proxima Centauri. This star is about 40 trillion km (24.8 trillion mi) away. Because stars are so far from Earth, writing their distance in kilometers or miles becomes awkward.

To simplify the writing of such large distances, astronomers use a unit called a light-year. A **light-year** is the distance light travels in one year, which is nearly 10 trillion km (6 trillion mi). Proxima Centauri is 4.2 light-years from Earth.

Fact Checker

A light-year is not a measure of time, but of distance.

When you observe a distant star, you are seeing what it looked like in the past. A star you observe today may have stopped glowing millions of years ago. However, its light is still making its way through space. The light we see from the Proxima Centauri system left there about 4.2 years ago.

Nearest Stars to Earth

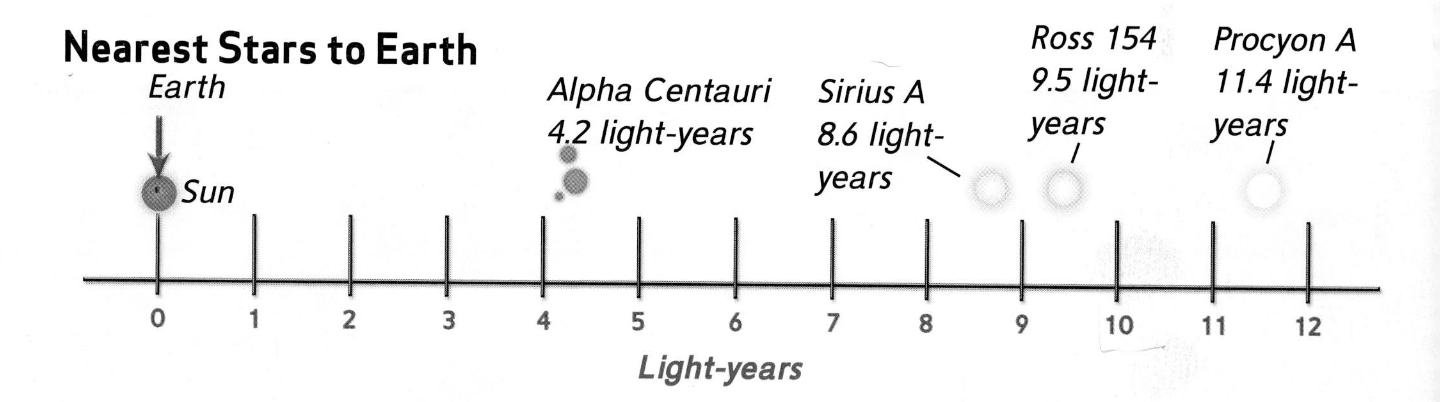

Skill Builder **Read a Diagram**

In this diagram, Earth is right next to the Sun at 0. Count the light-years from Earth-Sun to Alpha Centauri.

Star Cycles

Stars form when matter comes together and starts to give off energy. Stars go through stages, or cycles, between their beginning and ending. Different kinds of stars have different cycles. The cycle of a star depends on how much hydrogen the star contains. A star's cycle ends when it stops giving off energy.

A star forms out of a nebula. A **nebula** is a huge cloud of gases and dust. Gravity pulls the mass of the nebula, most of which consists of hydrogen gas, closer together. As hydrogen atoms move closer, they collide with one another. These collisions produce heat, and the temperature in the cloud rises. When the temperature reaches at least 10,000,000° Celsius (18,000,000° Fahrenheit), hydrogen atoms begin combining to form a new gas, helium. This process gives off tremendous amounts of heat and light. The nebula becomes a *protostar*, or beginning star. The protostar continues to gain mass because of its gravitational pull. Its heat makes it glow.

Make Connections

Jump to the Matter section to learn about how atoms and molecules make up all matter.

1 *nebula*

2 *protostar*

The Sun, and other stars like it, started with a medium amount of hydrogen. That hydrogen is the fuel that produces energy in the Sun. For a few billion years, hydrogen atoms continue combining to form helium, and the star increases in temperature.

Eventually the heat forces the hydrogen on the edge of the star to expand into space. As the expanding hydrogen moves farther from the center of the star, it cools and turns red. At this stage in its cycle, the star has become a *red giant*. A red giant is many times larger than the original star. In the star's core, the temperature has risen to about 100,000,000°C (180,000,000°F). Helium atoms now combine to form atoms of carbon.

When all the helium is gone, the star can no longer combine helium to form carbon. Now the star begins to cool and shrink, becoming a white dwarf. A **white dwarf** is a small and very dense star that shines with a cooler white light. The white dwarf stage is the end of a medium-sized star's cycle.

About 10 billion years pass during this cycle. Because the Sun is approximately 5 billion years old, it is about halfway through the cycle.

Skill Builder

Read a Diagram

Follow the numbers to better understand the stages of a medium-sized star. The Sun is in stage 3 of this cycle.

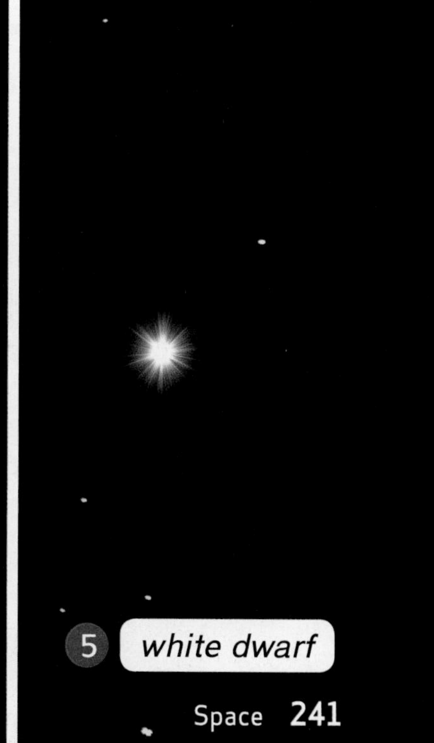

Star Cycles - Cycles of Larger Stars

Stars that start off with greater amounts of hydrogen end their cycle differently. After they become red giants, the temperature of the core of these stars increases to about 600,000,000°C (1,080,000,000°F). At this temperature, their atoms combine to form atoms of iron.

Eventually the iron core produces more energy than gravity can hold together, and the star explodes. The exploding star is called a **supernova**. Supernovas shine brightly for days or weeks and then fade away. A supernova will form a new nebula.

If a star is very massive, it may end its cycle as a black hole. A **black hole** is an object that is so dense and has such powerful gravity that nothing can escape from it, not even light.

Skill Builder

Read a Diagram

Look for a pattern between temperature and color in these different stars.

Star Classification

Stars are characterized by their size, color, and temperature. The Sun is a medium-sized yellow star with a surface temperature of about 6,000°C (11,000°F). Giant stars have diameters that are 10 to 100 times that of the Sun. Super giants may have diameters that are 1,000 times that of the Sun. Neutron stars are the smallest stars and are 60,000 times smaller than the Sun.

Color and Surface Temperatures of Stars

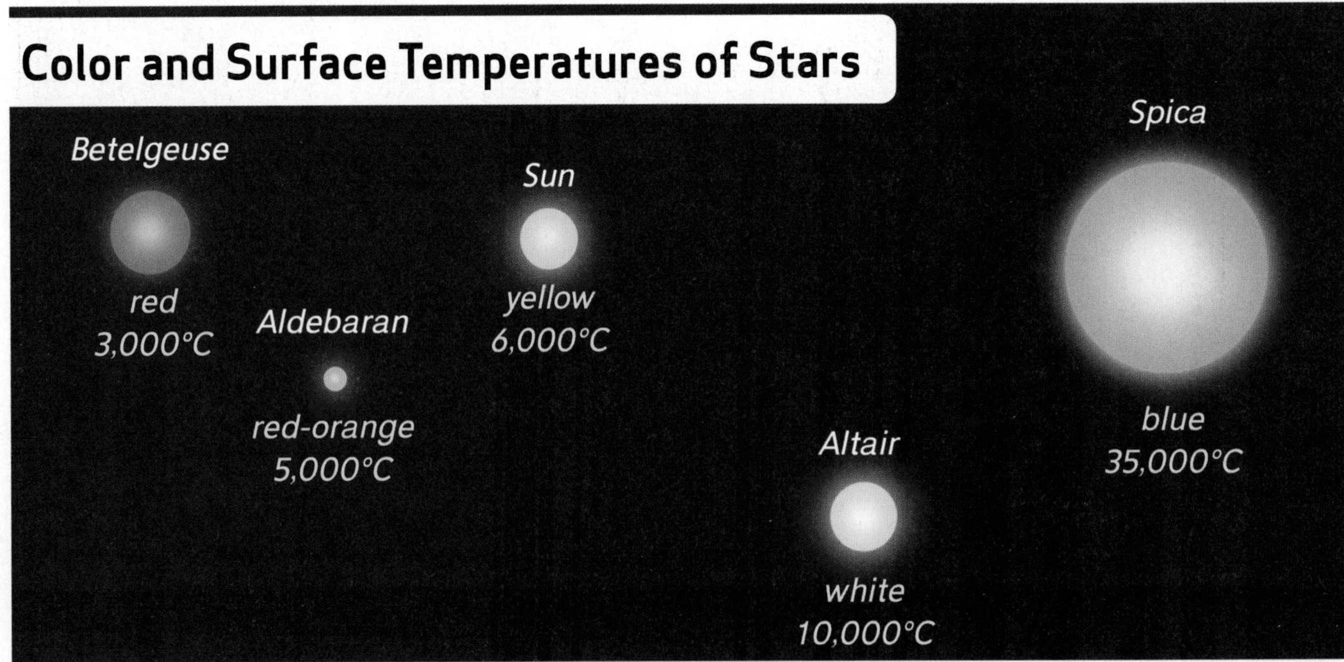

Distant Stars with Solar Systems

Planets around distant stars are too dim, small, and far away to be observed—even through the most powerful telescopes. How are these planets discovered? Remember that gravity causes all objects to pull on other objects. When scientists observe a star whose motion is not smooth, they infer that another source of gravity is present.

By measuring the motion of the star, astronomers can calculate the mass of the planet and its distance from the star. Using such methods, astronomers have discovered what may be more than 160 planets beyond our solar system.

Most of these distant planets are probably gas giants. However, scientists have reported finding what may be a rocky planet orbiting a red dwarf. Scientists calculated that the planet was five times more massive than Earth and three times farther from its star than Earth is from the Sun. Temperatures on its surface were thought to be about -220°C (-364°F). Astronomers found this planet by analyzing data of the star's brightness for changes that indicated that a planet passed in front of the star, which is called *gravitational microlensing.*

Scientists have theorized that distant solar systems might look like this.

Constellations

When people in ancient cultures looked at the night sky, they saw patterns in the stars. These patterns are called **constellations**. They were named after animals, fictional characters, or objects.

Star patterns have been useful to ancient and modern travelers. For example, if you can see either the Big Dipper or the Little Dipper in the night sky, you can use them to easily find Polaris, the North Star. If you travel in the direction of Polaris, you will be moving north.

The ancient Greeks divided the sky into 12 sections and named some constellations after characters from Greek myths, such as the hunter Orion and the hero Hercules. The ancient Chinese divided the sky into four major regions. The name of each region included a color, an animal, and a direction. For example, the western region was called the White Tiger of the West.

Today, astronomers divide the sky into 88 constellations. Many of the ancient names for constellations are still used today. Modern astronomers have named constellations visible in the Southern Hemisphere, which could not be seen by Ancient Greeks and Romans.

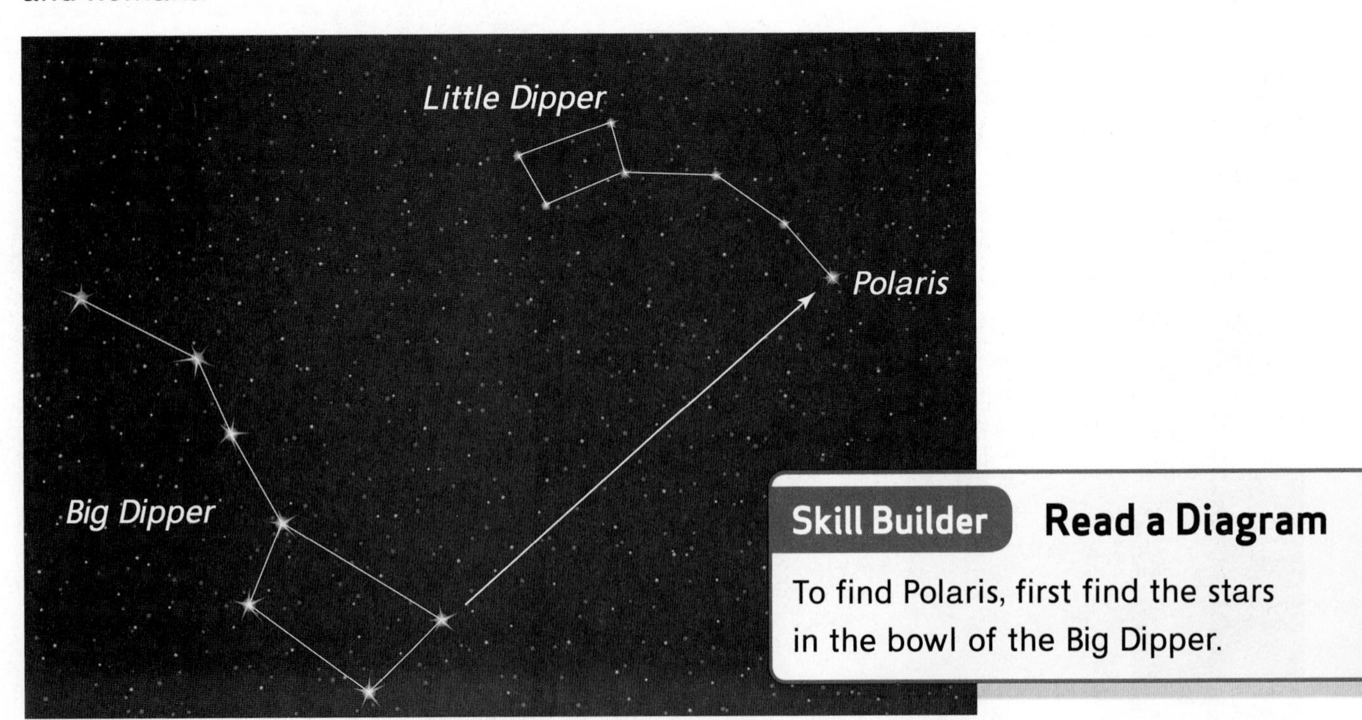

Skill Builder **Read a Diagram**

To find Polaris, first find the stars in the bowl of the Big Dipper.

Apparent Motion

The stars in the northern sky seem to circle around Polaris. The stars appear to move because of Earth's rotation. Although the stars appear to change position, their positions within constellations do not change.

As Earth revolves around the Sun, different constellations are visible to an observer on Earth. For example, Orion is a winter constellation in the Northern Hemisphere. It can be seen rising in the eastern sky on winter evenings. As the season changes, Orion sets earlier and earlier each night. In May, Orion disappears from the night sky in the Northern Hemisphere. In June, the constellation Scorpius, the scorpion, becomes visible.

These seasonal changes are caused by Earth's orbit around the Sun. Each night, the position of most stars shifts slightly to the west. Soon the stars once visible in the west cannot be seen, and other stars appear in the east.

There are a few constellations that are visible all year long. These are the ones closest to the North Star, Polaris. As Earth rotates, these constellations seem to circle the North Star. They never appear to rise or set. These constellations are known as *circumpolar* constellations.

Word Study

The word *circumpolar* comes from two words. *Circum* means "around," and *polar* means "of the poles."

Constellations make sense only to observers on Earth. Stars that seem close together are actually far apart. If you looked at the location of constellations from a different part of the universe, those patterns would change.

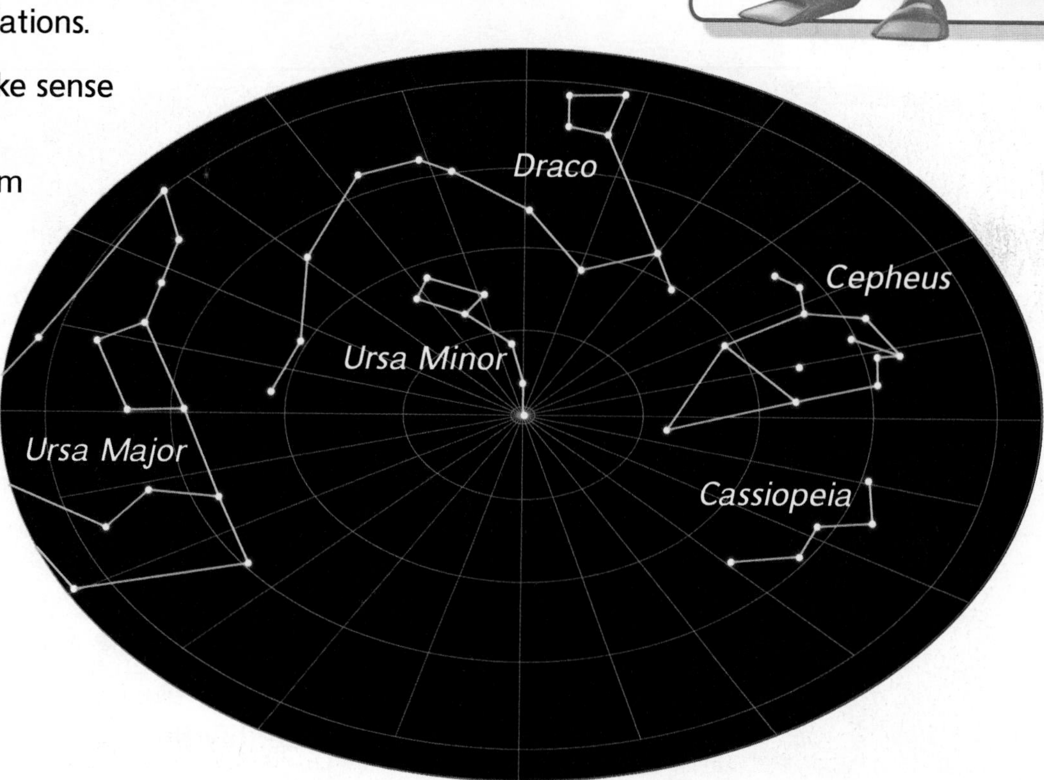

Our Star, the Sun

The Sun is the closest star to Earth. If you could travel to the Sun by car at highway speed, it would take more than 160 years. Despite its distance from Earth, the Sun's energy powers everything from living systems to Earth's weather. It provides our planet with heat energy and light.

The Sun contains 99.9 percent of the solar system's total mass. Most of the Sun—92 percent—is made up of hydrogen gas.

Like Earth, the Sun is made of layers. It has a core where hydrogen is being turned into helium. These reactions produce all the heat and light we receive on Earth. Energy moves outward from the core. It takes a long time, up to 170,000 years, for energy produced in the core to move through the *radiation zone* to the convective zone. In the *convective zone*, large bubbles of hot plasma move upward toward the surface of the Sun.

Sun Facts	
diameter	1.39 million km (865,000 mi)
period of rotation	25.4 Earth days
average distance from Earth	149.6 million km (93 million mi)
surface temperature	6,000°C (10,800°F)
core temperature	15,000,000°C (27,000,000°F)
size relative to Earth	1.3 million times larger

The Sun

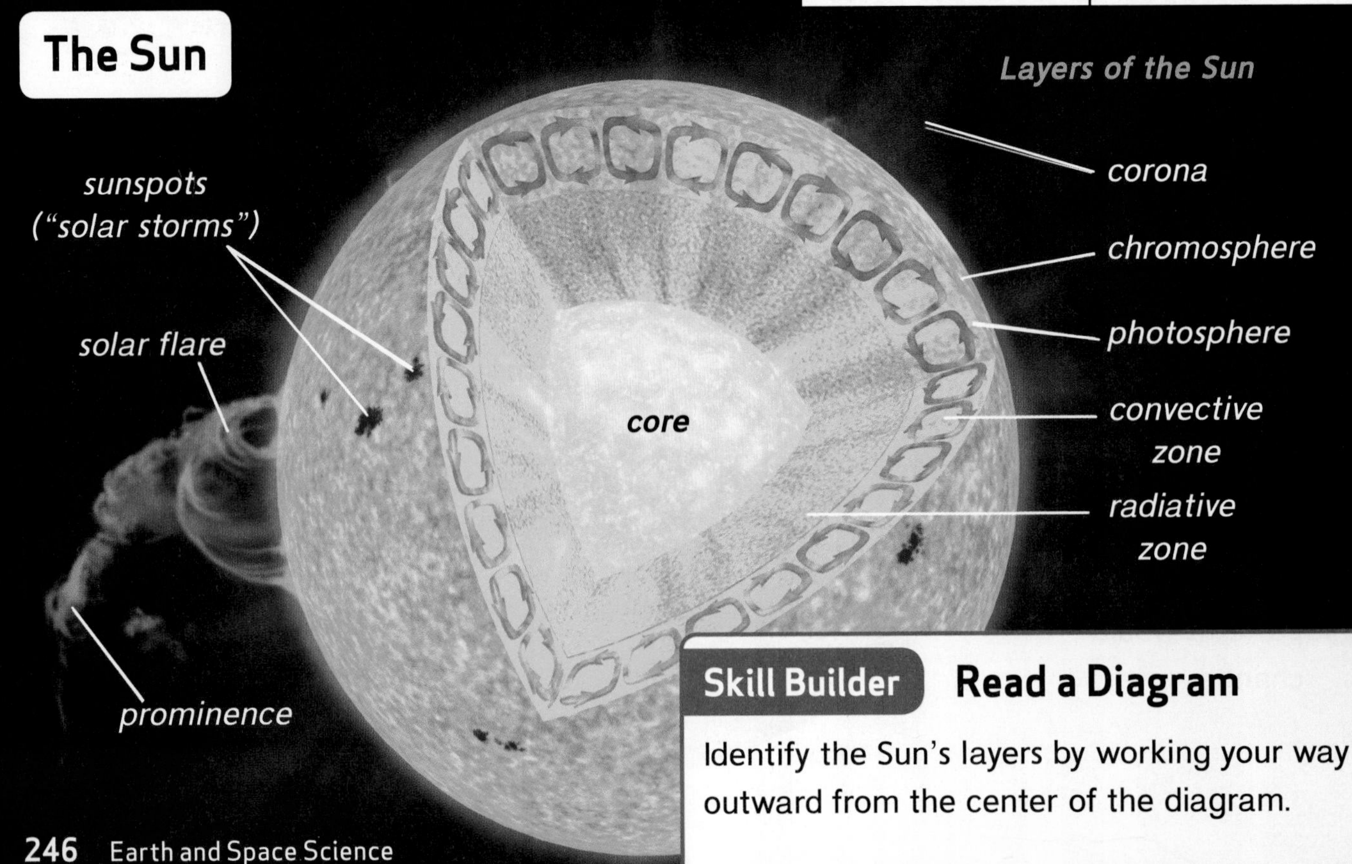

Skill Builder **Read a Diagram**

Identify the Sun's layers by working your way outward from the center of the diagram.

The Sun's surface is called the **photosphere**. It is from this region that most of the Sun's energy escapes outward to become the sunlight we observe on Earth. This energy takes about eight minutes to travel from the Sun to Earth.

Sunspots in the photosphere are areas with strong magnetic fields. These areas are cooler, and appear darker, than the material that surrounds them. Sunspots appear and disappear in an 11 year cycle. *Solar flares* are the sudden release of energy from the Sun's atmosphere. Both solar flares and sunspot activity can interfere with Earth's radio transmissions and satellites.

The thin *chromosphere* and *corona* lie above the photosphere. Light given off from these regions is usually too weak to be seen on Earth. However, when the Moon is aligned with the Sun as it moves through its orbit, these layers can be seen. *Prominences*, or dense clouds of gas, sometimes project from the chromosphere into the corona. These events are also associated with magnetic forces.

Make Connections

Jump to Heat Transfer in the Physical Science section to learn about convection and radiation.

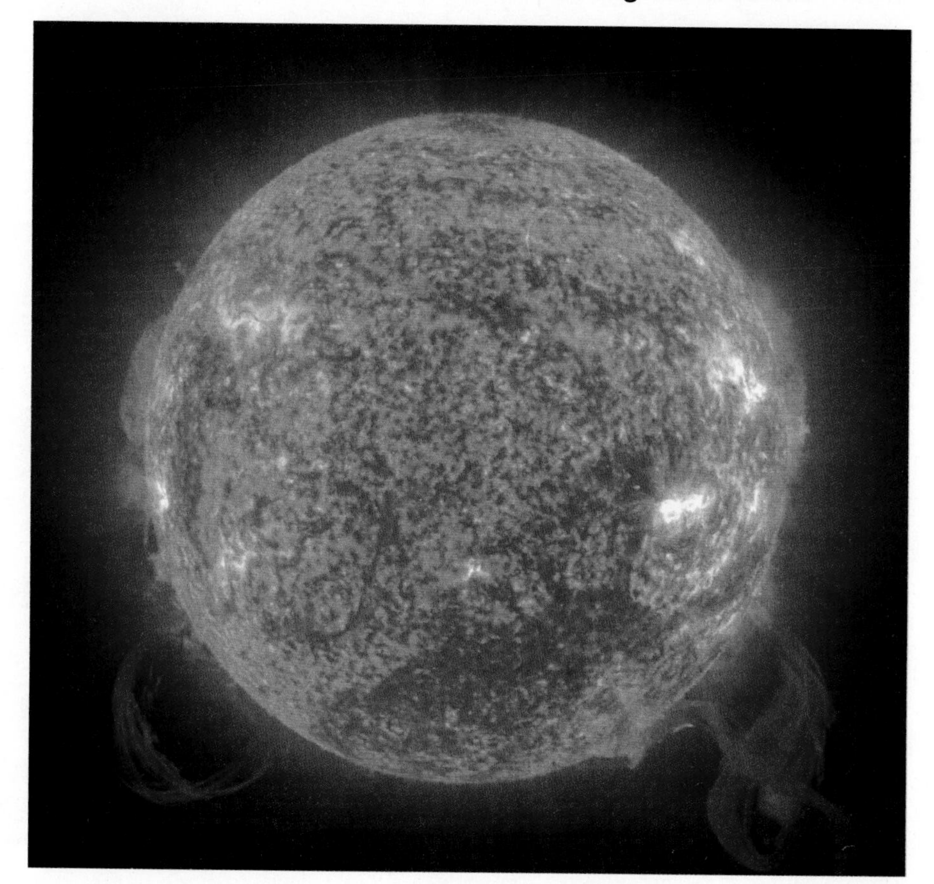

SOHO (ESA & NASA)

This image was taken with a special lens. It allows you to see parts of the Sun that are not normally visible from Earth.

Space Exploration

Until the early 1600s, people observed the night sky with only their eyes. Then early astronomers, including the Italian astronomer named Galileo Galilei, began looking at the sky through telescopes. Galileo observed things in space that no one had seen before.

Optical Telescopes

Galileo used an optical telescope, which uses lenses or mirrors to gather visible light. Among the objects Galileo saw were four moons revolving around the planet Jupiter. At that time, most people believed that all the objects in the solar system revolved around Earth.

an example of an optical telescope

Telescopes in Space

Clouds and city lights can make it hard to see through optical telescopes. For this reason, many telescopes are located in clear, deserted areas or on mountaintops. One good place for a telescope is outside Earth's atmosphere, in space. In 1990, the Hubble Space Telescope was placed into orbit around Earth. This telescope can show objects that are billions of trillions of kilometers from Earth.

Radio Telescopes

Back on Earth, radio telescopes record radio waves given off by objects in space. Groups of dishes focus the radio waves so the data can be recorded. Computers turn the data into images. Radio waves can pass through Earth's atmosphere without interference.

radio telescopes

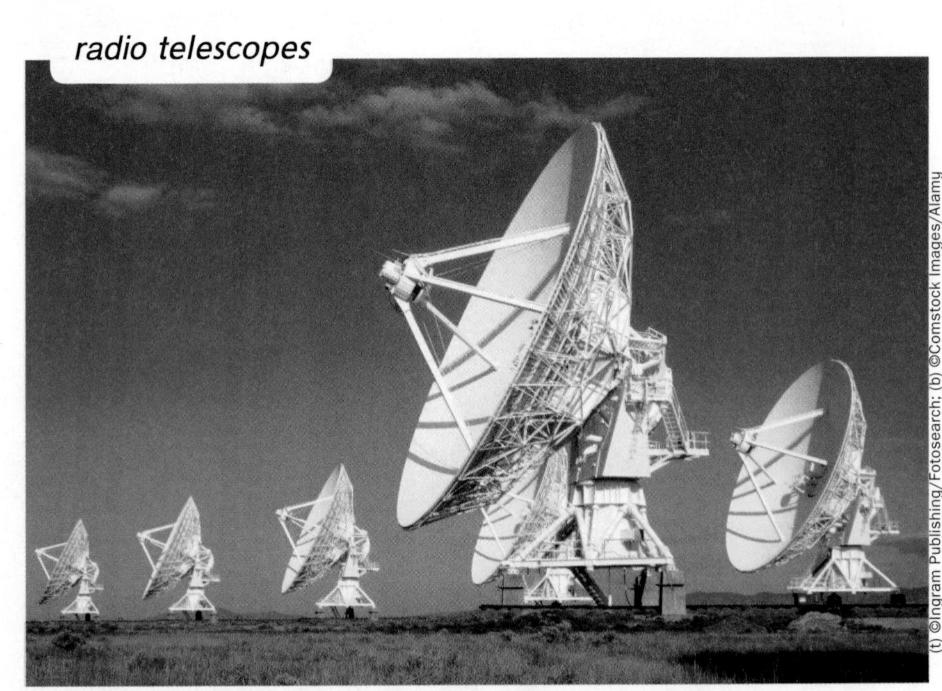

Astronauts

In the 1960s, NASA launched rockets that took people into space. Those people were the first astronauts. In the 1980s space shuttles carried astronauts to conduct experiments and launch satellites. Now, many countries, including the United States, share the *International Space Station* (ISS). Unlike rockets and the shuttles, the space station has stayed in orbit around Earth for a long time.

Space Probes

Space probes are unpiloted spacecraft that leave Earth's orbit. NASA has launched probes to planets, moons, and other objects. The probes send pictures and other data from space to Earth. Using space probes is safer and less expensive than sending astronauts into space.

In 2004 a space probe landed on Mars. Two robot explorers, called rovers, studied the surface and recorded data. These Mars rovers are named *Spirit* and *Opportunity*. On August 5, 2012, NASA's newest rover, *Curiosity*, landed on Mars. Its mission is to determine whether Mars ever had an environment able to support living organisms.

Because the solar system is so vast, some probes need many years to reach their targets. Probes have been launched to explore Saturn and its moons and even Pluto.

What Could I Be?

Aerospace Engineer

Do you ever think about building vehicles that could soar into outer space? As an aerospace engineer, you could make flights into space possible by helping build high-speed spacecraft. Turn to the Careers section to learn more.

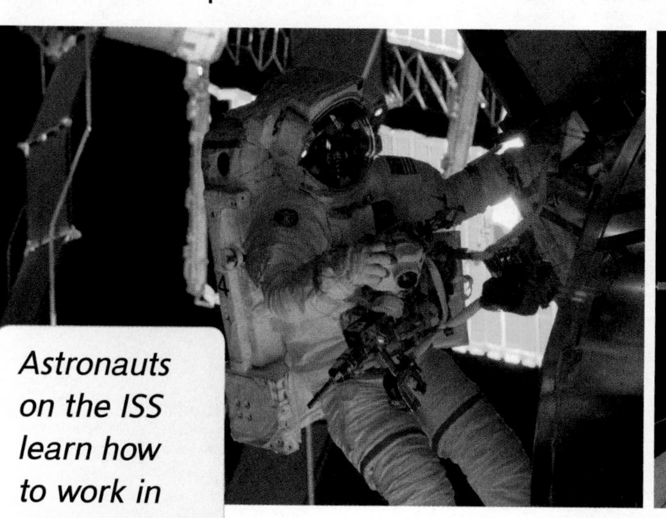

Astronauts on the ISS learn how to work in microgravity.

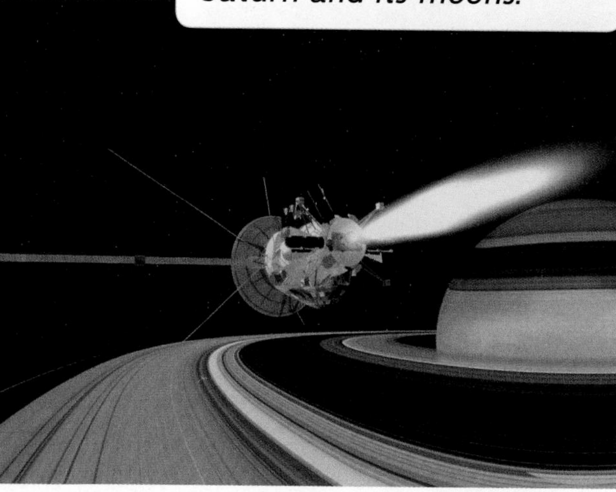

The Cassini spacecraft is exploring the planet Saturn and its moons.

(t) Monty Rakusen/Getty Images; (bl) NASA; (br) Purestock/SuperStock

Future Career

Photonics Engineer What do you think of when you hear the word laser? A photonics engineer thinks about a world of opportunity. Photonics is the science of light, and an engineer in this field builds and uses light in many different forms. Lasers are used in very detailed surgeries, such as in the human eye. They are also used to read movie, music, and computer disks that hold a lot of information. Photonics engineers are good at working with systems of many different sizes. One type of laser and light technology called fiber optics works to send information all over the world!

Physical Science

Physical Properties - Solubility

The ability of matter to dissolve in a liquid is called **solubility.** When a substance dissolves, the tiny particles it is made of become too small to be seen. Think about dissolving sugar in a glass of iced tea. It looks like the sugar disappears, but it is easy to tell it is still there because the tea tastes sweet.

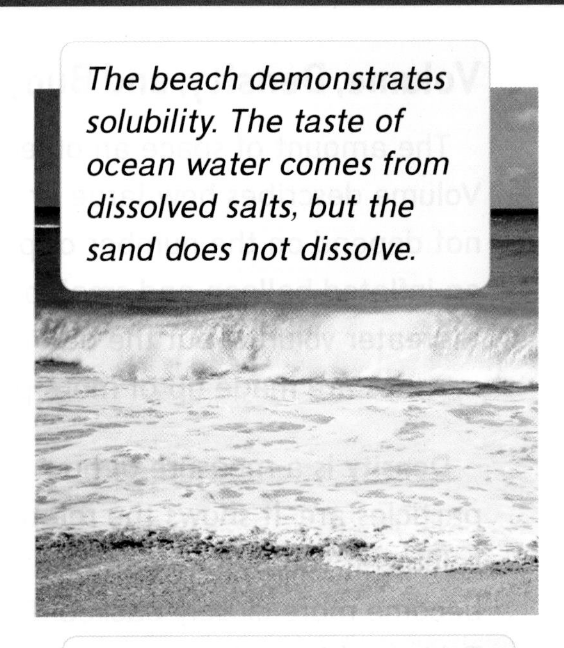

The beach demonstrates solubility. The taste of ocean water comes from dissolved salts, but the sand does not dissolve.

Conductivity

Electrical and thermal conductivity are two more properties of matter. *Electrical conductivity* describes how well electricity can move through a material. Metals, such as copper, gold, iron, and silver, are good conductors of electricity. A **conductor** is a material that transfers heat or electricity easily. Wood, glass, and plastic stop or slow the flow of electricity. These materials are called *electrical insulators.* An **insulator** is a material that stops or slows the flow of energy, such as electricity or heat.

Electrical wires are often covered with plastic to protect people and objects from the flow of electricity.

Thermal conductivity is a measure of how well a material conducts thermal energy. Thermal conductors allow thermal energy to easily flow through them. Metal pots and pans are good thermal conductors. They quickly transfer thermal energy from the burner to the food in the pan. Some materials, such as wood, rubber, glass, and plastic, slow or stop the flow of thermal energy. These materials are called *thermal insulators.*

Knowing which materials are conductors and which are insulators of thermal energy can keep you from being burned when cooking.

Flexibility and Elasticity

Objects that can bend without breaking are *flexible.* A thin wire that is bent into the shape of a paper clip is flexible. Objects that can stretch and then return to their original shape are *elastic.* Rubber bands are elastic.

Other Physical Properties

There are many other physical properties that scientists use to identify materials. Some of these are listed in the table below.

Other Physical Properties	
	**Color and Shape** *Color* helps you quickly identify which backpack is yours. The *shape* of the pack allows it to fit snuggly against your back.
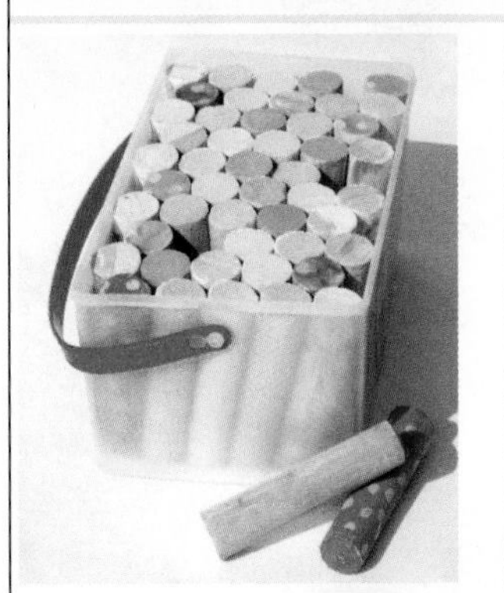	**Hardness** *Hardness* is a measure of how resistant a material is to scratching, bending, or denting. Chalk is very soft. It is easily rubbed onto the sidewalk. Other materials such as metal or stone are much harder.
	**Magnetism** **Magnetism** is the ability of an object or material to be pushed or pulled by a magnet. Some metals, including iron, cobalt, and nickel, are magnetic. Objects can be identified based on whether they are attracted to a magnet or not.
	**Reflectivity** **Reflectivity** is the way light bounces off an object. Smooth, flat surfaces, such as mirrors and shiny pans, reflect light evenly. You can see your reflection. Rough surfaces scatter light. They do not produce a reflection you can see.
	**Texture** *Texture* describes how an object feels. Imagine touching the balls shown here. The basketball would feel rough, or pebbly. The soccer ball would feel smooth. Both would feel rubbery.

(t-b) Tim Pannell/SuperStock; Corbis RF/Alamy; Ken Cavanagh/McGraw-Hill Education; Katie Weeks/Getty Images; D. Hurst/Alamy

Physical Properties - States of Matter

State is another physical property of matter. Solids, liquids, and gases are the common forms, or *states,* that matter can take. Each state has specific characteristics.

Ice is a solid. A solid has a definite shape and takes up a definite amount of space. A solid stays in a definite shape with a definite volume no matter what container it is in. The particles of matter in a solid are packed together tightly and vibrate in place. Often they are packed in a regular pattern.

Ocean water is a liquid. A liquid has a definite volume, but it does not have a definite shape. It can be poured from one container to another. Whatever shape the container is, the liquid fills that shape from the bottom up. In general, the particles of a liquid are less tightly packed than those in a solid. The particles can move and slide past one another.

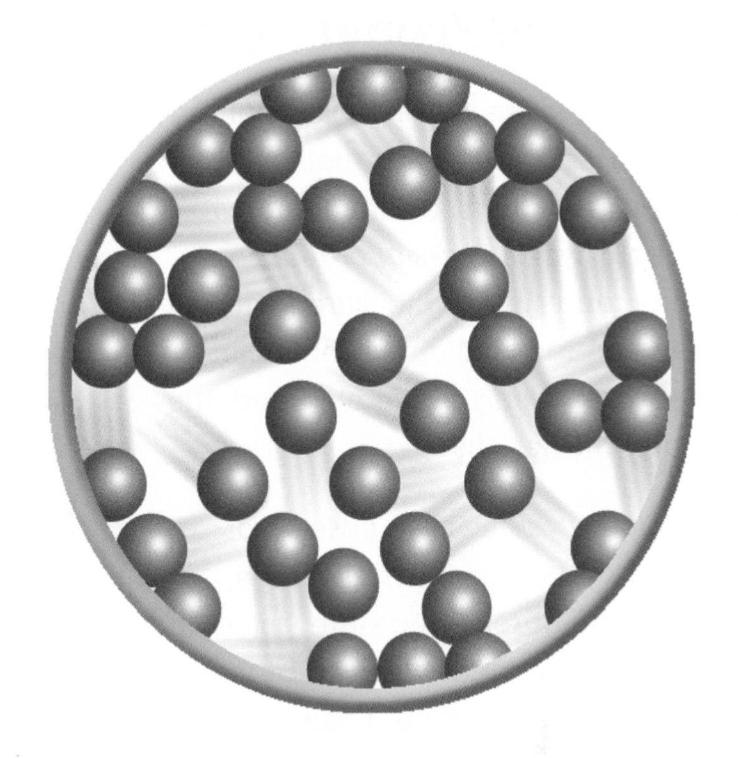

The air that makes up the atmosphere is a gas. Gases have no definite shape or volume. The particles in a gas are farther apart than those of solids or liquids, and they can move around each other very easily. Gases spread out and completely fill a closed container. If you make the container bigger, the gas will expand to fill it. That is why you can inflate a balloon with air.

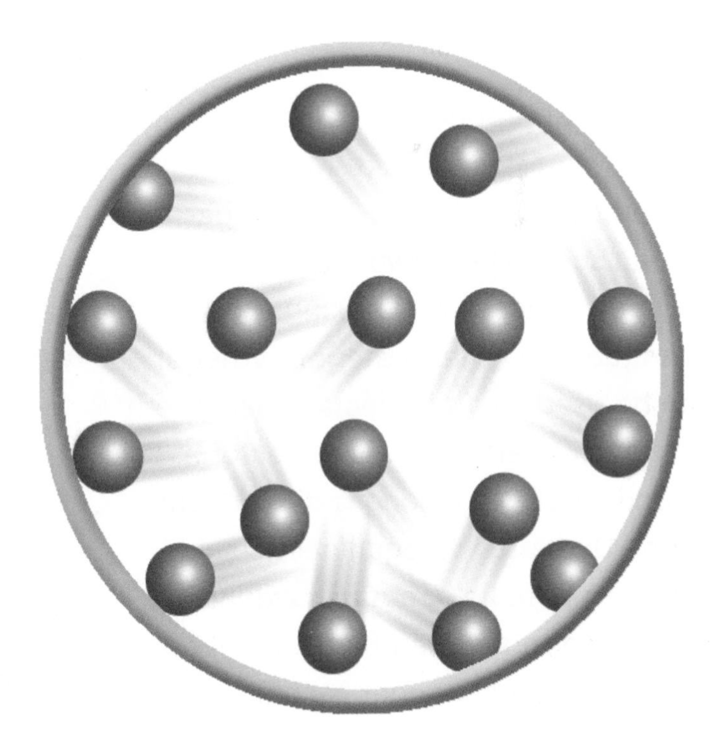

Chemical Properties

Properties are characteristics of matter that can be observed or measured. Properties can be classified into two different types. *Physical properties* can be observed or measured without changing the kind of matter. When *chemical properties* are observed, the kind of matter changes.

Suppose you want to observe or measure the properties of vinegar. When you observe the state, color, and odor of the vinegar, the kind of matter in the vinegar does not change. You can also measure the mass of a gallon of vinegar without changing the kind of matter. These properties are physical properties of vinegar.

Chemical properties of vinegar include that it is an acid. To observe this property, you could add vinegar to baking soda. These two materials react, and both vinegar and baking soda are changed into other materials. Vinegar can be used as a disinfectant because it reacts with certain types of bacteria. The vinegar used in cooking is diluted with water. Undiluted vinegar, acetic acid, can burn skin. Its fumes can explode if exposed to an open flame. In all of these examples, vinegar is changed into other kinds of matter. Its acidity, ability to burn, and its ability to kill bacteria are chemical properties of vinegar.

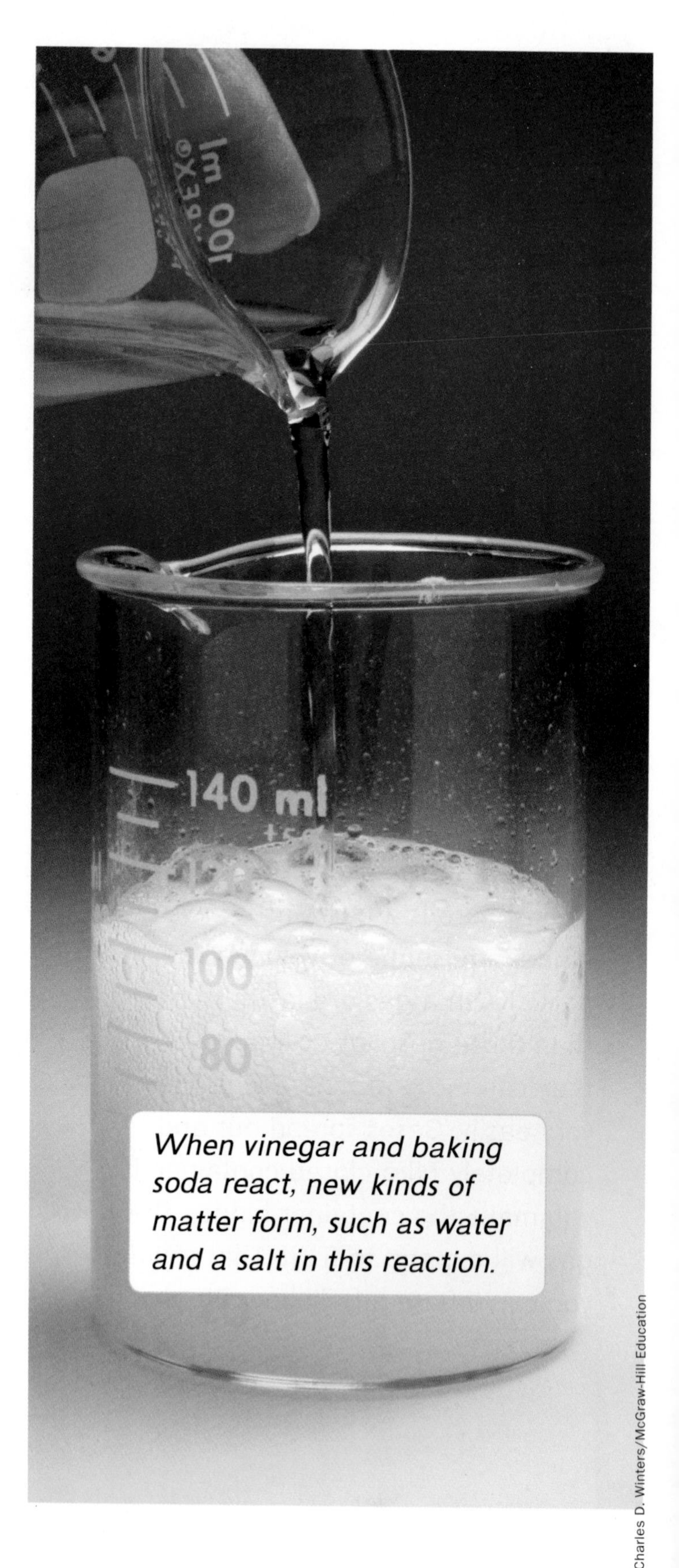

When vinegar and baking soda react, new kinds of matter form, such as water and a salt in this reaction.

Chemical properties can be observed only when the type of matter changes. The table lists several other chemical changes you may have observed.

Ability to Burn
Ability to burn is a chemical property of certain materials, such as wood. Some metals, such as magnesium, also will burn. Being able to burn means the material is *combustible*.

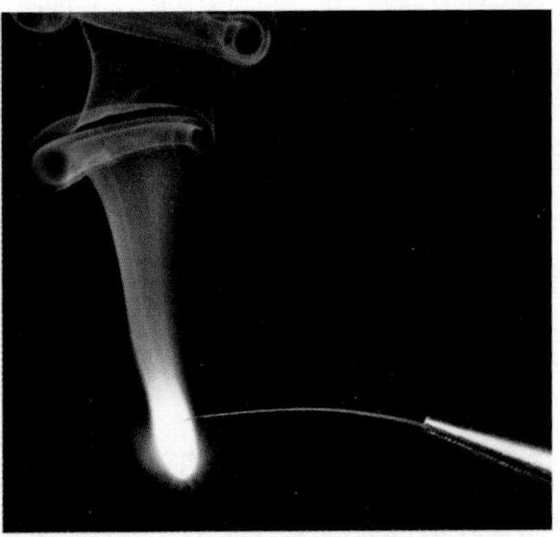

Reaction with Air
Some metals will react with materials in the air, causing the metal to rust or tarnish. Iron rusts. Silver and copper both tarnish. These processes are also called **corrosion**.

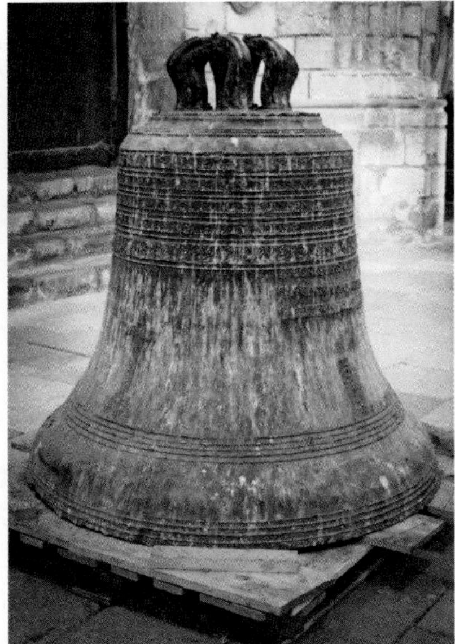

Change When Heated
Some matter has the ability to change when heated. Most foods show this chemical property when they are cooked.

Observing and Measuring Matter

Observing and measuring are different ways to identify properties. You can use your senses to observe many properties. You use your ears to hear the sound a harmonica makes as air blows through it. Your senses of taste and smell can be used to tell the difference between a grape and a piece of pineapple, even if you cannot see or feel them. Your sense of sight is used to observe the color of a leaf. Touch helps you tell whether a spoon is warm or cool.

Measuring is a way to compare sizes or amounts. People use tools marked with standard units to measure certain properties of matter. A *standard unit* is a unit of measurement that people agree to use. Systems of standard measurement include customary measurement units, such as feet, miles, cups, and quarts. You probably use these units every day, but they are not used by scientists. The system of standard measurement used by scientists is the **metric system**. The metric system is used every day in most countries of the world.

Make Connections

Jump to the Science Guide section to learn more about units in the metric system.

Skill Builder

Read a Table

Follow each row of the table from left to right to learn about each type of measurement.

Measuring Properties

Measurement	Property of the Object	Sample Units		Sample Measuring Tools
		Customary System	Metric System	
length	how long it is	inch, foot, yard, mile	millimeter, meter, kilometer	ruler, meterstick, yardstick, tape measure
volume	how much space it takes up	cup, pint, quart, gallon	milliliter, liter, cubic centimeter	measuring cup, graduated cylinder
temperature	how warm or cold it is	degrees Fahrenheit	degrees Celsius	thermometer
mass	how much matter it contains	slug	gram, kilogram	balance
weight	how strongly gravity pulls on it	pound	newton (N)	spring scale

Measuring Matter

Measuring Mass Mass is the amount of matter in an object. Mass can be measured on an equal pan balance. Gravity pulls on the standard pieces and on the sample. When the pans are level, the amount of matter in the sample and in the standard pieces is the same.

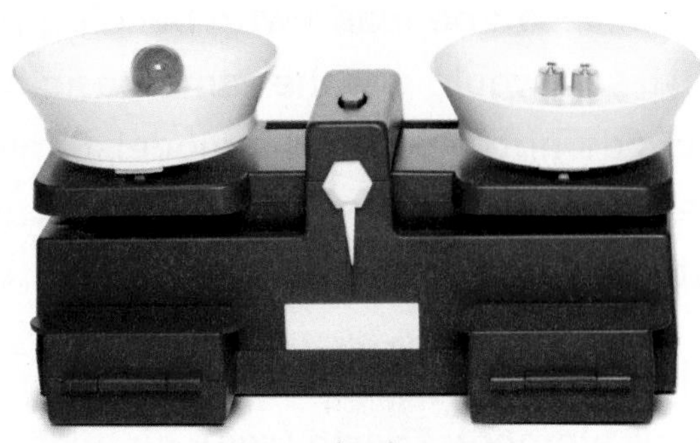

pan balance

Measuring Weight Weight is how strongly gravity pulls on an object. If an object has more mass, it will also have more weight. Weight is measured in newtons (N) using a spring scale. One **newton** is equal to 0.225 pounds (lb) in the customary system.

scale

Measuring Volume Volume is a measurement of how much space an object takes up. Volumes of liquids are often measured in milliliters (mL) by using a graduated cylinder, a beaker, or a measuring cup. The volume of a solid is usually measured in cubic centimeters (cm^3). A volume of 1 cm^3 equals 1 mL. A solid that has an irregular shape or is difficult to measure in some way can be measured using displacement. A measured amount of water is placed in a cylinder. The object is placed in the water. The object pushes water out of the way, and the water level rises. The original water level is subtracted from the new water level. The difference is the volume of the solid.

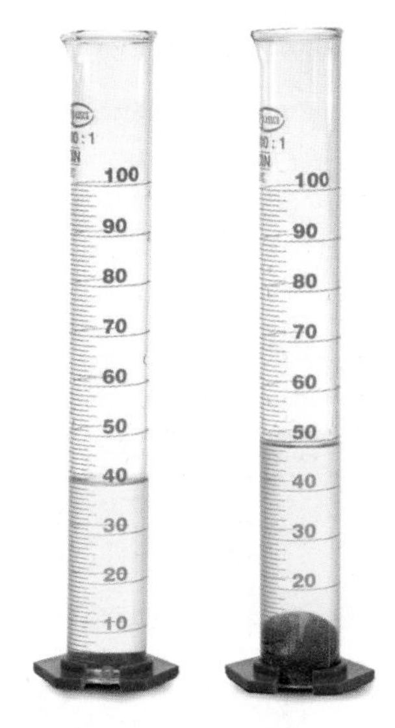

graduated cylinders

Elements

Models can be made with building blocks. However, if you take the models apart, you would get the same basic blocks. When mixed together, the blocks do not look like the model. Similarly, most matter on Earth is made of a set of "building blocks," the chemical elements. An **element** is a material that cannot be broken down into anything simpler by chemical reactions.

The ancient Greek philosopher Aristotle believed that all matter was made of four elements: earth (soil), air, water, and fire. These elements are not true elements. Fire is not matter. Air and soil are made of many different materials. Water is made up of hydrogen and oxygen. Hydrogen and oxygen, however, cannot be broken down into simpler substances. Hydrogen and oxygen are elements.

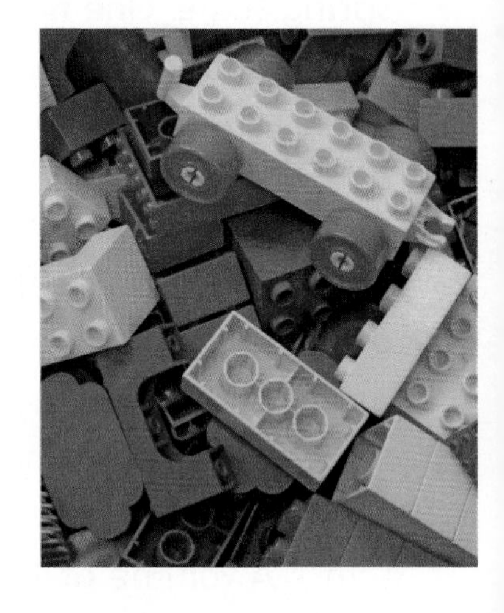

Scientists know of 118 elements, which have different properties. Three important properties of elements are the state of matter at room temperature, the way they mix with other elements, and whether they are metals, nonmetals, or metalloids.

Most elements are solid, some are gases, and two are liquid at room temperature. Some elements are more likely than others to combine with other elements to form new substances. These elements are said to be more chemically reactive.

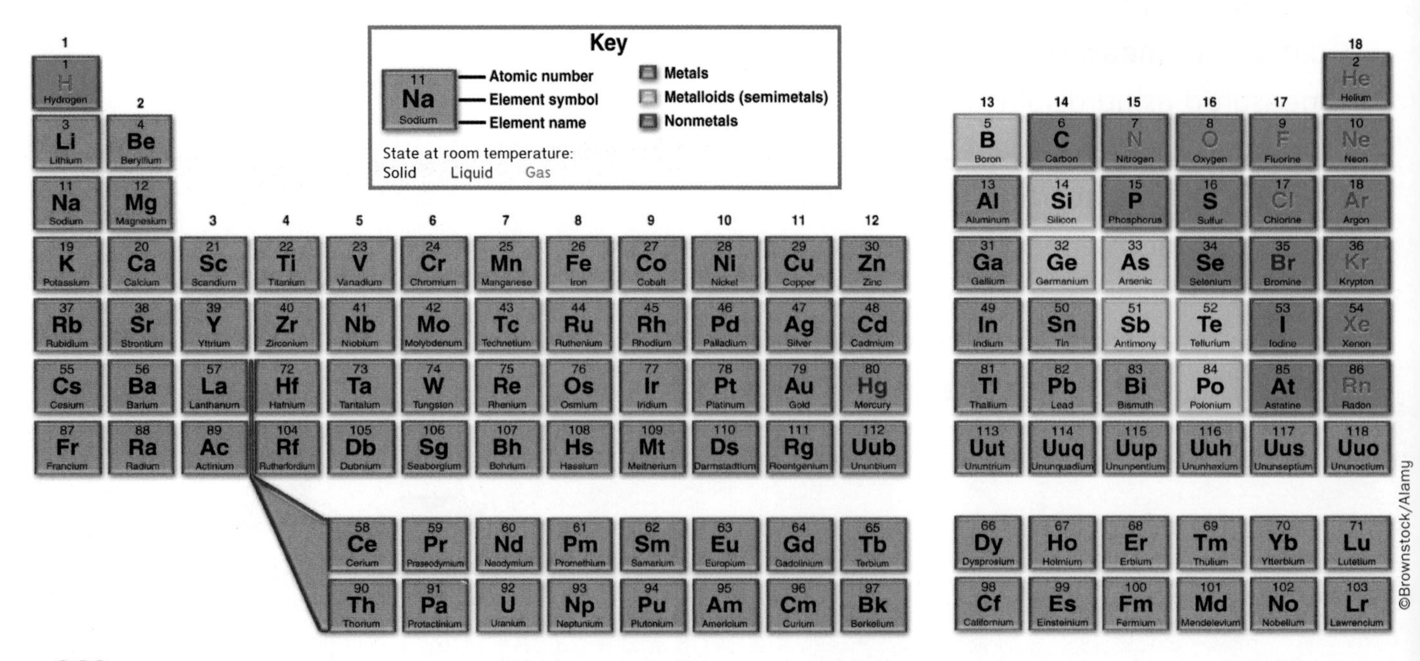

Atoms

If you cut a piece of an element in half, the two halves have the same properties as the original element. If you kept cutting it in half, again and again, eventually you would have the smallest piece of the element possible. In 1803, John Dalton proposed that elements are made of tiny particles. He believed that these particles could not be cut into smaller pieces. Today we know that these particles do exist, and we call them atoms. An **atom** is the smallest unit of an element with the properties of that element.

All atoms are made of even smaller particles. The *nucleus* is the center of an atom. It is made of small particles called protons and neutrons. The *proton* is a particle with one unit of positive electric charge. The number of protons in an atom is called the *atomic number* and determines which element it is. A *neutron* is a particle with no electric charge—it is neutral.

Atoms also contain *electrons,* which are smaller particles, each with one unit of negative electric charge. Electrons move within the space outside the nucleus. Most of an atom is empty space. Usually, the numbers of electrons and protons are equal, so atoms have no overall charge.

Did You Know?

An atom is almost completely empty space. Picture an atom as a baseball field. The nucleus is like a pebble in the middle of the field.

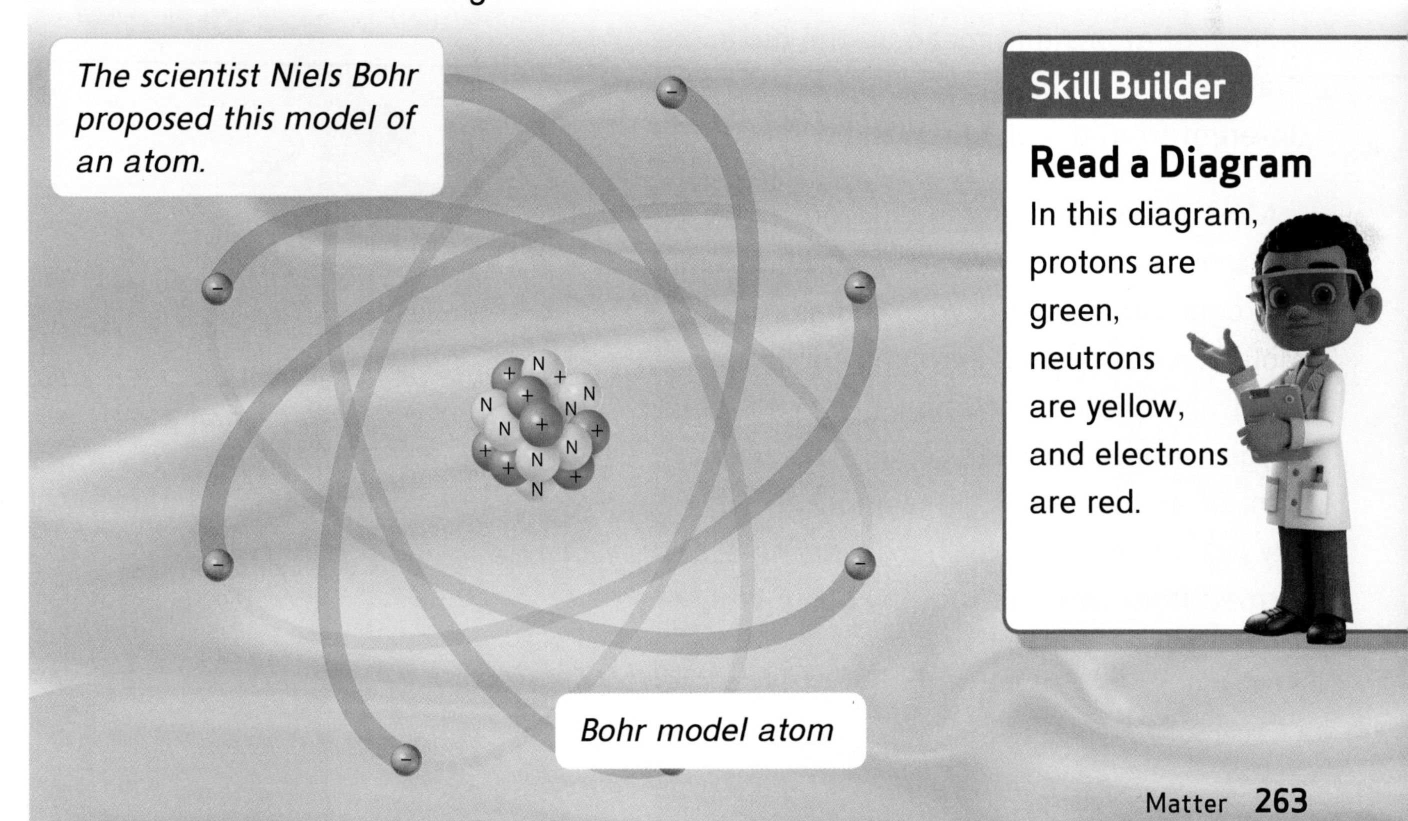

The scientist Niels Bohr proposed this model of an atom.

Bohr model atom

Skill Builder

Read a Diagram

In this diagram, protons are green, neutrons are yellow, and electrons are red.

Molecules and Compounds

Few elements exist by themselves. Most elements are found as combinations of one or more elements.

Molecules

Molecules are particles with more than one atom joined together. Most of the atoms in the world do not exist on their own but as part of a molecule. Most objects in the world are made from many molecules grouped together.

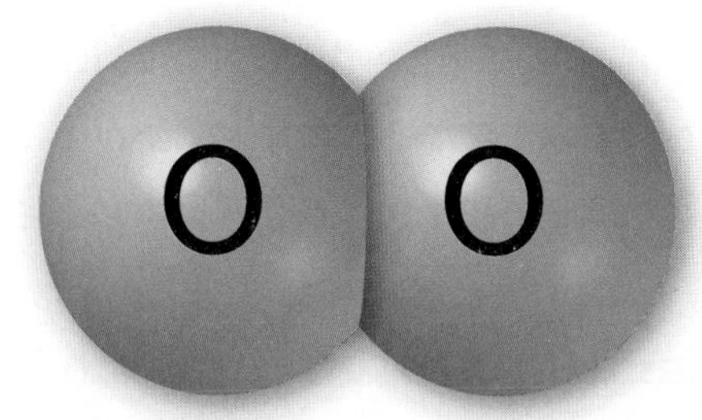

An oxygen molecule is made of two oxygen atoms that are joined together.

When a molecule forms from elements, atoms link together through their electrons. Each atom in a molecule shares electrons with at least one other atom in the molecule. In general, the arrangement of electrons determines the properties of the substance. Because electrons are arranged differently in a molecule than they are in the individual atoms that make up the molecule, the molecular substance has properties different from the elements that make it up.

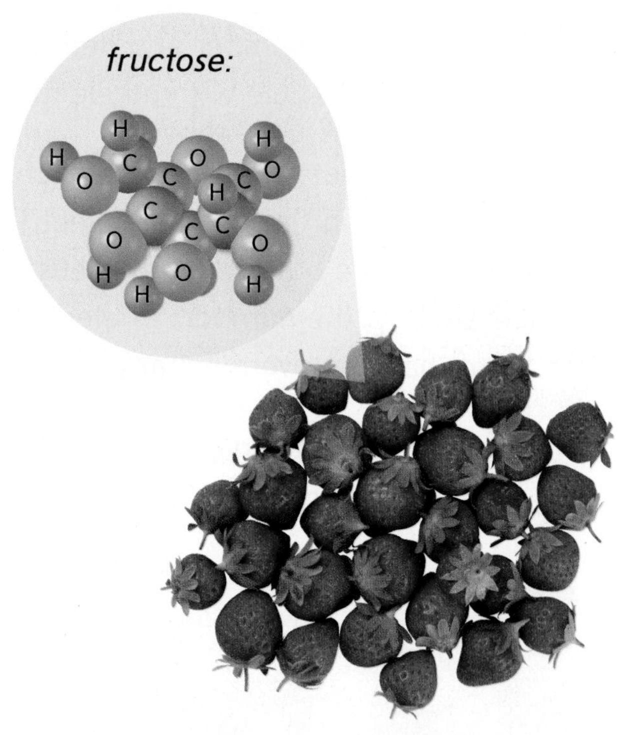

Fruit contains fructose molecules. Each molecule contains 6 carbon atoms, 12 hydrogen atoms, and 6 oxygen atoms.

Most molecules contain different types of atoms. A water molecule is formed from hydrogen atoms and oxygen atoms. A molecule of heptane, the main compound in gasoline, contains carbon, hydrogen, and oxygen atoms. However, some molecules form of atoms of a single element. In air, oxygen and nitrogen both exist as molecules formed from two atoms of the same type joined together.

Compounds

Compounds are formed by the chemical combination of two or more elements. For example, the shells of birds' eggs are made of a compound containing the elements calcium, carbon, and oxygen. Citric acid, which is found in oranges, lemons, and other citrus fruits, is a compound composed of the elements carbon, hydrogen, and oxygen.

Compounds have properties that are different from the properties of their individual atoms. For example, a common compound is table salt. This compound is composed of the elements sodium and chlorine. The properties of table salt are quite different from the properties of the elements that are part of it. Sodium is a soft, silvery, and highly reactive metal. Chlorine is a poisonous green gas. When these two elements combine, they form the compound sodium chloride, which is the white, brittle, solid that is known as table salt.

Most of the matter in and around you is made up of compounds. Most of your body is made up of water. Water is not an element. It is a compound made up of two different elements, hydrogen and oxygen. Hydrogen and oxygen are both colorless, odorless gases. The properties of water are quite different from the properties of hydrogen and oxygen.

Forming a Compound

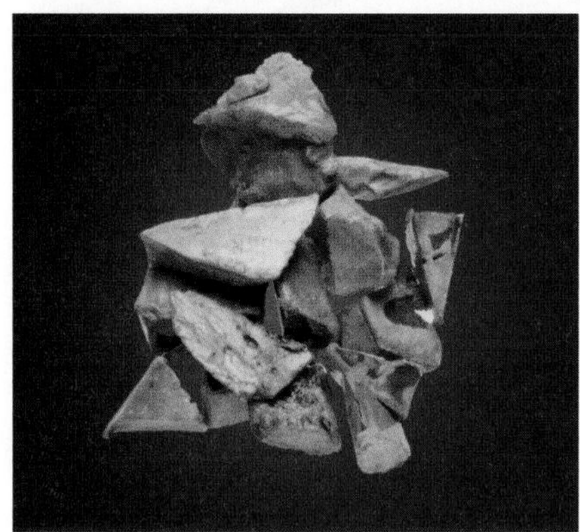

Sodium is a soft, reactive metal that explodes on contact with water.

Chlorine is a poisonous, yellow-green gas. Putting sodium in chlorine causes a fiery reaction.

The reaction produces sodium chloride, also known as table salt.

Changes in Matter

Matter can be changed in many ways. A **physical change** begins and ends with the same kind of matter. A **chemical change** begins with one kind of matter and ends with another.

Changing State

The three common states of matter are solid, liquid, and gas. When matter changes from one state to another, a physical change occurs. When liquid water boils and becomes water vapor, a physical change takes place. When liquid water freezes and becomes ice, another physical change occurs. These changes do not change the kind of matter. Water vapor, liquid water, and ice may look and feel quite different, but they are all water.

Particles in objects move. In solids, particles vibrate in place. In liquids, particles vibrate as they move past one another. In gases, particles move quickly and away from one another. The average movement of particles in an object is measured by its temperature. Changes in temperature occur when an object gains or loses heat energy. When energy is added to the particles, they move faster. When energy is lost, particles in matter move more slowly. When enough energy is gained or lost, there is a change of state.

Skill Builder

Read a Diagram

In this diagram, temperature is indicated by color. Blue arrows indicate colder temperatures. Red arrows indicate warmer temperatures.

Changes of State

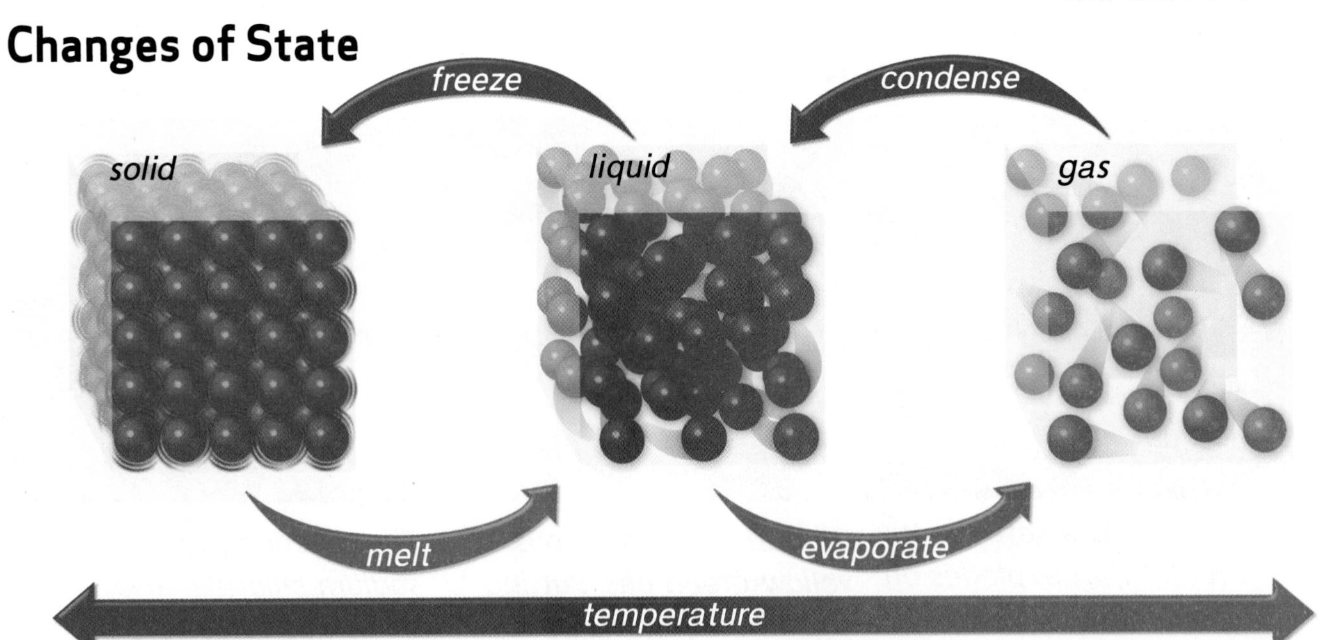

As heat energy is absorbed, particles move faster and become less organized.

How Water Changes State

Adding Heat Energy Heat is energy that flows between objects with different temperatures. When energy is added to a solid, the particles start to move more quickly. When they move quickly enough that they slide by each other, the solid changes to a liquid in a process called *melting.* If enough energy is added to this liquid, the particles in the liquid move away from each other. The particles spread apart, and the liquid boils (a quick change) or evaporates (a slower change), becoming a gas.

Removing Heat Energy Changes of state also occur when energy is removed from a substance. If a gas loses energy, its particles slow down and move closer together. They start to cling together and slide by each other. A liquid forms in a process called condensation. If the liquid loses enough energy, freezing occurs and a solid forms.

When energy is added to ice, the water molecules move faster. The ice melts.

As energy is added to liquid water, the particles move faster. Some turn to gas.

Skill Builder **Read a Diagram**

Examine the circles showing the arrangement of particles in a solid, liquid, and gas to better understand state changes.

Water vapor is a gas. Its particles move very fast.

Mixtures

At first glance, a garden salad and fog seem to have little in common. However, both are mixtures. A **mixture** is a physical combination of two or more substances. These substances remain the same, even though they are close together. You can separate mixtures using physical properties. For example, you could pick all the tomatoes and olives out of a garden salad by picking out different shapes and colors.

What Could I Be?

Biochemist

Researching how materials combine is one of the jobs of a biochemist. Some biochemists develop new drugs to fight diseases. Others work in food and agricultural fields looking for ways to improve products and crops. Learn more by turning to the Careers section.

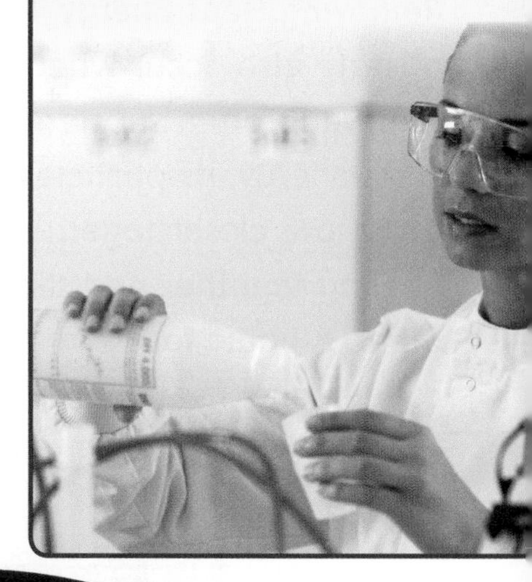

Heterogeneous Mixtures

Mixtures that have parts that you can plainly see are called *heterogeneous mixtures.* You can see the individual parts of a garden salad. It is a heterogeneous mixture. Gravel in a driveway and trail mix are other examples of heterogeneous mixtures. Mixtures can have different combinations. A garden salad can have many tomatoes or only a few. There are no mixing rules, and there can be more of one ingredient in some parts than in others.

Suspensions The parts of heterogeneous mixtures are not always easily seen with the unaided eye. If you look at peanut butter, it might look as if it is the same throughout. However, looking at it through a microscope shows pieces of peanuts mixed with oil. This kind of heterogeneous mixture is called a *suspension.*

This salad is a heterogeneous mixture.

Over time, one or more parts of a suspension will settle to the bottom, like mud in a stream. Stirring or shaking the mixture, however, will make it look the same again. You might have suspensions around your home. If you use salad dressing that has instructions to "shake well before using," you are probably using a suspension.

Colloids A **colloid** is a heterogeneous mixture in which the parts are so small that they do not settle out. In a colloid, undissolved particles or droplets stay mixed in another substance. Fog is a liquid-in-gas colloid. Smoke is a solid-in-gas colloid. However, the parts are large enough that they make the mixture look cloudy or creamy. Smoke, mayonnaise, and colored glass are all examples of colloids.

Skill Builder

Read a Photo

Note how the water changes from one photo to the next. This change gives you a clue that the mixture shown is a suspension, not a colloid.

Suspension in Water

The appearance of muddy water under a microscope shows that it is a heterogeneous mixture.

Most of the clay has settled out of the muddy water.

Mixtures - Homogeneous Mixtures

Not all mixtures are heterogeneous. *Homogeneous mixtures* have parts that are so small they cannot be seen, even with a microscope. Another name for a homogeneous mixture is a **solution**. Sugar dissolved in water is a solution. If you take samples from different parts of a glass of sugar water and look at them under a microscope, they would all look the same. The individual parts of the solution cannot be seen.

A carbonated beverage is a solution of carbon dioxide gas in liquid water under pressure. When the pressure is released, the carbon dioxide gas bubbles out of the solution.

The material present in the greatest amount in a solution is the *solvent.* Any other part of the solution is a *solute.* In sugar water, water is the solvent, and sugar is a solute. Solvents and solutes can be in any state. Saltwater is a solution made with both solids and gases mixed in a liquid. Air is a solution made from different gases. Rubbing alcohol is a solution of two liquids. Brass is a solution of two solid metals. Brass is an *alloy,* which is a solution of a metal and another solid.

Water vapor is a solute in air. Nitrogen gas is the solvent. When water vapor condenses, clouds form.

The greatest amount of solute that can dissolve in a certain amount of solvent is called the **solubility** of the solute. Usually, solubility of a solid in a liquid increases when the temperature of the solvent increases. For example, you can dissolve more sugar in hot tea than you can in iced tea.

The brass in this instrument is a solid solution of the metals copper and zinc.

Separating Mixtures

Making mixtures requires a physical change. Physical changes are also needed to separate mixtures. Parts of a mixture with different properties act differently when changed in the same way.

You can use a physical change to push, pull, lift, or otherwise separate one part of a mixture from another. Density, solubility, particle size, magnetism, melting points, and boiling points are all properties used to separate mixtures. Depending on the mixture, the parts could be separated by picking the parts out by hand, melting a solid, or boiling off a liquid. The methods used to separate a mixture should be based on the properties of the parts of the mixture.

Separating a Mixture

Several methods can be used to separate a mixture of sand, iron filings, sugar, and sawdust.

Adding water dissolves the sugar in the mixture. The iron, sand, and sawdust do not dissolve in water.

The sugar water can pass through the filter. Sand, iron, and sawdust cannot. If the water is evaporated or boiled out of the sugar solution, the sugar is left behind.

Iron is attracted to a magnet, but sand, sugar, and sawdust are not.

Sawdust floats on water, but sand, iron, and sugar do not.

Chemical Changes

Physical changes, such as tearing paper or making mixtures, do not result in new substances. A chemical change occurs when atoms link together in new ways to produce substances different from the original substances. A chemical change is also known as a **chemical reaction**. You can tell that a chemical change occurs because the properties of the starting materials are different from the properties of the ending materials.

If you combine baking soda with vinegar, atoms in the baking soda and vinegar link together in new ways. During the chemical reaction, carbon dioxide bubbles form. A new white solid, sodium acetate, is left behind. The liquid from the reaction is water. The new substances that have formed have different properties from the vinegar and the baking soda.

Acids and Bases

Acids and bases are compounds that react easily with other substances. You can identify them using *litmus paper*. Litmus is a dye that changes color when it touches acids or bases. The pH scale measures the strengths of acids and bases. *Acids* turn blue litmus paper red, and they have a pH below 7. Lemon juice is an acid. *Bases* turn red litmus paper blue, and they have a pH above 7. Soaps are basic. A substance with a pH of 7, such as distilled water, is neutral.

When an acid and a base of equal strength are mixed, they react. This process, known as *neutralization*, produces water and a salt.

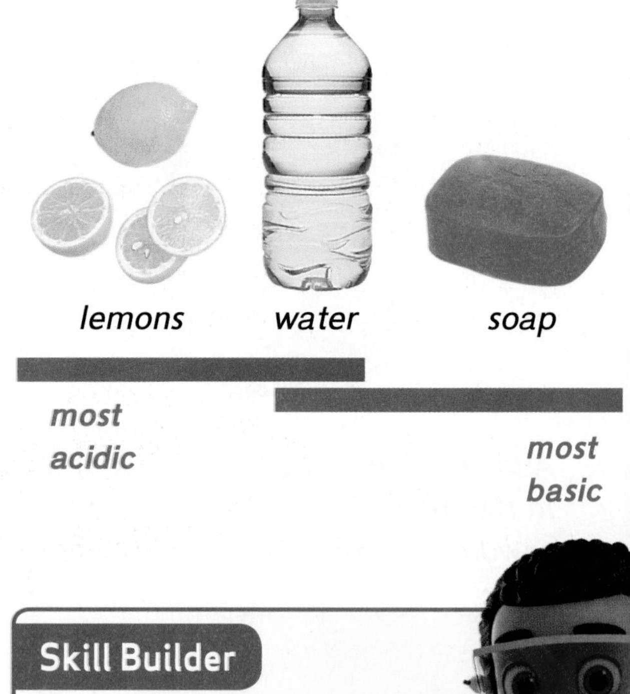

Skill Builder

Read a Diagram

The strips below the items are litmus paper. Look carefully at how each substance changes the paper strips.

Signs of Chemical Change

Chemical changes produce new substances. Often you can see, hear, or smell the formation of new substances as a chemical change occurs.

Change in Color The change from shiny metal to rust and tarnish shows that certain metals have chemically changed. The color change of green leaves to orange and red also shows that a chemical change occurred in the leaves.

Change in Odor A change in odor is another sign of chemical change. Have you ever toasted marshmallows over a campfire? An uncooked marshmallow has little odor. While your marshmallow is toasting, it gives off a pleasant smell. This change in odor shows that a chemical change happened.

Energy Change Many reactions cause matter to become warm. Others cause the matter to become cold. Some reactions give off light. Fireworks undergo chemical changes that give off both heat and light.

Formation of a Gas Bubbles often indicate the release of a gas. For example, dropping antacid tablets into water will release bubbles that are the result of chemical change.

Formation of a Precipitate Chemical changes can produce more than just gases. A *precipitate* is a solid formed from the chemical reaction of some solutions. One example of a precipitate is soap scum. Soap scum forms when dissolved soap reacts with minerals dissolved in tap water.

Fireworks release so much energy that they light up the sky.

Gas bubbles are evidence of chemical change.

Chemical Changes - Names and Formulas

All compounds have chemical names, and many have common names as well. The chemical name indicates which elements make up a compound. For example, the chemical name for rust, a compound of iron and oxygen, is iron oxide. The chemical name for table salt, a compound of sodium and chlorine, is sodium chloride. Chemical names use the names of elements. Often, the name of the second element is used in a different form. You can see this change in iron *oxide* and sodium *chloride*.

The chemical formulas for compounds have more than one elemental symbol because they are made from more than one element. For example, the chemical formula for iron oxide is Fe_2O_3. This formula indicates that two iron atoms combine with three oxygen atoms to form iron oxide.

Chemical Equations

In studying math, you probably have seen equations such as 2 + 6 = 8, or 7 + 3 = 6 + 4. In an equation, both sides equal the same amount. Chemists write equations for chemical changes similar to math equations. For example, hydrogen gas and oxygen gas combine to form water. A chemical equation keeps track of which substances are used, and in what ratio.

There are 2 iron atoms and 3 oxygen atoms in the compound rust.

Fuse/Getty Images

The chemicals on the left side of a chemical equation are the **reactants**. The chemicals on the right side of the equation are called **products**. For the water-forming reaction, the reactants are hydrogen and oxygen, and the product is water. Note that an arrow separates the reactants and products. The same atoms are present on both sides of the arrow.

The reactants and the products are made of the same elements, but the elements are rearranged. Equal numbers of each atom are on both sides of the arrow. These numbers show that the chemical equation is balanced.

A chemical reaction does not create new matter. Instead, it forms new bonds among existing atoms. The total mass of the reactants equals the total mass of the products. Scientists call this fact the **law of conservation of mass**. According to this law, all the atoms present before a reaction begins will also be present after the reaction ends. However, the atoms may bond with other atoms in different ways to form different substances.

Math equations make sense whether you read them from right to left or left to right. Chemical equations are similar. When a chemical reaction is reversed, the products break apart or combine to form the original reactants. For example, water can be broken down into hydrogen and oxygen gas.

What Could I Be?

Chemical Engineer

Do you like to work with others to solve problems? Are you interested in using chemistry to develop new materials or find new uses for old ones? You might be interested in a career in chemical engineering. Learn more by turning to the Careers section.

Water-Forming Reaction

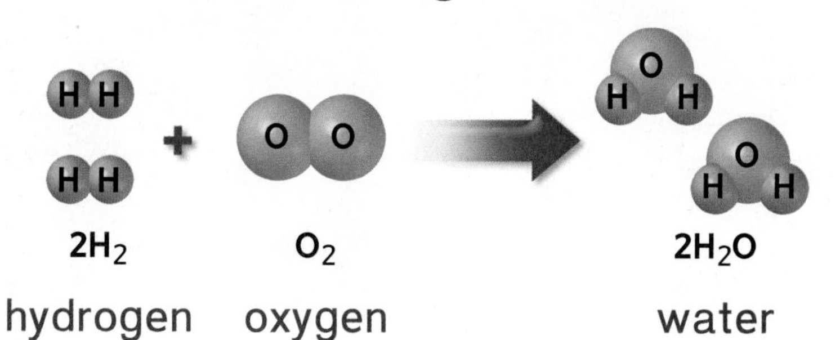

This equation reads:
"2 hydrogen molecules +
1 oxygen molecule yields
2 water molecules".

Forces and Motion

Position and Motion

The *position* of an object is its location. The position of an object or place can be shown on a grid. A grid is series of lines that show units of distance in two directions. The map on this page shows a grid. Vertical lines show positions north and south. Horizontal lines show positions east and west.

When an object changes position on a grid, you can draw an arrow between the old position and the new position. The arrows on the grid below show changes in the positions of two cars. The arrows represent two details used to describe changes in position—distance and direction.

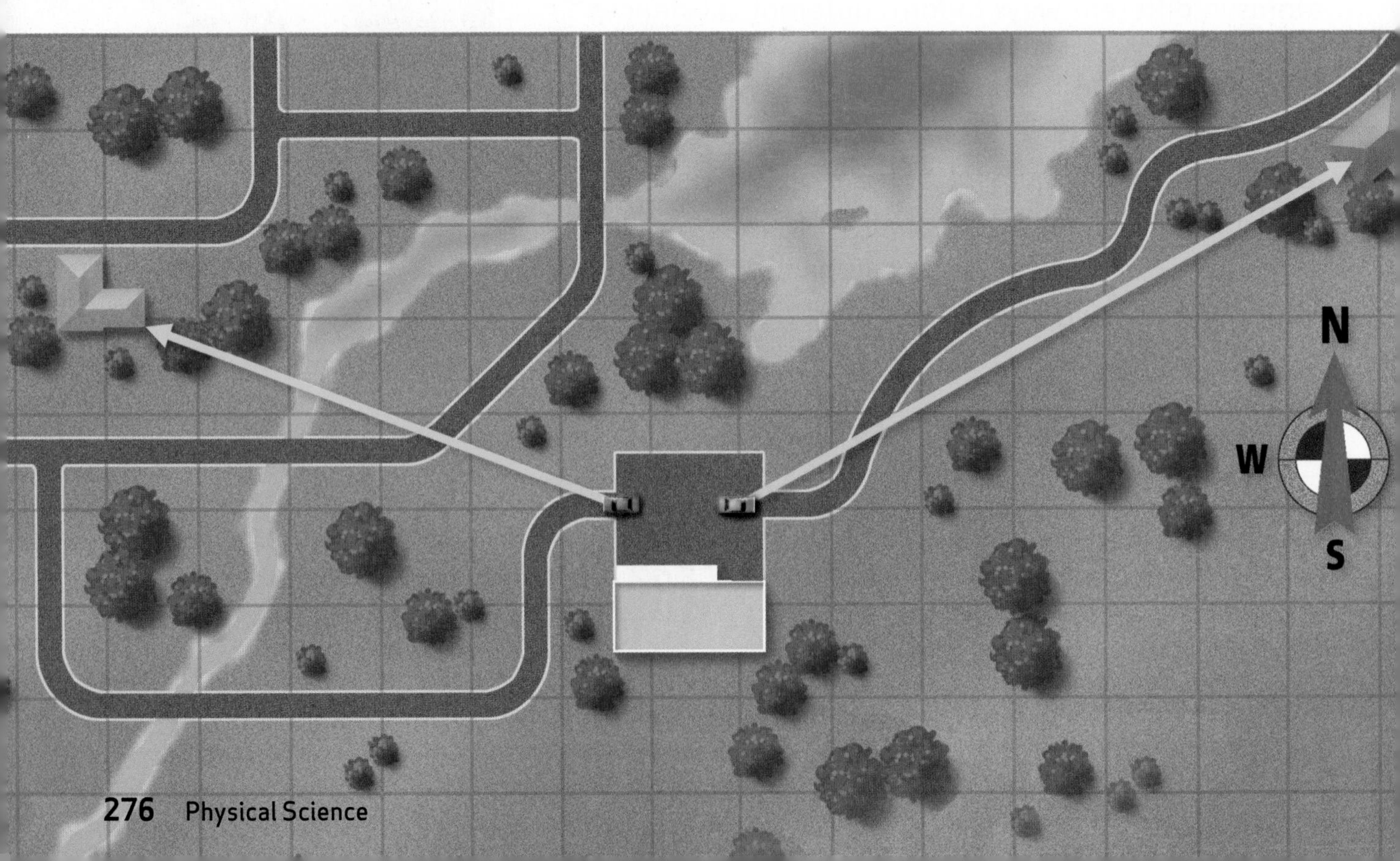

Distance is the length of the space between the starting point and ending point. You can measure the length of the arrow on the grid with a ruler. Units such as meters (m), kilometers (km), feet (ft), or miles (mi) describe distance. *Direction* is the way the arrow points. Words such as *left, right, north, south, east,* and *west* describe direction.

When an object is changing position, it is in motion. **Motion** is the process of an object changing position over time.

Frame of Reference

Position and motion make sense only if you have a frame of reference. A **frame of reference** is a group of objects from which you can measure a position or motion. Your classroom and the objects inside it are a frame of reference. If your friend told you he was moving three meters to the right of his desk, you would know his new position compared with his desk.

Almost anything can be a frame of reference: a baseball field, the solar system, or a grid placed on a map. Even moving objects can be a frame of reference. For example, the inside of a moving car is a frame of reference. Inside a moving a car, if you move your hand to the right, other passengers see you moving your hand to the right compared with the inside of the car. But if the road is the frame of reference, your hand is moving forward, along with the car and all the other passengers.

How you perceive motion depends on your frame of reference. To anyone outside the car, you appear to be moving very fast. When you are in motion and look outside the car window, the ground seems to be moving very fast, even though you know it is really not moving at all.

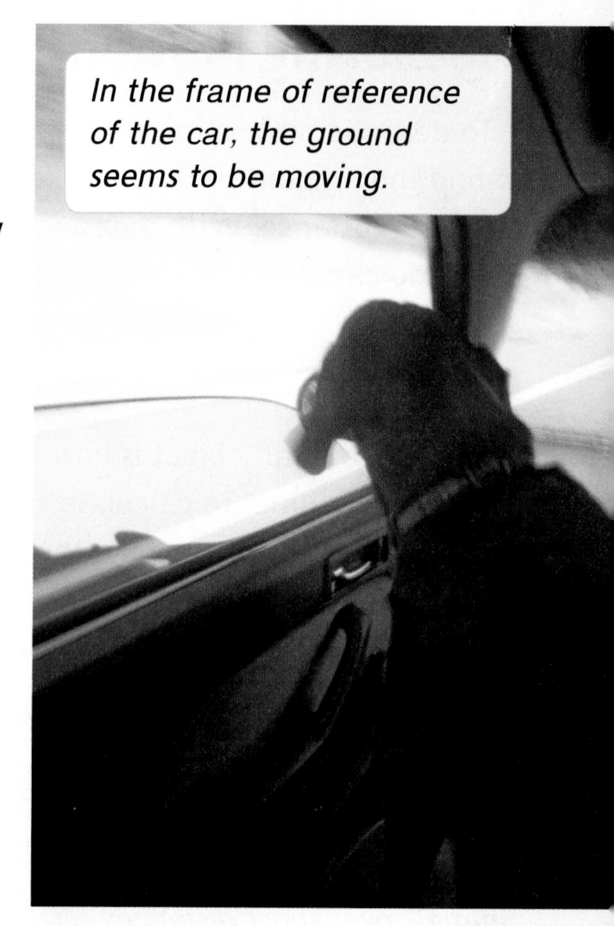

In the frame of reference of the car, the ground seems to be moving.

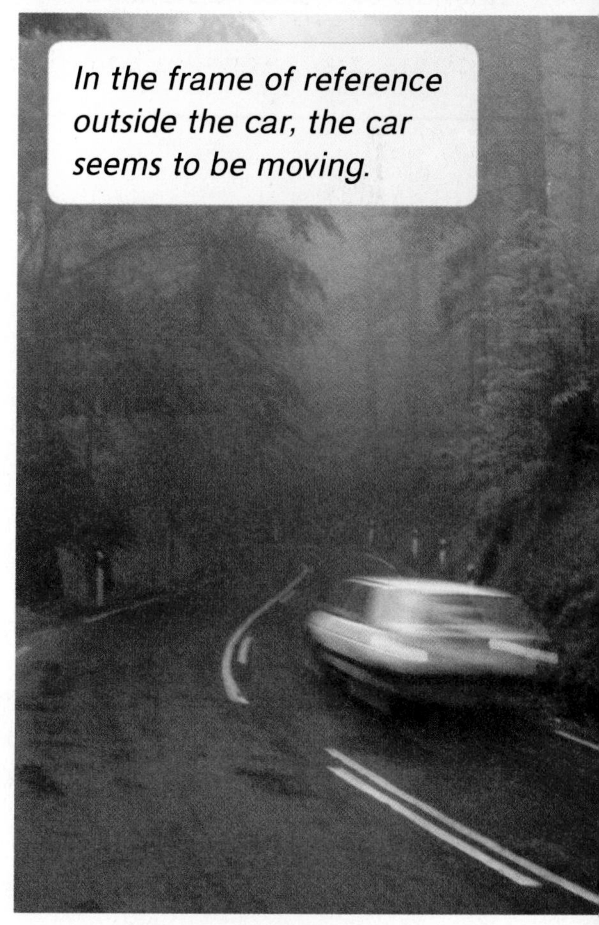

In the frame of reference outside the car, the car seems to be moving.

Measuring Motion

To describe motion more completely, you also need to find the amount of time it takes an object to move a certain distance. With measures of distance and time, you can describe motion and how it changes.

Speed

The *speed* of an object is how fast its position changes over time. To calculate speed, divide the distance traveled by the time the object travels. Units of speed are units of distance per unit of time, such as meters per second (m/s), kilometers per hour (km/h), or miles per hour (mph).

Calculating Speed

distance = 100 m

time = 10 s

speed = distance ÷ time

= 100 m ÷ 10 s

= 10 m/s

These are the fastest speeds of animals over short distances.

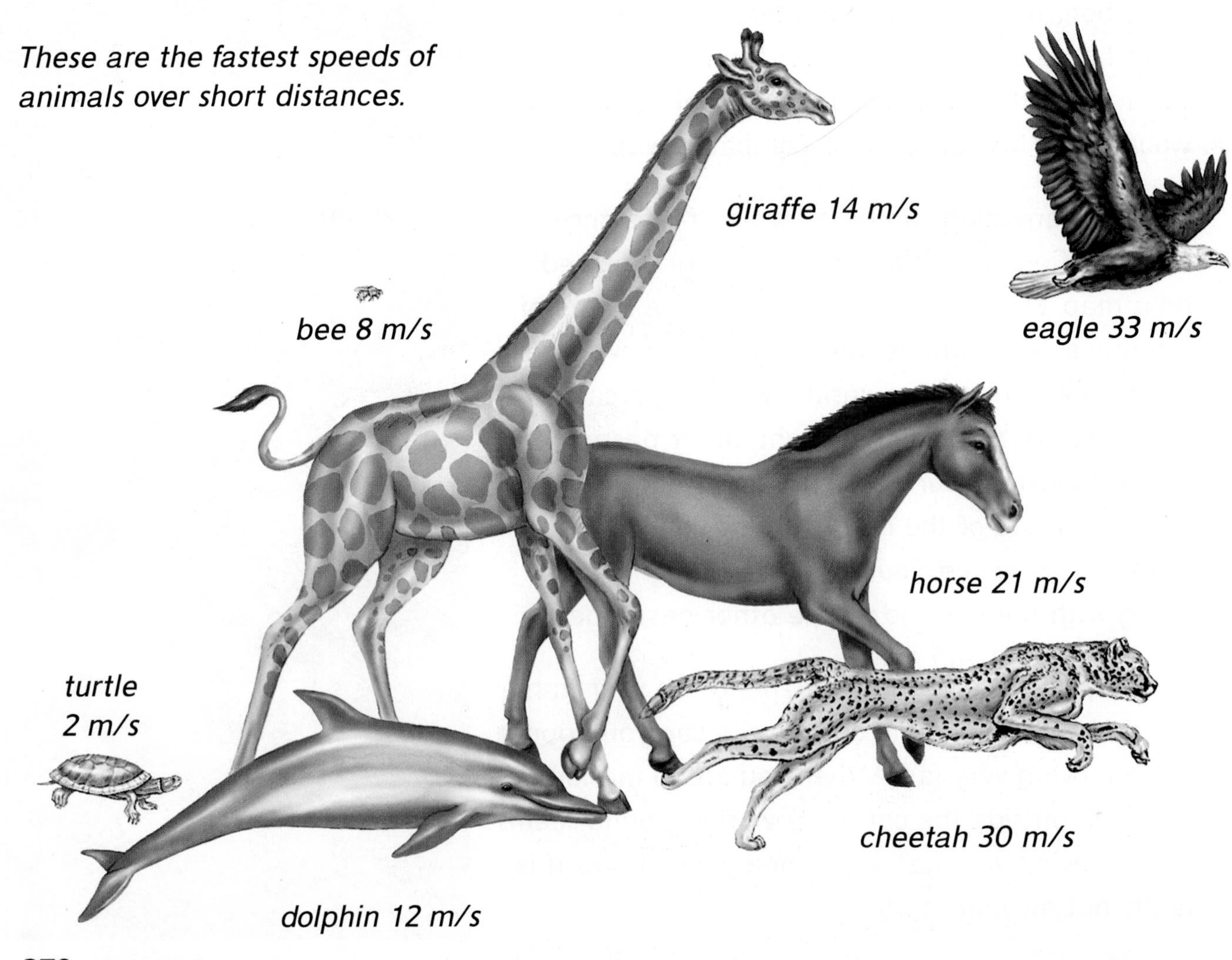

Velocity

Speed describes how fast an object is moving, but it does not include the direction. A pilot flying a plane, for example, needs to know both the speed of the plane and the direction the plane is flying.

Velocity is the speed and direction of an object. The units of velocity are the same as the units of speed, distance per unit of time. Velocity, however, must also include a direction. A plane that flies 640 kilometers in 2.0 hours has a speed of 320 km/h. If the plane is traveling south, then its velocity is 320 km/h south.

Two objects can have the same speed but different velocities. A car might travel 20 m/s (about 45 mph) east along a straight road. Another car might travel 20 m/s west along the same road. The speeds of the cars are the same. Their directions, however, are different, so they have different velocities.

The plane's velocity depends on both its speed and its direction.

Cars moving in different directions have different velocities, even if their speeds are the same.

Make Connections

Velocity does not apply only to smaller moving objects. Jump to Earth's Changing Surface to learn how an understanding of velocity is important in the study of Earth's plate tectonics. Jump to Measuring Weather and Climate to learn how understanding velocity affects scientists' ability to forecast weather.

Measuring Motion - Acceleration

When a moving object speeds up, slows down, or changes direction, it accelerates. A change in velocity over time is *acceleration.* Units of acceleration are units of velocity divided by units of time: meters per second per second ((m/s)/s). Just like motion and velocity, acceleration has a direction.

Suppose a car is facing north at a red light along a straight road. The light changes to green, and the driver steps on the gas pedal (also known as the "accelerator"). When the car reaches 12 m/s (about 27 mph), the driver keeps the car traveling at a constant speed. You note that it took 4 seconds for the car to go from 0 m/s to 12 m/s.

The acceleration is 3 (m/s)/s north. The speed increases by 3 m/s each second. After 4 seconds, the car has reached a final speed of 12 m/s. When the driver maintains a constant velocity, the car is no longer accelerating.

A car also accelerates when it slows down. Acceleration for decreasing speed is a negative number. For example, a stopping car might accelerate at –4 (m/s)/s. Another way of saying this is that the car *decelerates* at 4 (m/s)/s.

Calculating Acceleration

change in speed = 12 m/s

time = 4 s

acceleration = change in speed ÷ time

= 12 m/s ÷ 4 s

= 3 (m/s)/s

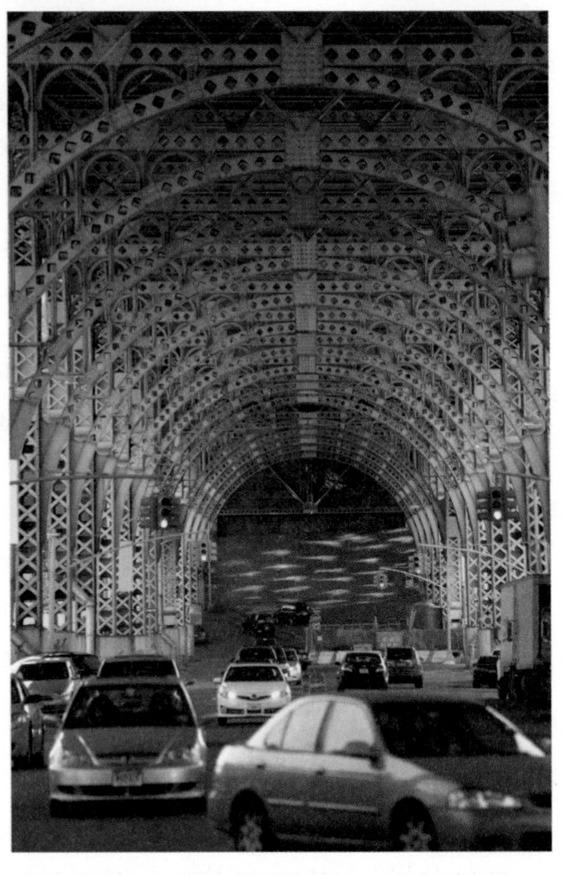

A car accelerates when its speed is increasing.

Slowing down is also acceleration.

(t) fotog/Tetra images/Getty Images; (b) Big Cheese Photo/Getty Images

Fact Checker

Acceleration always has direction, but the calculation shows only the change in speed.

Changing Direction

Suppose you are traveling forward in a canoe. As long as your speed does not change, you are not accelerating. If you begin to paddle faster, however, your speed increases and you accelerate. You can decelerate by paddling backward. In both of these cases, you accelerate by changing speed. With enough backward paddling, you start moving backward. In that case, your changing acceleration changes the direction of your velocity too.

Acceleration is a change in velocity, not just a change in speed. You also accelerate by changing only direction. When you travel around a curve, the direction of your motion changes. It doesn't matter whether your speed is constant or changing. As long as your direction of motion is changing, you are still accelerating.

Velocity and acceleration can both be represented by arrows. As shown in the diagram, the directions of an object's velocity and acceleration are not always the same.

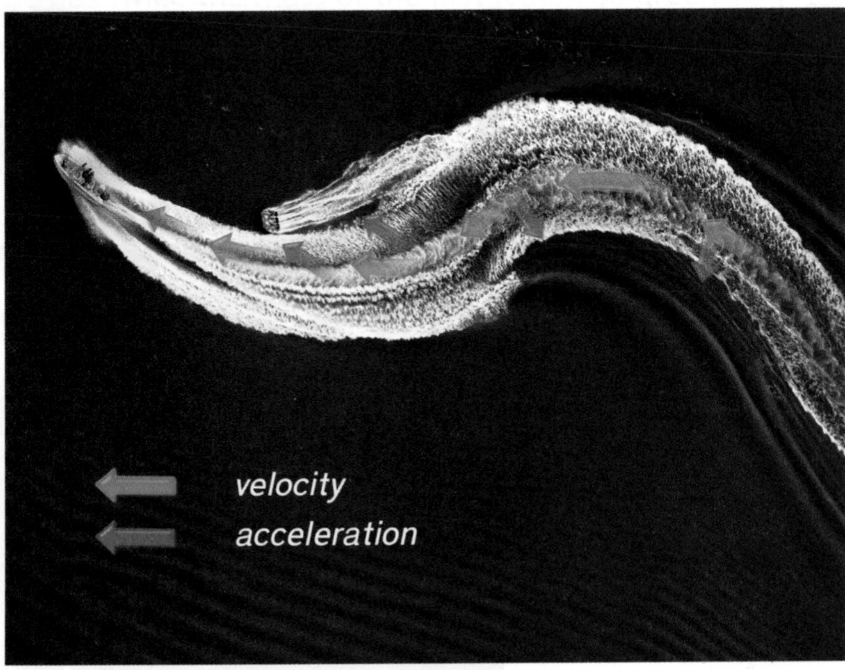

Each red arrow shows how the direction part of the velocity changes during acceleration.

What Could I Be?

Barge Pilot

Controlling the speed and direction of a giant barge with heavy cargo is not easy! A barge pilot understands velocity and acceleration and uses that knowledge to move huge boats that are hard to maneuver. Learn more about becoming a barge pilot in the Careers section.

Force - Balanced and Unbalanced Forces

The forces that act on an object can combine in different ways. Forces that act in the same direction add up to produce a stronger force. Forces that act in opposite directions can cancel each other or produce a weaker force. The total force is the sum of all the forces acting on an object.

Combining forces

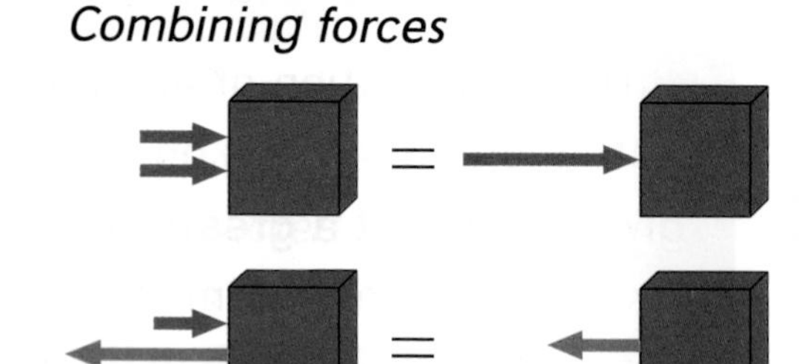

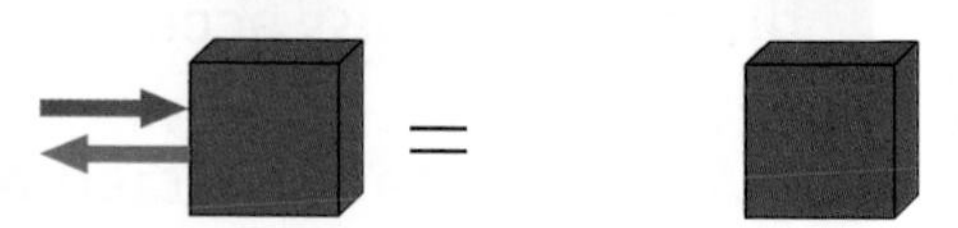

Balanced Forces When forces act on an object without changing its motion, they are called **balanced forces.** Balanced forces often point in opposite directions, and they always add up to zero. There may be more than one pair of forces acting on an object, but as long as the object is stationary, the forces are balanced.

If an object is not moving, the forces on it are balanced. However, the forces on a moving object can be balanced too. Think of a bus moving at a constant speed down a straight road. The force of the engine pushes the bus forward. As long as this force is balanced with other forces from air against the bus and the road against the tires, the bus will move at the same speed in a straight line. If the total of the forces on an object equals zero, the object will not accelerate, or change velocity.

Fact Checker

Forces do not cause motion. They cause *changes* in motion. An object that is in motion with no force acting on it will remain in motion.

The motion of the bus will continue without changing while the forces on it are balanced.

PhotoSpin, Inc/Alamy

Unbalanced Forces Forces that do not add up to zero are unbalanced. **Unbalanced forces** change the motion of an object. If two forces that are not equal act on an object, the greater force determines the direction of motion. Unbalanced forces can affect an object's direction, speed, or both. If the total force on an object is not zero, then the object will accelerate in some way.

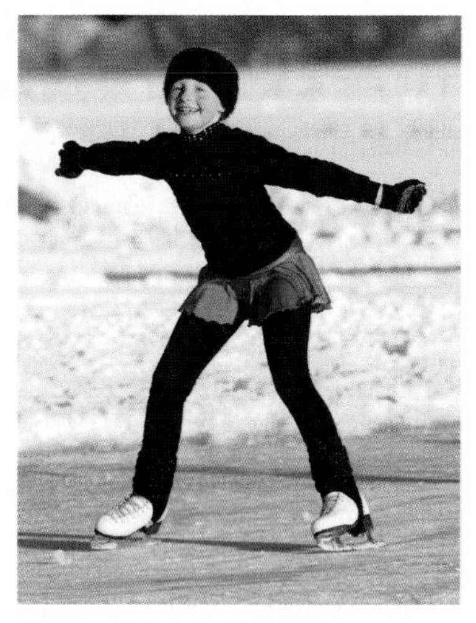

Unbalanced forces cause this skater's speed and direction to change.

Unbalanced forces change motion. All objects have a property called inertia. **Inertia** is the tendency of an object in motion to stay in motion or of an object at rest to stay at rest. A bicycle, for example, will not start to move unless an unbalanced force acts on it. A rider changes the speed and direction of a bike by applying forces to it. When a bicycle rider is moving fast and puts on the brake, the bike slows down. The rider can feel his or her weight shift toward the front of the bike as the body is still trying to move forward. This weight shift is the effect of inertia.

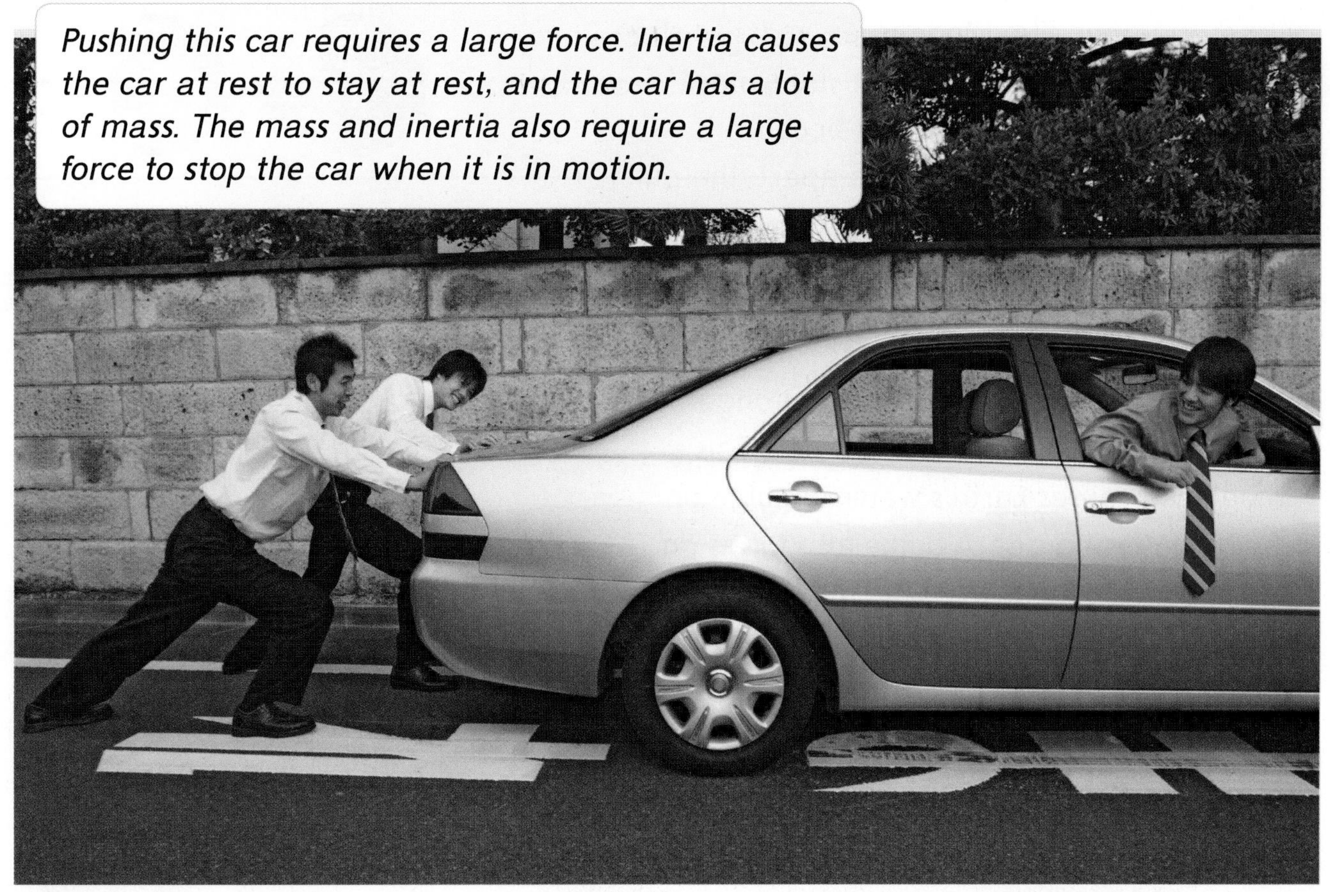

Pushing this car requires a large force. Inertia causes the car at rest to stay at rest, and the car has a lot of mass. The mass and inertia also require a large force to stop the car when it is in motion.

(t) Michi B/Getty Images; (b) Holos/Getty Images

Force - Gravity

Gravity is a noncontact force that acts over a distance and pulls all objects toward each other. The pull of gravity between two objects depends on two factors. One is the amount of matter in the objects. The other is the distance between the objects.

Objects with more mass have a stronger pull. For instance, the mass of Earth is huge. Its strong gravity pulls on all the smaller objects near it, holding everything around us to Earth's surface. The direction of the force pulls toward the center of Earth. We exist completely within Earth's gravity, so we sense that pull and identify the direction of it as "down." This force is what keeps your feet on the ground.

Gravity is also stronger when objects are closer together. The Moon is less massive than Earth, so its pull is weaker. But you cannot sense the gravity of the Moon because it is also far away. As objects move apart, the pull of gravity between them is weaker.

The pull of gravity between Earth and the Moon causes tides. The pull of the Moon's gravity acting on Earth's ocean is stronger on the side of Earth facing the Moon. This imbalance causes Earth's water to bulge on the Moon-facing side. A bulge also forms on the side facing away from the moon. The water level rises where the bulge is, causing high tide, and falls where it is not, causing low tide.

The force of gravity between Earth and the Moon is about 200 billion newtons!

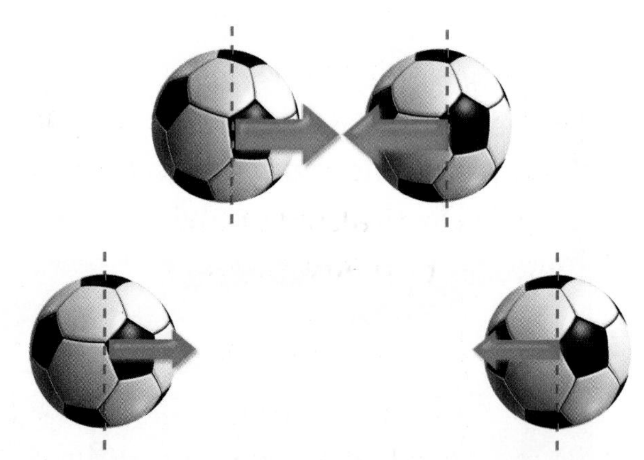

Gravity decreases as distance increases.

Make Connections

Jump to Space - Solar System to learn more about how gravity affects the movement of objects in our solar system.

NASA/JSC

Friction

Jogging on an icy sidewalk would be dangerous because the ice is so slippery. Jogging on a rough sidewalk is easier because there is more friction between your shoes and the sidewalk. *Friction* is a force that opposes the motion of one object moving past another. Friction keeps your shoes from sliding out from under you on a regular sidewalk. In this case, friction is helpful.

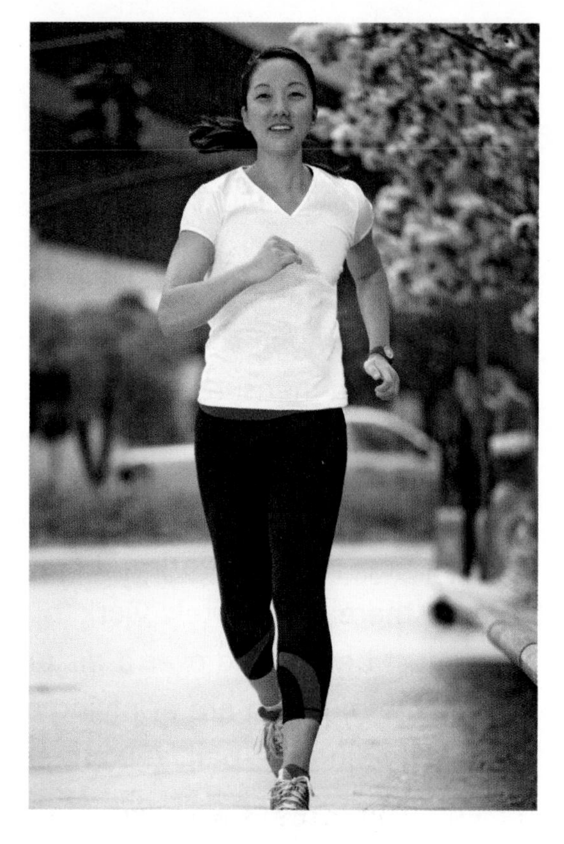

Friction between the shoes and the grounds keep the runner from slipping.

Friction is not always helpful though. Friction causes machine parts to wear out when they rub against each other. It causes tires to wear out from rubbing against the road. Vehicles that produce a lot of friction rubbing against the road and the air require more fuel for energy to keep them moving.

Friction depends on the surfaces of the two objects and how hard the objects are pushed together. Smooth surfaces usually have less friction than rough surfaces. Friction increases when surfaces are pressed together with greater force. Friction also increases with the weight of one object against the other.

Friction between two surfaces produces heat. You can feel this heat if you rub your hands together very fast. Friction explains why it helps to rub your hands together to warm them on a cold day.

Friction usually increases with the roughness of a surface.

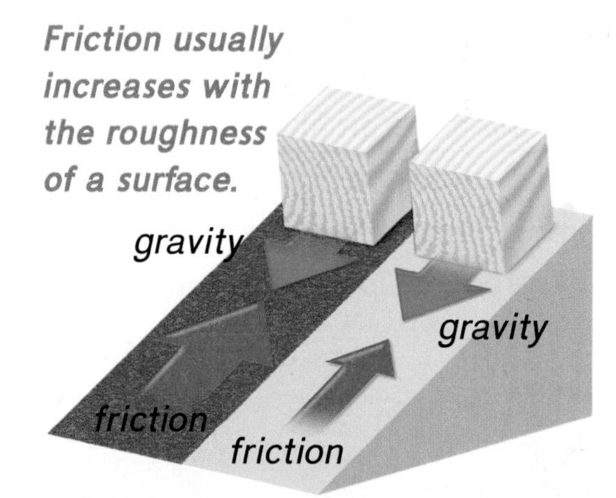

Friction increases with weight.

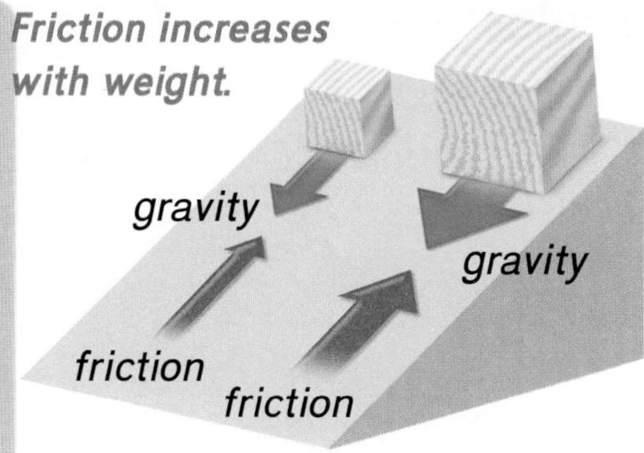

Skill Builder **Read a Diagram**

Look at the size of the red arrows (friction) to determine which block experiences the strongest friction force.

Simple Machines

Work

You might think picking up all the things in your room and putting them away is a lot of work. It is more work to pick up bigger things than it is to pick up the smaller ones. Work has a specific meaning in science. *Work* is the use of a force to move an object. You measure work as the force used to move an object multiplied by the distance the object moves.

There are many things that seem like they are work but are not. For instance, it is not work to hold a ball over your head. Lifting the ball there is definitely work, but just holding it there is not. A force must be applied over a distance to qualify as work. When you lift the ball, you are applying a force over a distance. When you are holding the ball, you are still applying a force, but the ball is not moving, so the distance equals zero.

Calculating Work The unit of work is the unit of force times the unit of distance, known as newton-meters (N m) or joules (J). If you lift a box that weighs 10 N onto a shelf that is 1 m high, you are performing 10 J of work.

If the force on an object and the distance the object moves are in the same direction, the work is positive. If the force and distance are in opposite directions, the work is negative. Lifting a box off the ground is positive work. Lowering the box is negative work.

Types of Simple Machines

Machines are tools that make work easier. Some have a motor and are made of many parts, but others are made of just one or two parts. A *simple machine* is a device that changes the direction, the distance, or the strength of one force. A simple machine can increase either the force or the distance in an example of work, but the total measure of work stays the same.

The force you apply to a machine is called the effort, or *input force.* The force the machine supplies, in turn, to do work is called the *output force.* The object moved by the output is called the *load.* So a simple machine changes your effort to move a load. It changes your effort in one of two ways. Simple machines either change the direction or the amount of effort force.

Deciding which machine is helpful in performing a task depends on whether you are more interested in decreasing the amount of force you need to use or the amount of distance you need to move to get a job done.

What Could I Be?

Mechanical Engineer

A mechanical engineer designs products, devices, and manufacturing systems. Understanding how machines change force or direction is important to a mechanical engineer's job. Learn more about becoming a mechanical engineer in the Careers section.

lever

wheel and axle

pulley

inclined plane

wedge

screw

Simple Machines - Types of Simple Machines

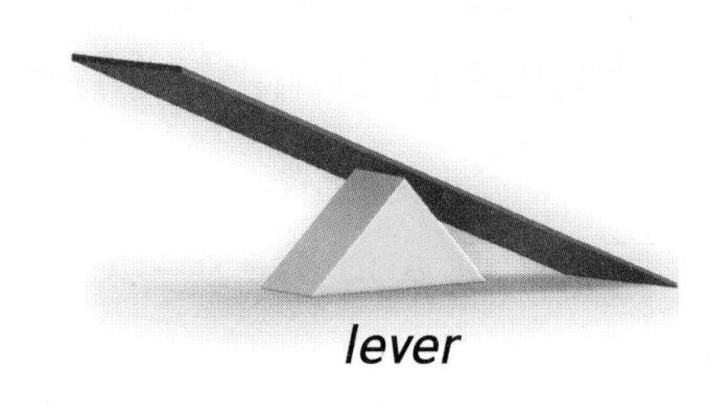
lever

Lever A *lever* is a type of simple machine made up of a bar that rotates on a pivot point called a *fulcrum.* Levers are divided into three classes, depending on the location of the fulcrum, the effort force, and the output force along the bar. Levers can multiply effort, distance, and speed. Some also redirect effort.

Word Study

The word *lever* comes from the Latin word *levare*, which means "to lift."

The position of the fulcrum determines how a lever does work. When a load is closer to the fulcrum, less effort force is needed to move the load. When a load is farther from the fulcrum, more effort force is needed to move it, but the load moves farther or faster.

A *first-class lever* has a fulcrum between the load and the effort force. The effort force and the output force are in opposite directions. The ratio of the forces depends on the ratio of the lengths of the sides of the lever on either side of the fulcrum. If the arms on both sides of the fulcrum are the same length, the effort force and the output force are the same.

examples of first-class levers

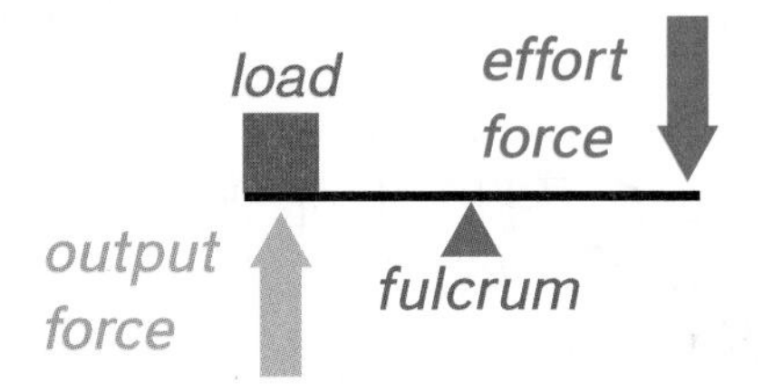

A *second-class lever* has a fulcrum at one end and the load in the middle. The distance of the effort is longer, but it makes it easier to move the load.

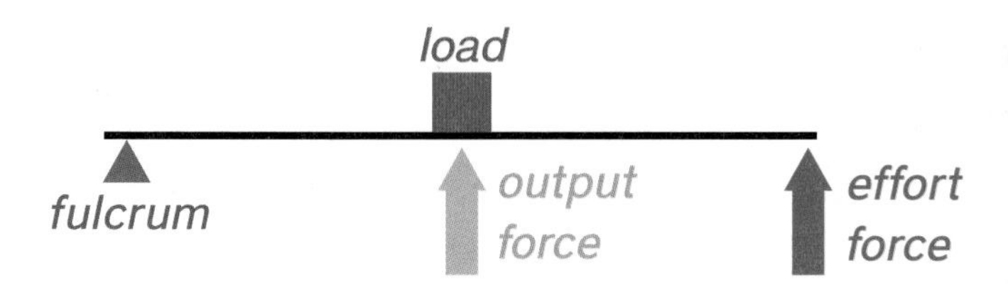

example of second-class levers

A *third-class lever* has a fulcrum at one end and its effort force in the middle. The effort and the load move in the same direction. The load moves a greater distance than the effort.

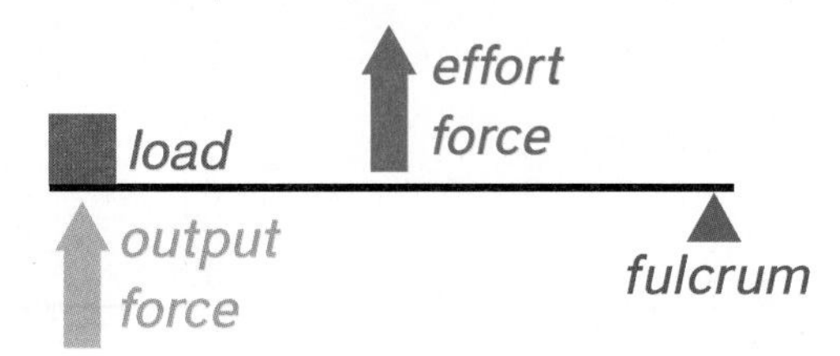

example of third-class levers

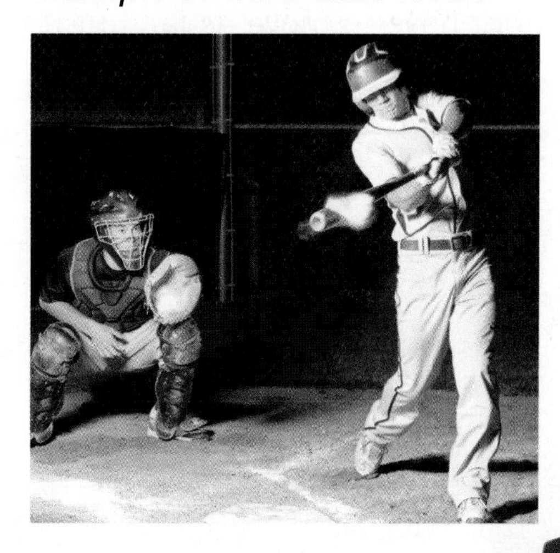

What Could I Be? Physical Therapist

The motion of the human body requires various forces. Sometimes an injury can cause a person to have to re-learn how to move the load of his or her body. Physical therapists use knowledge of forces and motion. They use exercises that apply pushes and pulls to help people regain the ability to move around. Learn more about becoming a physical therapist in the Careers section.

Did You Know?

The human body contains many levers. Your head and neck act as a first-class lever. Your foot acts as a second-class lever. Your forearm acts as a third-class lever.

Simple Machines - Types of Simple Machines

Wheel and Axle A *wheel and axle* is a type of simple machine made of an axle, or rod, through the center of a wheel. A wheel and axle is similar to a lever. The axle acts like the fulcrum of a lever, and the wheel acts like the two arms of a lever. Like a lever, a wheel and axle can multiply force, speed, or distance, depending on how it's used. Turning the wheel causes the axle to turn. Turning the axle causes the wheel to turn.

wheel and axle

Inclined Plane A ramp is a type of simple machine called an *inclined plane.* Comparing the effort distance to the output distance tells how much the machine will multiply effort. A longer ramp means the load has to be moved farther, but less effort force is required to move it. A shorter, steeper ramp means the load does not have to move as far, but it takes more effort to move it.

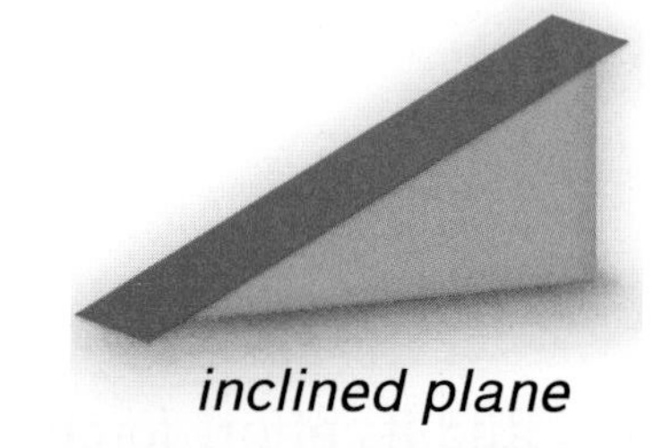

inclined plane

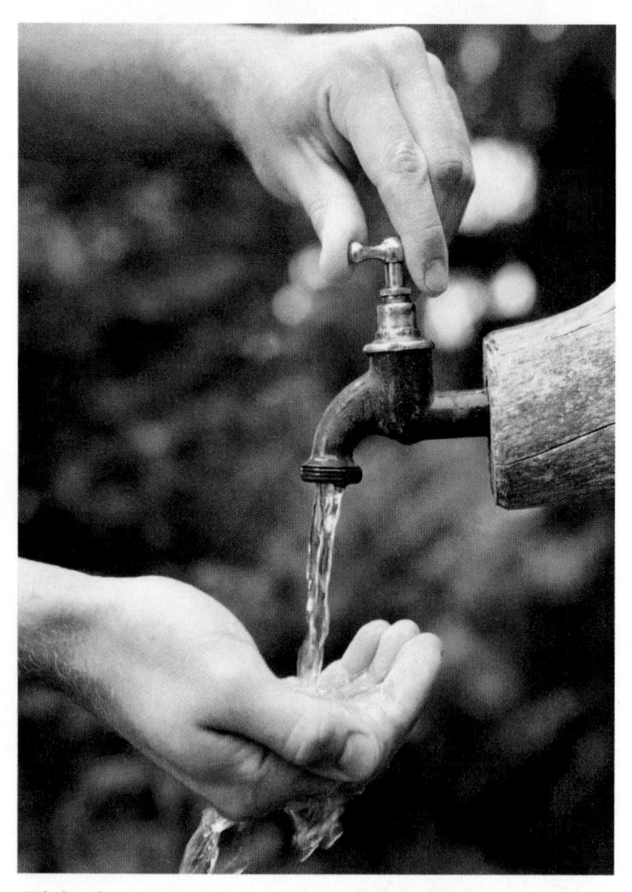

This faucet is a wheel and axle. A small effort force is applied to a faucet handle. A greater output force opens the valve.

Using a ramp reduces the force needed to move a load.

Wedge If an inclined plane is pushed into an object, the effort of pushing it in will cause a spreading force and push apart the two halves of the load. Chopping wood spreads force. When an inclined plane is used to separate two objects, it is called a *wedge.* Wedges may have one or two sloping sides. They may also be driven under an object to lift it. In this way, a wedge acts very much like a ramp.

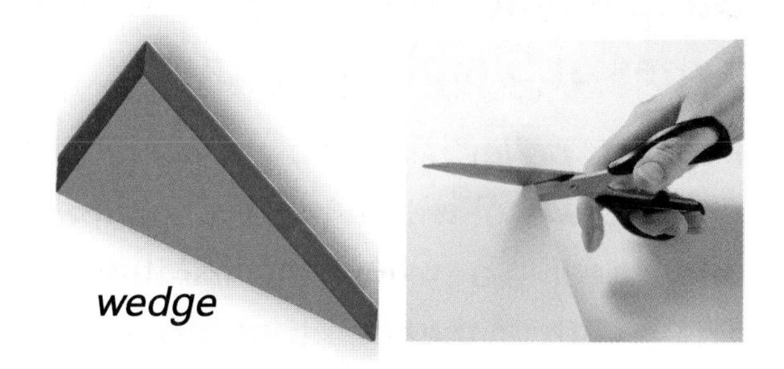

Screw A *screw* is an inclined plane wrapped around a cylinder. The farther apart the threads, the faster the screw moves into the material when you turn it. However, the farther apart its threads, the more effort it takes to turn the screw. Friction acts along the entire length of the threads of the screw. The more threads, the greater the friction force that acts on the screw.

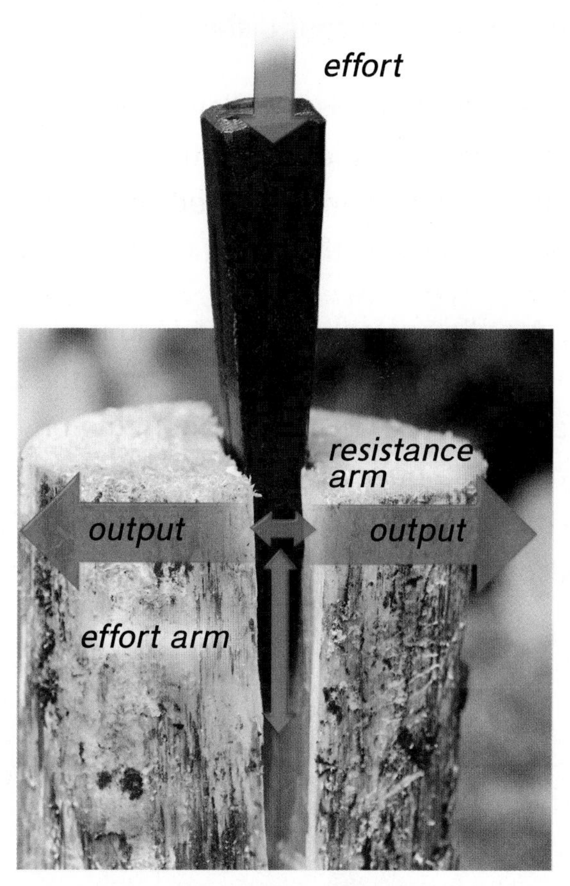

The spreading force of the wedge splits this log.

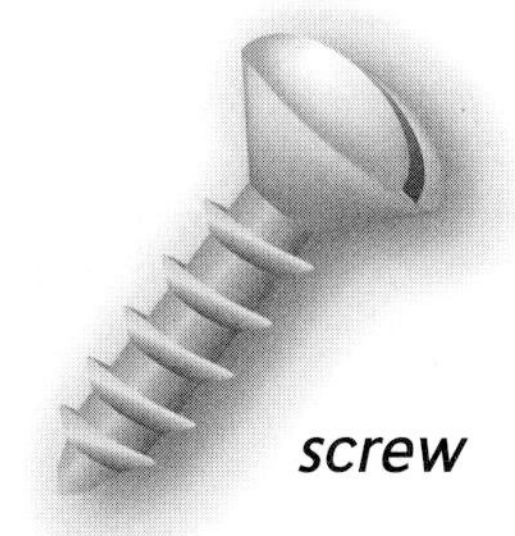

Screws used to build objects have inclined planes wrapped around a cylinder. The screws use friction to keep two parts together.

Simple Machines - Types of Simple Machines

Pulley A *pulley* is a simple machine made of a grooved wheel with a rope running along the groove. The wheel in a pulley acts like the fulcrum of a lever. The length of the rope that you move while applying the effort is the effort distance. The distance the load travels is the output distance.

pulley

A *fixed pulley* is a pulley attached to an unmoving object. Pulling the end of the rope in one direction moves the load in the other direction. A fixed pulley simply changes the direction of a force.

A pulley attached to the load is called a *movable pulley.* A movable pulley changes the amount of force need to do work. Depending on how the moveable pulley is used and how many pulleys are attached to the rope, the direction of the force might also change.

Skill Builder

Read a Diagram

The arrows show the effort force and output force of the three pulley systems.

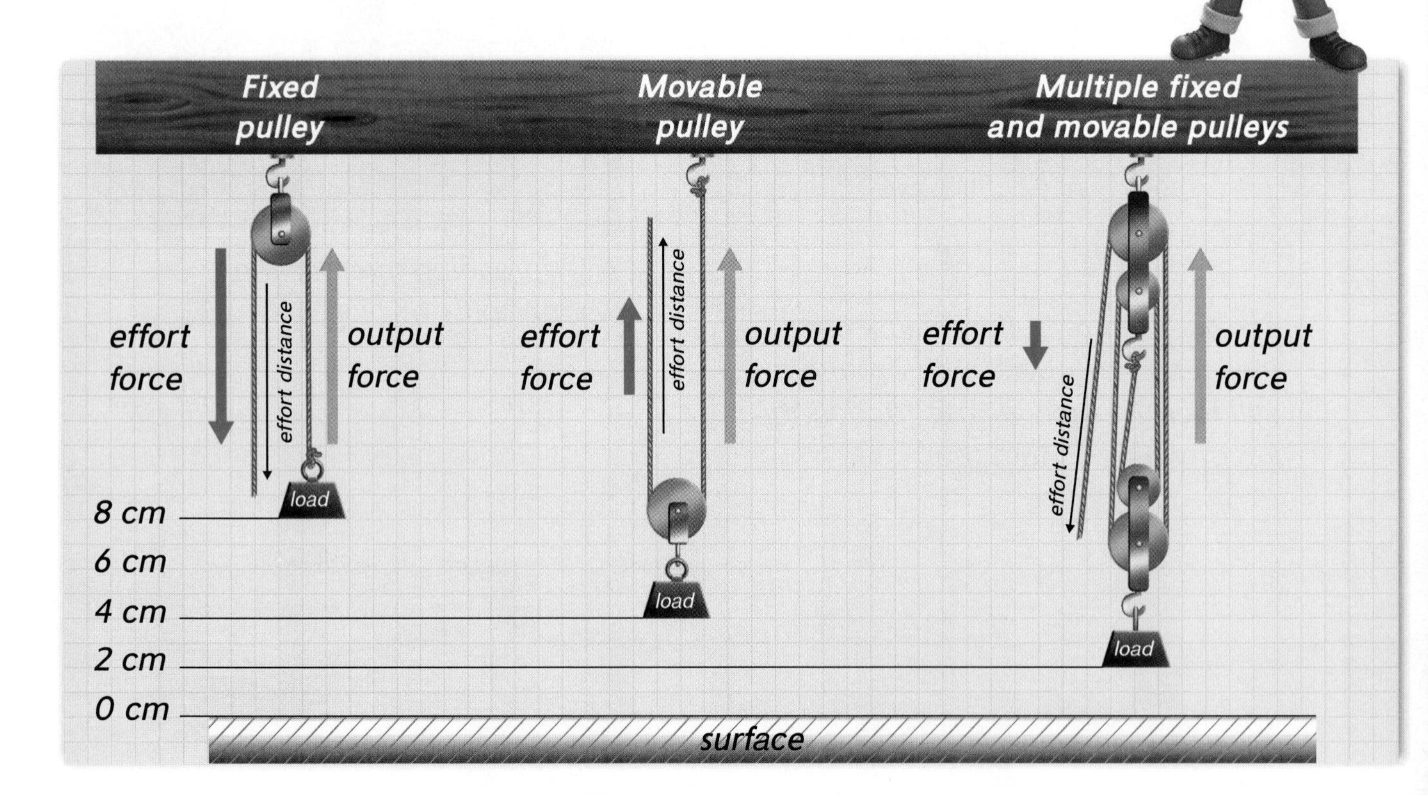

Compound Machines When two or more simple machines are combined, they form a *compound machine.* Some compound machines use just one type of simple machine repeatedly. Others include different types of simple machines. The figures shown here identify some of the simple machines included in each compound machine.

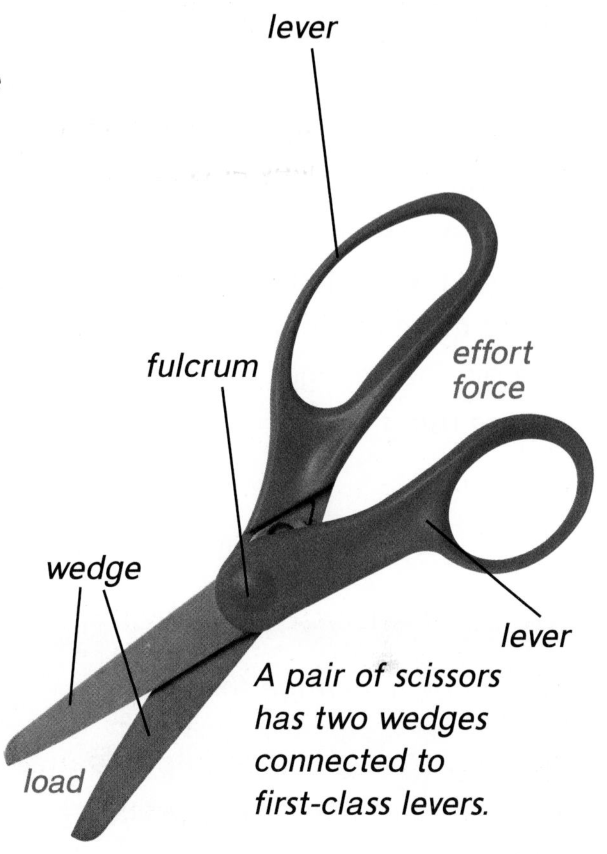

A pair of scissors has two wedges connected to first-class levers.

(l) Darren Mower/E+/Getty Images; (tr) Siede Preis/Getty Images; (br) Nikada/Vetta/Getty Images

Energy

Energy is the ability to do work or to change an object. The units of energy are expressed the same way as the units of work—*joules (J)*. A volcano uses energy to erupt and change the land around it. Plants use energy to grow. Ocean waves move and crash against rocks because they have energy. The changes that you see happening around you involve using energy.

Using Energy

energy to change things

energy to grow

energy to move

energy to do work

(tl) Jim Vallance/USGS; (tr) Jaap Hart/Getty Images; (bl) Design Pics/Don Hammond; (br) Steve Allen/Brand X Pictures

Forms of Energy

Potential Energy

Energy that is stored in the position or condition of an object is called **potential energy**. An object with potential energy is not using the energy now. Instead, the energy is stored in the object and can be used later. One form of potential energy is present when objects are above a surface. The higher the object is above the surface, the more potential energy it has.

Objects can also store energy because of their elasticity. For example, when a person jumps down on a trampoline, the trampoline gains potential energy because it is stretched. The trampoline uses this energy to push the person back up. A spring has potential energy if you compress it. When you release the spring, it will use the stored energy to expand. A rock on the edge of a cliff stores energy because of its position and the force of gravity.

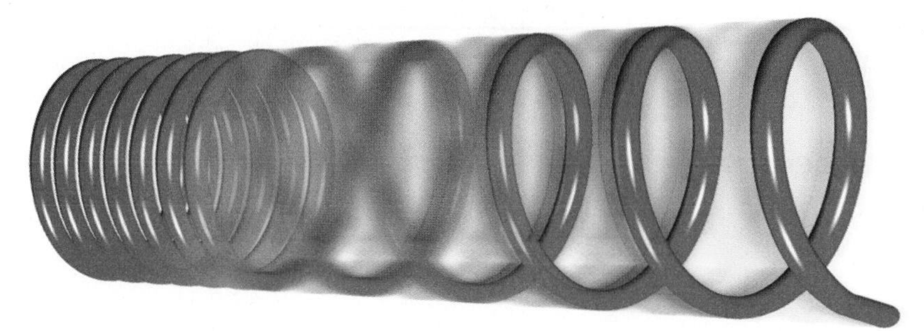

Releasing a compressed spring releases stored energy.

Increasing Potential Energy

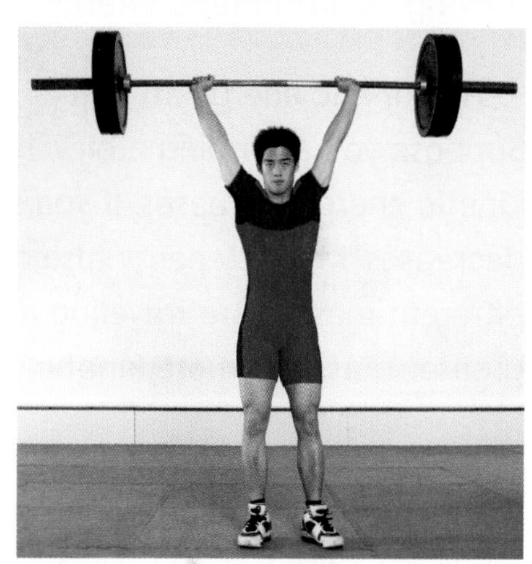

As the weightlifter raises the weights higher, the weights gain potential energy.

Word Study

The word *potential* means "possible." Potential energy makes it possible to do work or cause a change.

Forms of Energy - Kinetic Energy

Kinetic energy is the energy of motion. All moving objects have kinetic energy. A car traveling along a road has kinetic energy because it is moving. If the car stops, however, it no longer has kinetic energy.

Unlike potential energy, kinetic energy does not change if an object moves higher or lower above a surface. A car has kinetic energy as it moves up and down hills, but it also has kinetic energy when it moves along a flat highway. As long as the car is moving, it has kinetic energy.

The kinetic energy an object has depends on the object's speed. Suppose you are riding a bicycle. If you increase your speed, your kinetic energy increases. If you slow down, your kinetic energy decreases. Kinetic energy also depends on mass. If two objects with different masses are traveling at the same speed, the object with the greater mass has more kinetic energy.

You give objects energy by doing work on them. A pitcher does work on a baseball by throwing it. This work causes the ball to start moving, so the ball has kinetic energy as it moves toward the batter. Gravity also does work on the ball, so the ball also has potential energy. A player does work on the ball by catching it. This work decreases the speed of the ball, so its kinetic energy decreases.

Word Study

The word *kinetic* comes from a Greek word meaning "to move." Kinetic energy is always related to motion.

The airplane has potential energy because it is above the ground. It also has kinetic energy because it is moving.

Mechanical Energy

The energy that an object gains when work is done on it is called *mechanical energy.* Potential energy and kinetic energy are the two forms of mechanical energy. An object gains kinetic energy if the work causes the object to start moving. Work that changes the speed of a moving object changes its kinetic energy.

If work causes an object to move higher above a surface, then the object gains potential energy. The engines of an airplane, for example, do work that causes the airplane to take off. If you lift a backpack, you do work on the backpack that gives it potential energy. Work that stretches or compresses a spring or elastic object also gives the object potential energy. If you stretch a rubber band, for example, you do work on the rubber band, and it gains potential energy.

Skill Builder

Read a Diagram

The numbers on the diagram relate to the captions.

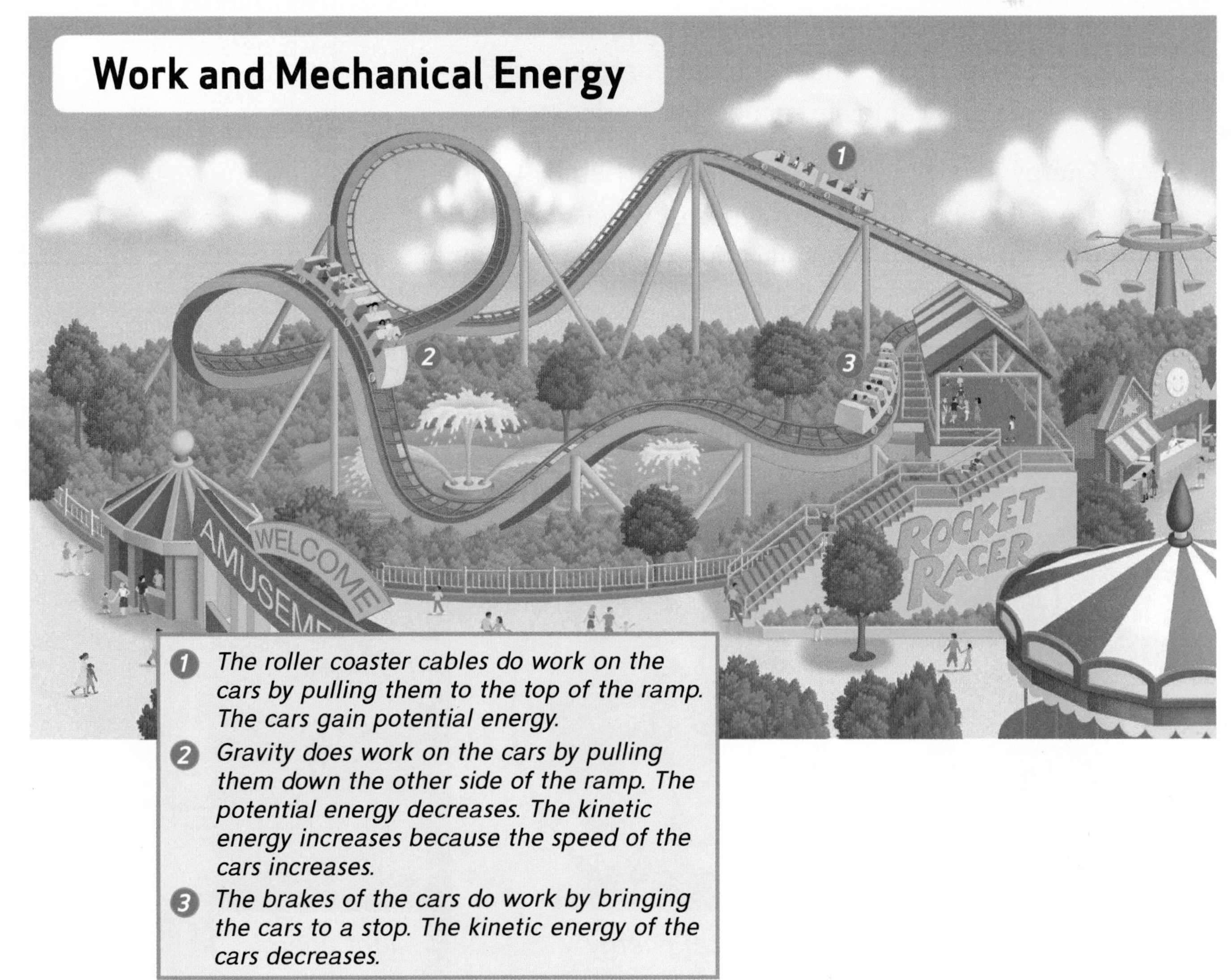

1. *The roller coaster cables do work on the cars by pulling them to the top of the ramp. The cars gain potential energy.*
2. *Gravity does work on the cars by pulling them down the other side of the ramp. The potential energy decreases. The kinetic energy increases because the speed of the cars increases.*
3. *The brakes of the cars do work by bringing the cars to a stop. The kinetic energy of the cars decreases.*

Other Forms of Energy

There are many forms that potential and kinetic energy can take.

Chemical Energy *Chemical energy* is stored in links between the atoms and molecules that make up food and other fuel. These links are energy that can be released by a chemical reaction, such as burning wood and digesting food. Chemical energy is potential energy. A chemical reaction is needed to change this potential energy into kinetic energy.

Electrical Energy Energy that comes from the movement of charged particles is *electrical energy.* This energy is a type of kinetic energy because it involves the motion of the particles. Some electrical energy comes from batteries. Most of it, however, comes from power plants that burn fuel to make electricity. The electrical energy is sent through wires and cables to homes and businesses.

Light Energy *Light* is a type of energy carried by waves. Stars are visible because these types of waves carry light energy through space. The Sun is the main source of light energy on Earth. Plants use light energy for photosynthesis. Light energy can be collected with solar cells. Some scientists use light energy from lasers. A laser is a form of light that can cut through some metals.

What Could I Be?

Machinist

Lasers are not just for scientists. Many machinists use lasers to precisely cut metal parts. These machinists program and monitor the work of machines, read and interpret blueprints, and maintain the precision cutting instruments. Learn more about the work of machinists in the Careers section.

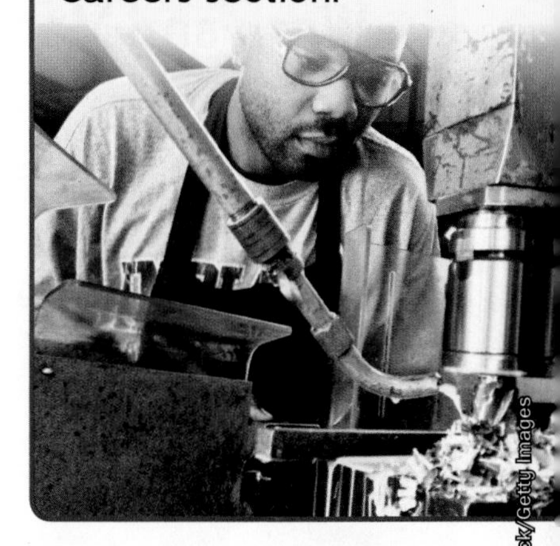

Forms of Energy

chemical energy

electrical energy

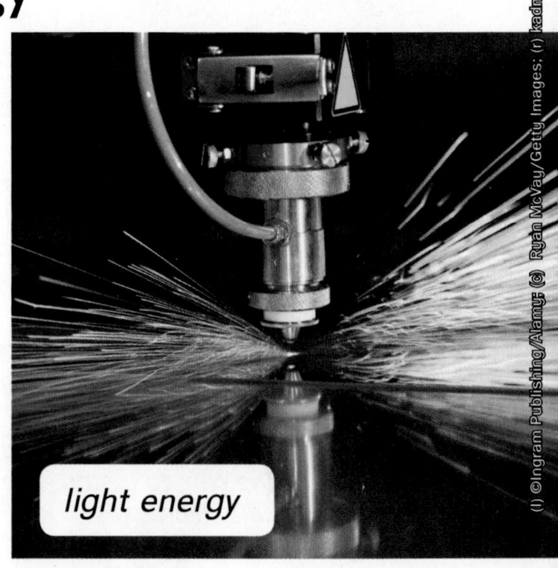

light energy

Thermal Energy A stove, a heater, and a match have thermal energy. They give off heat. *Thermal energy* is a type of kinetic energy. It depends on the motion of the tiny particles in matter. The faster these particles move, the warmer a substance gets. The warmer the substance, the more thermal energy it has.

Nuclear Energy Protons and neutrons make up the center of an atom, called the nucleus. Energy stored in the links between these tiny particles is called **nuclear energy.** Like chemical energy, nuclear energy is a type of potential energy. It takes a nuclear reaction to release the energy stored in a nucleus. At nuclear power plants, atoms are split apart. This is the process that releases the energy of the atoms.

Sound Energy *Sound* is a type of energy produced by vibrations of matter. A *vibration* is a fast back and forth movement. When a drummer beats a drum, the drum vibrates. The vibrations of the drum make the air vibrate. The vibrations travel away in all directions. Because sound depends on the movement of the particles that make up matter, it is a type of kinetic energy.

Make Connections

Jump to the end of the Physical Science section to learn more about sound energy.

thermal energy

nuclear energy

sound energy

Energy Changes

Energy does not always keep its form. It also does not always stay in one place.

Transforming Energy

Energy is *transformed* when it changes from one form to another. A light bulb transforms electrical energy to light and thermal energy. A lawn mower transforms chemical energy to mechanical energy. When walking, you are transforming chemical energy in your body to kinetic energy in your leg muscles.

Transferring Energy

Energy is *transferred* when it passes from one object to another. Newton's cradle is a toy that shows how energy can be transferred. Pull one ball up and release it. When it hits the other balls, the ball on the opposite end swings out. Energy is transferred from the first ball to the last and back again.

Skill Builder

Read a Diagram

Look at the numbers in the diagram, and then read the captions to identify some of the ways energy is transformed inside a home.

Transforming Energy

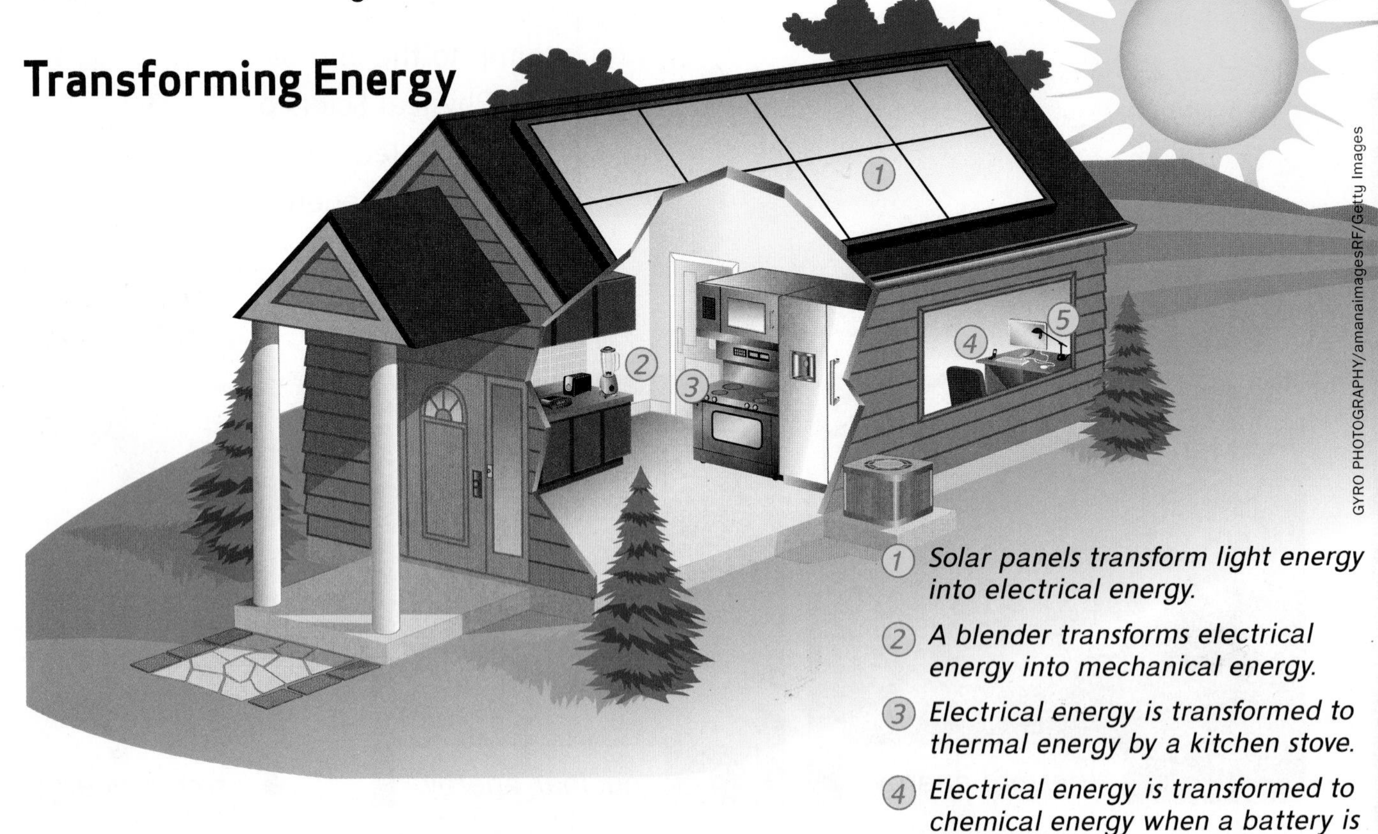

1. *Solar panels transform light energy into electrical energy.*
2. *A blender transforms electrical energy into mechanical energy.*
3. *Electrical energy is transformed to thermal energy by a kitchen stove.*
4. *Electrical energy is transformed to chemical energy when a battery is charged.*
5. *A lamp transforms electrical energy into light and thermal energy.*

Conservation of Energy

Energy cannot be created or destroyed. It can only change form. This idea has been observed so many times that it is called the **law of conservation of energy.**

A roller coaster, for example, cannot gain kinetic energy without losing potential energy. As the cars of a roller coaster move up a hill, they gain potential energy. The decrease in speed, however, means that they lose kinetic energy. The kinetic energy transforms into potential energy. When the cars start to roll down the hill, the potential energy changes to kinetic energy.

You might think that a roller coaster loses energy. After all, the roller coaster steadily slows down. This lost energy, however, is not destroyed. It changes to heat and sound through the work of friction.

Whenever energy is used to do work, that energy changes. Electricity does work in an oven by moving particles around and changes into heat. Heat does work on a loaf of bread and changes into chemical energy. Chemical energy in the bread does work and changes into kinetic energy in your muscles.

Fact Checker

Conserving energy is not the same as the law of conservation of energy. Conserving energy means to use energy resources carefully in order to extend their supply.

A roller coaster is a good example of conservation of energy.

Shenval/Alamy Images

Heat Energy

Energy is needed for animals to stay warm. Whether it is from the Sun or your body, thermal energy keeps you warm. **Thermal energy** is the energy of the moving particles of matter. The faster the particles move, the more thermal energy the matter has.

When particles of one substance come in contact with particles of another substance that are moving at a different rate, thermal energy is transferred. **Heat** is the flow of thermal energy from one substance to another.

Thermal energy moves from an object with a higher temperature to an object with a lower temperature. *Temperature* is a measure of the average kinetic energy of particles in an object. All the particles in an object are vibrating with kinetic energy. Objects with a higher temperature have particles that are vibrating faster. Objects with a lower temperature have particles that do not vibrate as much. When a hot object touches a cold object, their particles bump into each other. During these collisions, the particles from the hot object pass on some of their energy to the particles in the cold object. The cold object becomes warmer, and the hot object becomes colder.

A lizard can warm itself by sitting in sunlight. Heat flows from sunlight into the lizard. It also flows from the rock into the lizard.

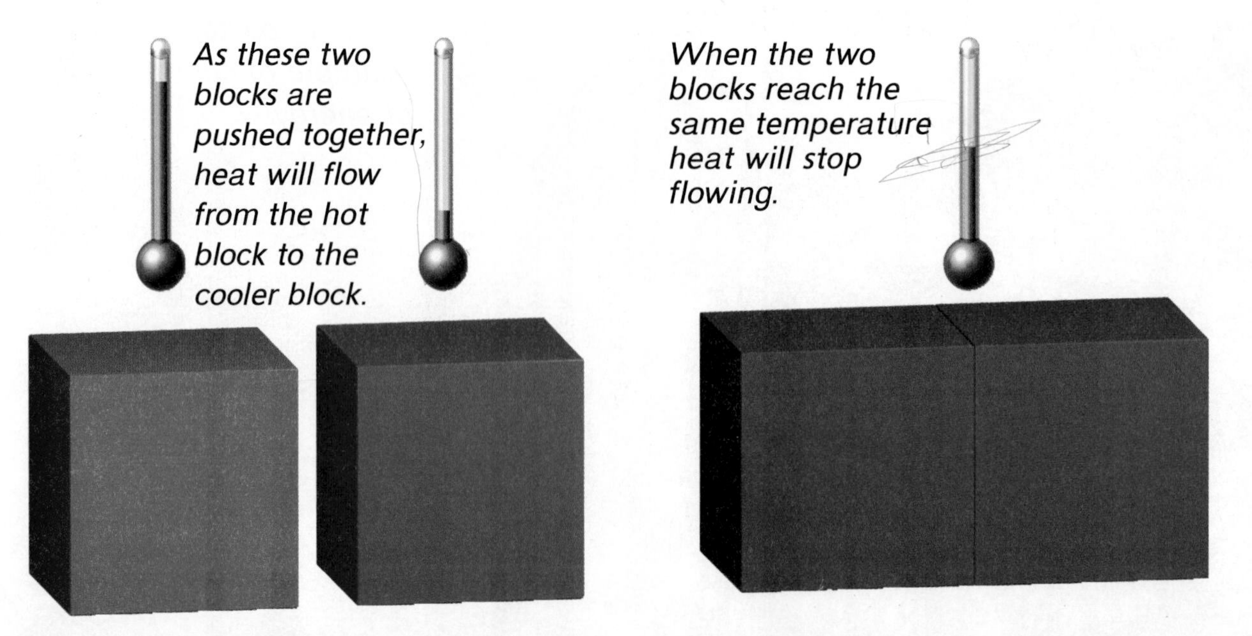

As these two blocks are pushed together, heat will flow from the hot block to the cooler block.

When the two blocks reach the same temperature heat will stop flowing.

Tom Horton, Further To Fly Photography/Getty Images

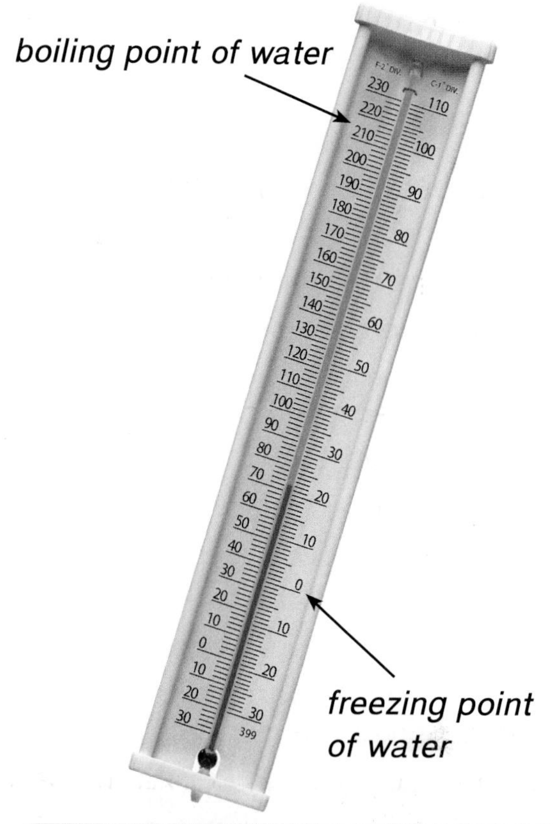

Temperature is measured with a thermometer. When a **thermometer** touches a material, heat will flow either into or out of the thermometer. When the thermometer and the material are at the same temperature, heat stops flowing and the thermometer shows the temperature. The thermometer displays the temperature in units of degrees Celsius (°C) or degrees Fahrenheit (°F).

Temperature and Mass

Keep in mind that temperature and heat are not the same things. Temperature measures the average kinetic energy of particles. The total amount of thermal energy in a substance depends on temperature and mass. A cup filled with boiling water has a high temperature, but little mass. It has less thermal energy than a jug full of water that is warm, but not boiling. Overall a gallon of warm water has more energy than a cup of boiling water. Although the temperature of the jug is lower, the total amount of energy is higher because the jug has more mass.

Skill Builder **Read a Photo**

Read the temperature on the thermometer by finding the mark nearest the top of the red line.

Heat and Temperature

The soup has a higher temperature, but the lake's greater mass releases more heat than the soup.

Producing Heat

Like all energy, heat cannot be created or destroyed. It can be released as energy changes occur. If you rub your hands together very fast, you can feel your hands get warmer. Friction between your hands changes kinetic energy into thermal energy. Friction is a force that opposes the motion of an object. Friction occurs when two or more objects come in contact and at least one of the objects is moving.

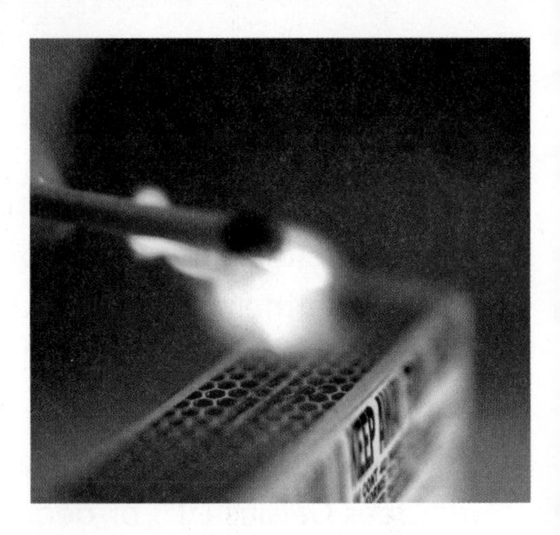

Friction between the match head and the surface produces enough heat to light the match.

Sometimes the heat produced by friction is useful, such as pushing the brake pedal in a car. However, friction often causes a problem. When the parts of a machine rub together, friction between the parts produces heat. The machine has less energy to do work because some of its energy changes to heat.

Mixing and burning are other ways to produce heat. Some of the energy in a chemical reaction is given off as heat. In a campfire, chemical energy stored in wood is released as heat when logs burn.

The flow of electricity can also produce heat. Electric charges moving through a wire release energy as heat. The thinner the wire, the more heat produced. Toasters, some electric lights, and space heaters produce heat in this manner.

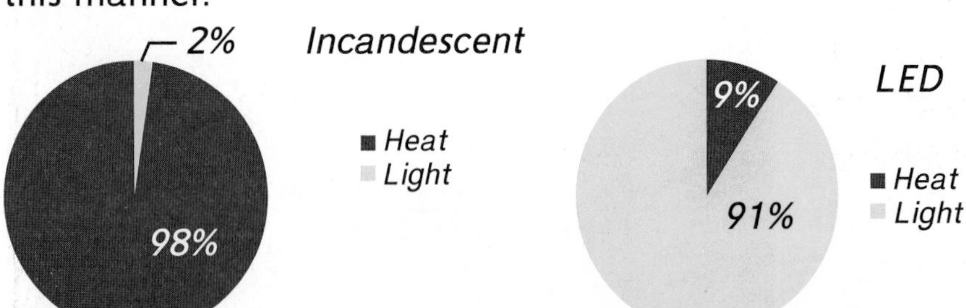

Skill Builder **Read a Graph**

Compare the amount of heat versus light produced by incandescent and LED bulbs.

Heat and Physical Properties

The particles that make up matter are always moving. By adding energy to those particles or by taking energy away, you can change matter.

Expansion and Contraction As heat flows into a substance, the energy gained by the substance can cause its particles to move farther apart. As the particles in a substance spread out, the substance usually increases in volume. An increase in volume caused by an increase in temperature is called *thermal expansion*. The opposite happens if you reduce thermal energy. Most matter shrinks when cooled. The particles move closer together. This movement is called *thermal contraction*.

Joints such as this one are used in bridges. As the pavement heats, it expands, and the joints move closer together. As the pavement cools, it contracts. The joints move farther apart.

Scientists, engineers, and architects consider the effects of thermal expansion when choosing materials to build houses and other structures. Not all materials respond to changes in temperature in the same way. Some expand more than others.

Specific Heat Not all materials change temperature at the same rate. If you apply the same amount of heat to a gram of water and a gram of cooking oil, the cooking oil will warm more. Oil has a lower specific heat than water. The *specific heat* of a substance is the amount of energy needed to raise the temperature of 1 g of the substance by 1°C. Specific heat is a physical property of matter that results from how a material's particles hold together. Materials with low specific heat change temperature quickly. Most metals have a low specific heat, so little energy is needed to increase their temperatures. Water has a high specific heat, so more energy is needed to raise its temperature.

Some saws and drills are cooled by water. Water absorbs the heat produced by the friction of the blade or bit cutting through a material.

Heat Transfer

Heat always flows from a higher temperature material to a lower temperature material. Heat can move in three ways.

Conduction The particles that make up matter are always vibrating. **Conduction** is heat transfer that occurs because of these vibrations. The material itself does not move. The vibrations, however, spread from the warmer part of an object to the cooler part of the same object. If two objects are touching, the vibrations in one object can make particles of the other object vibrate faster so that they have more energy. Heat travels from the warmer object to the cooler object.

Convection Inside a pot of heating water, heat spreads as warm and cool parts of the water move around. **Convection** is heat transfer through a liquid or a gas. Convection is caused by cool parts sinking and pushing the warmer parts up into the liquid or gas. As warm and cool parts of the gas or liquid move, they cause rotating currents. The currents spread thermal energy throughout the material. Convection occurs in gases and liquids but not in solids.

Make Connections

Turn back to the Earth Science section to learn how convection causes many of the weather patterns on Earth.

How Heat Moves

In conduction, heat moves directly from the stove to the pan to the eggs.

In convection, cold water falls, pushing the warmer water up through the liquid.

Radiation Earth's surface is warmed by radiation from the Sun. **Radiation** is the transfer of energy through *electromagnetic waves.* These waves include visible light, X-rays, and radio waves. As the waves travel from their source, they carry energy from one place to another. Matter is not needed to transfer heat by radiation. Electromagnetic waves can travel through empty space. Otherwise, heat from the Sun could not warm Earth!

Infrared Radiation Hot objects radiate heat. The electromagnetic waves they produce are called *infrared.* They are called infra "red" rays because they are close in color to red visible light. You cannot see infrared rays, but you can feel them when you stand in the Sun. Cooks use infrared radiation when they place food under a heat lamp to keep it warm.

Some snakes and other animals have special sensory organs that detect infrared rays. Scientists have built special instruments and cameras so that humans can do the same. Computers can artificially color the images that these cameras take. These cameras can "see" to take pictures of objects, even in the middle of the night.

Make Connections

Jump ahead in this Physical Science section to learn more about electromagnetic waves.

Skill Builder

Read a Diagram

Look at the arrows in each diagram to better understand heat transfer.

In radiation, electromagnetic waves carry energy from the hot wires to the toast.

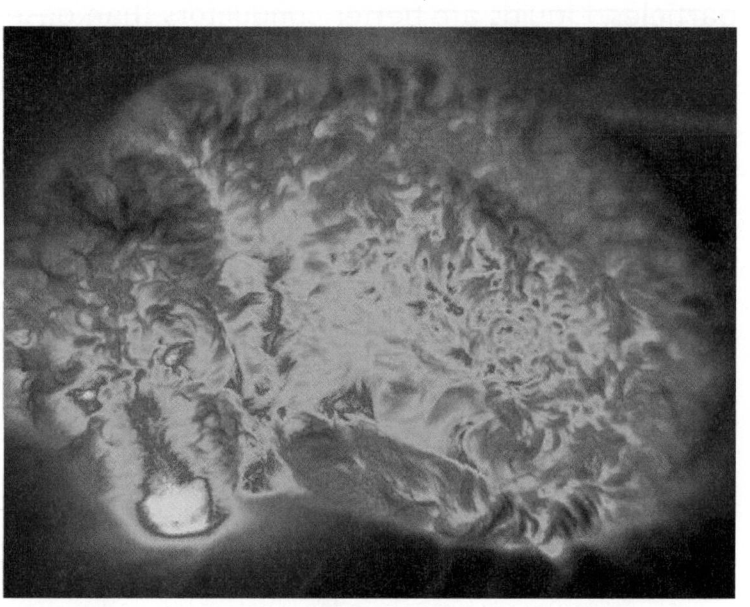

A special infrarred camera took this photo of a dog. Red shows where the most heat is produced. Black and dark blue areas are coolest.

Thermal Conductivity

Heat travels at different speeds. Radiation carries thermal energy through empty space at the speed of light. That is 300,000,000 meters each second (670,000,000 miles per hour)! Convection currents are rarely faster than about 56 meters per second (125 miles per hour). Conduction typically carries thermal energy more slowly than either radiation or convection. In conduction, energy is transferred as particles bump into other particles. Thermal energy can move only from one side of an object to another by moving through all the particles in between, which takes time.

This copper pot is a good thermal conductor.

Thermal Conductors The ability of a material to transfer heat is thermal conductivity. Materials that conduct heat easily are good *thermal conductors.* Most metals are thermal conductors. Thermal conductivity usually increases with density. The closer particles are together, the quicker heat can move through them. Solids make better conductors than liquids because of the closeness of the solids' particles. Liquids are better conductors than gases.

Thermal Insulators Materials that conduct heat poorly are *thermal insulators.* Air, for example, is a good thermal insulator. This fact explains why some winter coats are puffy. They contain pockets of air that keep heat from moving away from your body. The walls of homes usually have some type of thermal insulator inside them. This insulation slows down the transfer of heat into the home in summer or out of the home in winter. Other examples of thermal insulators are rubber and plastic.

Material	How Many Times Better Than Air It Conducts Heat
oak wood	6
water	23
brick	25
glass	42
stainless steel	534
aluminum	8,300
copper	15,300
silver	16,300
diamond	35,000 or more

Ways People Use Heat

Buildings are heated by systems designed to transfer heat energy. In a hot water system, water transfers energy from a boiler to the air in a room. The boiler heats water, which is forced through pipes. The pipes lead to radiators in the rooms, and the air around the radiators becomes warmer. Cool air is denser than warm air. Cooler air sinks to the floor and pushes the warmer air upward. Eventually the warmer air cools and sinks, forming a convection current that warms the room.

In a forced-air system, a room is heated with air alone. Hot air, forced up by fans from the furnace, heats the air in the room. Convection currents circulate the air in the room.

A thermostat can regulate themperature. Some thermostats contain a switch made of a metallic strip. The strip is made of two metals that expand or contract at different rates. This expansion causes the strip to bend. The bending of the strip turns the system off when the air in the room reaches a certain temperature. As the strip cools, it straightens and that turns the furnace on again.

Skill Builder

Read a Diagram

Follow the arrows to identify how air is warmed in each diagram.

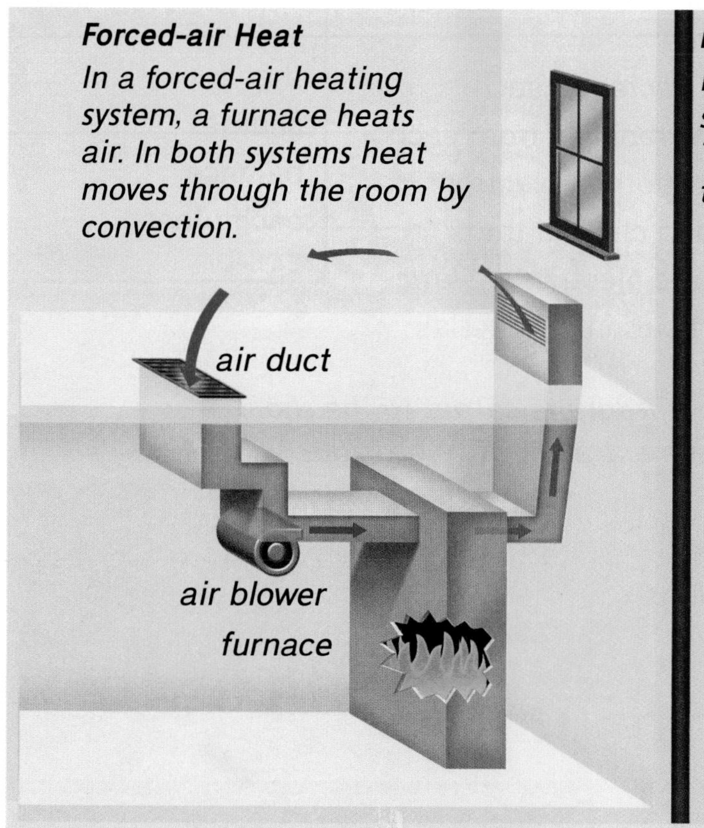

Forced-air Heat

In a forced-air heating system, a furnace heats air. In both systems heat moves through the room by convection.

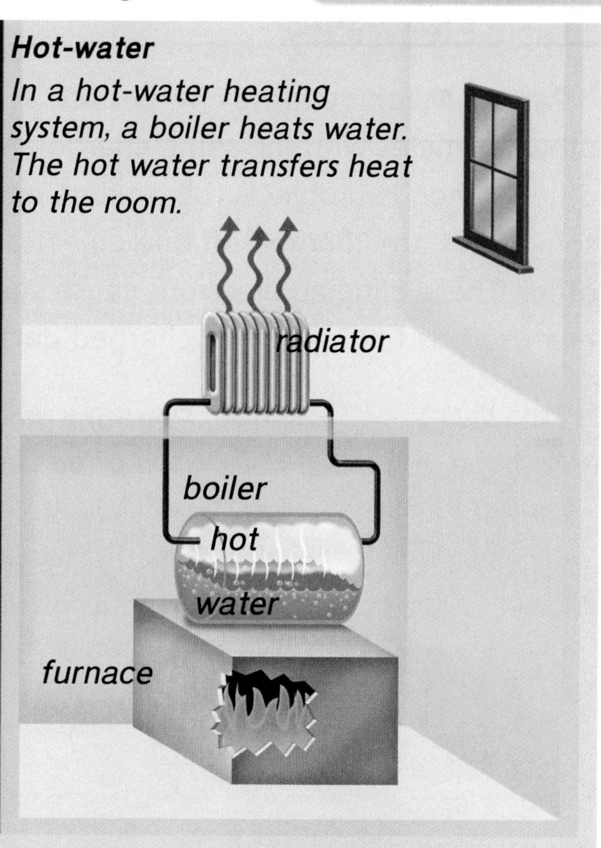

Hot-water

In a hot-water heating system, a boiler heats water. The hot water transfers heat to the room.

Electricity and Magnetism

Electrical Energy

There are two types of charged particles in an atom. Protons have a positive charge (+), and electrons have a negative charge (–). *Electricity* refers to electric charge and how it moves. In most materials, the electric charge that moves is the electron. Electrons can move from one object to another, causing the material to take on a temporary charge.

The energy carried by electric charge is called electrical energy. Like all energy, it is measured in joules. Units called volts describe how strongly electrons flow.

Fact Checker

An atom has charged particles in it, but it can still be neutral if the number of protons and electrons is equal. The tom itself has no overall charge, because the charges balance each other.

Static Electricity

Particles with opposite charges are *attracted* to each other. Particles with the same charge are *repelled* from each other. When two objects rub against each other, electrons sometimes are knocked off one object and cling to the other. These clinging electrons cause static electricity. *Static electricity* is the buildup of charged particles in a material.

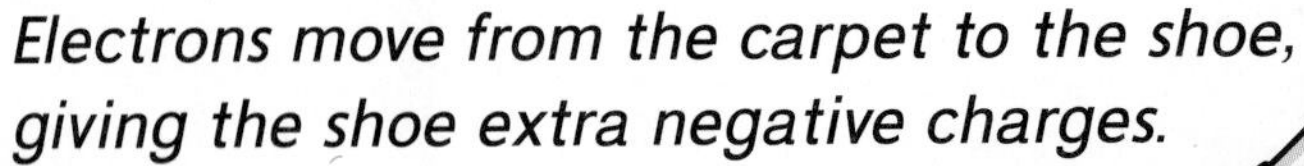

Electrons move from the carpet to the shoe, giving the shoe extra negative charges.

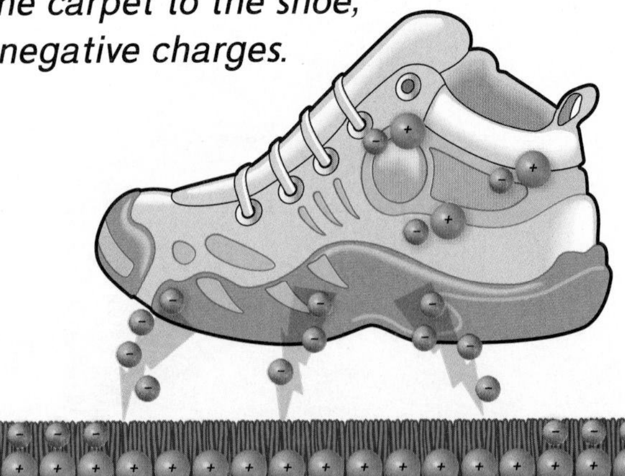

The attractive force between electrons and protons is strong. A buildup of electrons in one object will jump through the air toward nearby protons in another object. This movement forms a spark called *static discharge.* After the static discharge, the two objects have balanced charges. The objects are then *neutral.* A neutral object has equal numbers of protons and electrons.

Make Connections

Jump to the Earth Science section to learn more about lightning and grounding.

Since opposite charges attract, two oppositely charged objects can stick together. This is called *static cling.* It often happens when clothes rub together as they tumble in a dryer. Charged objects can also attract neutral objects. When a charged object nears a neutral object, the neutral object will act like it is slightly charged on one side and attract the charged object.

Electric charge moves easily through some materials, such as most metals. These materials are called *conductors.* Electric charge does not move easily through other materials called insulators. Wood, glass, and plastic are insulators.

Static electricity, including lightning, can cause damage. You can protect objects from static electricity by grounding them to Earth with a wire. **Grounding** occurs when a conductor shares its excess charge with a much larger conductor. Earth is a large conductor with a neutral charge.

This charged purple balloon has four extra electrons.

The neutral red balloon is attracted to the purple balloon.

The neutral green balloon is too far away to be attracted.

KEY
electrons
protons

Electric Current

When you are in the dark, a flashlight is useful. A flow of electric charge lights the bulb. **Electric current** is the flow of electric charge through a conductor.

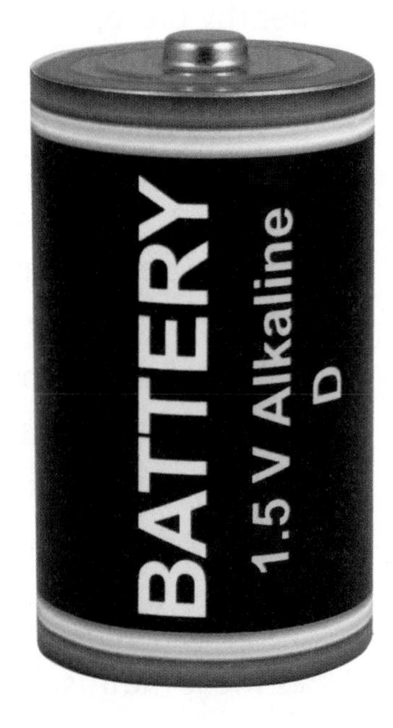

The chemicals inside a battery cause a buildup of extra electrons.

A **circuit** is formed when an electric current passes through an unbroken path of conductors. Often the path of a circuit consists of wires. Circuits must also have a device to move electrons along the path. These devices, called *voltage sources,* increase the volts of electrons flowing in the circuit. Batteries are a voltage source. Different sizes of batteries produce different quantities of voltage, or amounts of energy that move in the circuit.

A *switch* is a device that can open or close the path in the circuit. When the switch is closed, the electric current can flow around and around through the circuit. When the switch is open, the current cannot flow.

Electricity does not flow in the same way through every part of a circuit. An object in an electrical circuit that resists the flow of electrons is called a *resistor*. Electrons lose energy when moving through a resistor. This energy is transformed into a different form of energy, such as heat or light. A light bulb is a resistor. Resistance is measured in units called *ohms*.

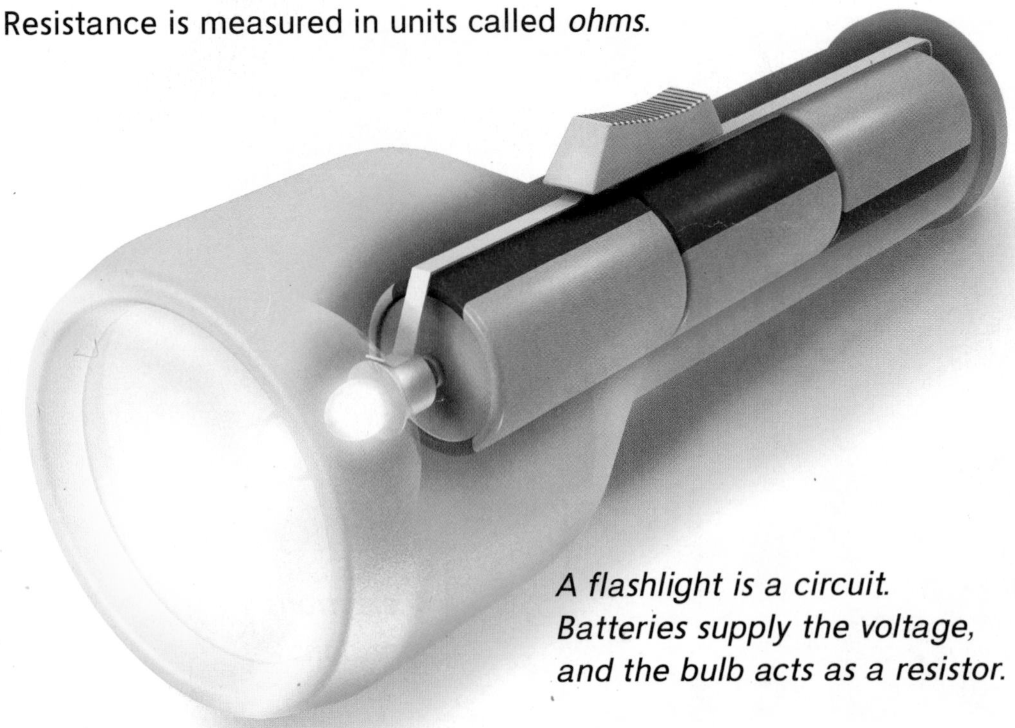

A flashlight is a circuit. Batteries supply the voltage, and the bulb acts as a resistor.

Electric charge travels fast in a circuit—almost as fast as the speed of light. And much more electric charge moves as electric current in a circuit than in the static discharge when a doorknob shocks your finger. The amount of electric charge moving in a circuit can be very powerful and very dangerous. Electric current is measured in units called *amperes*, or amps. There are about 6 billion billion electrons moving every second in 1 amp of current. Even currents as small as 0.05 amps can seriously harm you.

Your home has a collection of connected electric circuits hidden in the walls. When you plug a device that uses electricity into a wall outlet, you are making that device part of a circuit. The wires and outlets are covered with plastic insulators to keep you from coming into contact with the electrical current.

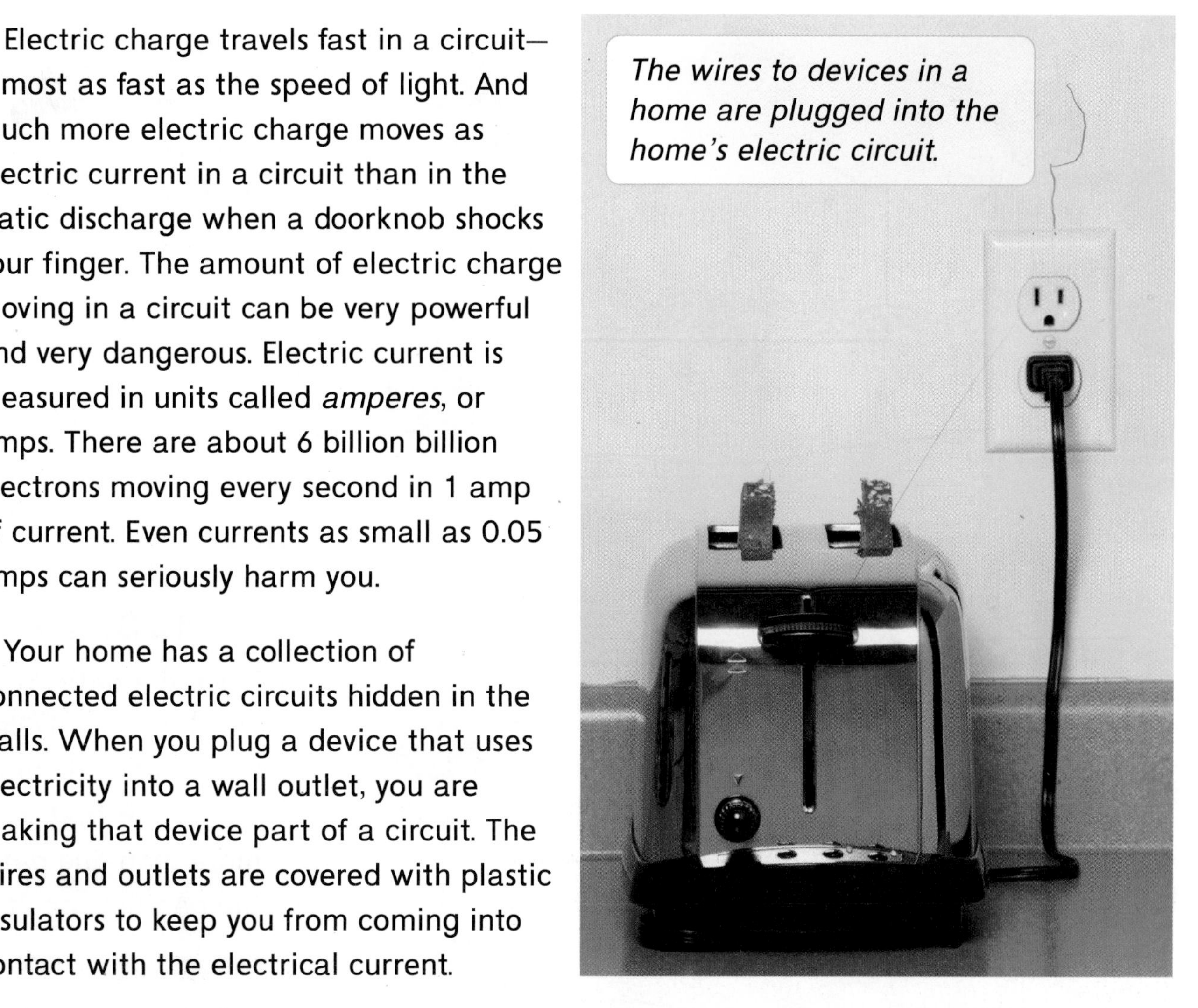

The wires to devices in a home are plugged into the home's electric circuit.

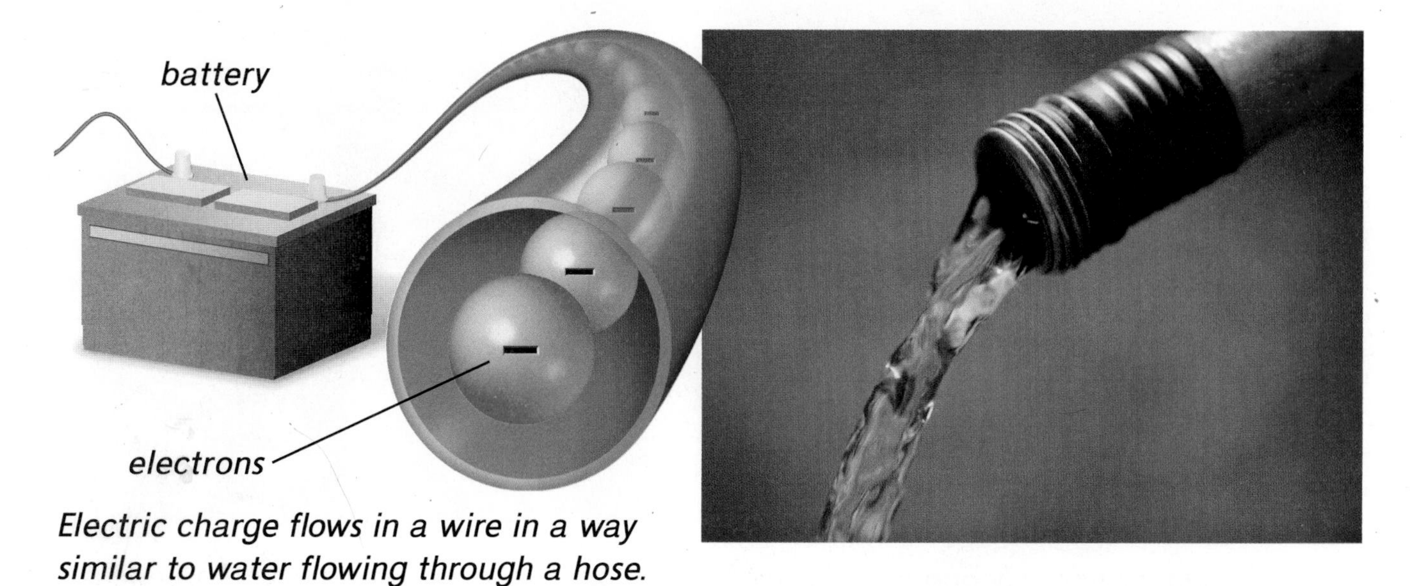

Electric charge flows in a wire in a way similar to water flowing through a hose.

Circuit Diagrams

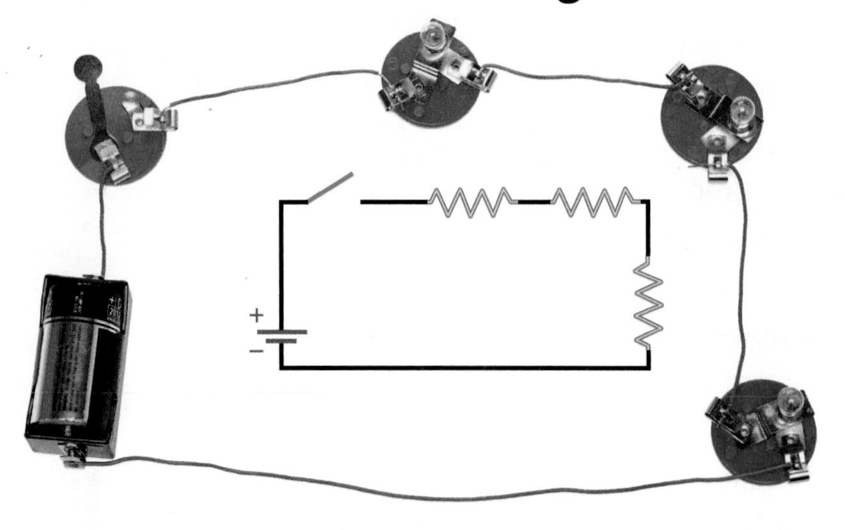

KEY

- *conducting wire*
- *resistor*
- *switch*
- *battery*

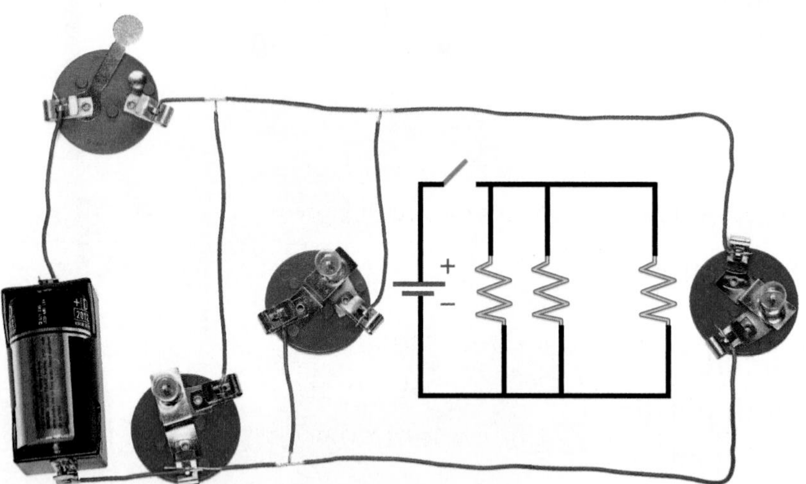

Skill Builder

Read a Diagram

Trace the paths through which electricity can flow in each diagram. The circuit that has more than one path is the parallel circuit.

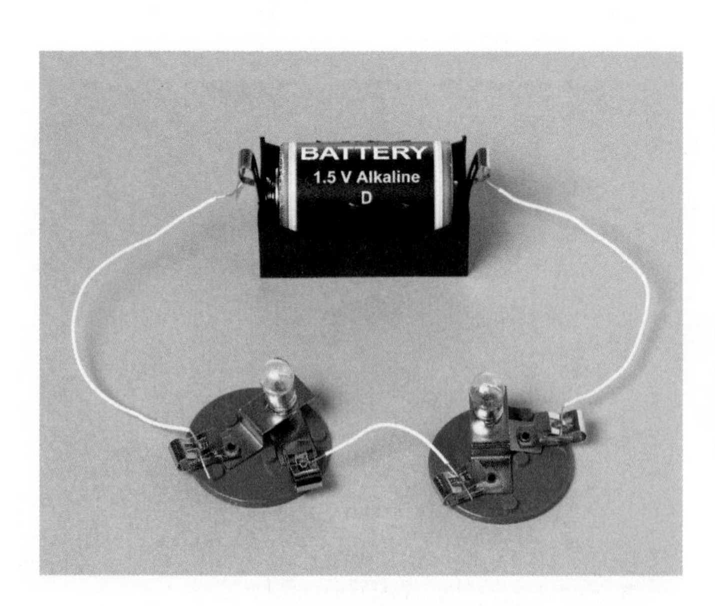

A series circuit has only one path through which electricity can flow.

A parallel circuit has more than one path through which electricity can flow.

Electric Current - Types of Circuits

A *circuit diagram* uses symbols to show parts of an electric circuit. The key at the left shows symbols used in circuit diagrams.

Series Circuit A *series circuit* has only one conductive path. In this type of circuit, the resistance increases with each resistor added. Remember that a resistor works against the flow of energy in a circuit. Electric charge travels through all the resistors one after another. As resistors are added, the energy each resistor receives is decreased.

Holiday lights are usually wired in series. If one bulb is taken out, the other bulbs will go out too. Using a series circuit in homes would cause problems. Turning off one appliance would turn off all the others!

Parallel Circuit Circuits in your home are parallel. A *parallel circuit* has more than one conductive path. The overall resistance of the circuit is smaller, so more current will flow.

Electric charge flows through all paths in a parallel circuit at the same time. The smaller the resistance of the path, the more current flows through it. If one path is broken, the current flows through the remaining paths.

Short Circuit A *short circuit* is a path with little or no resistance that connects the ends of an electrical source. The low resistance in short circuits causes currents large enough to damage appliances or start fires. Frayed wires are a common cause of short circuits.

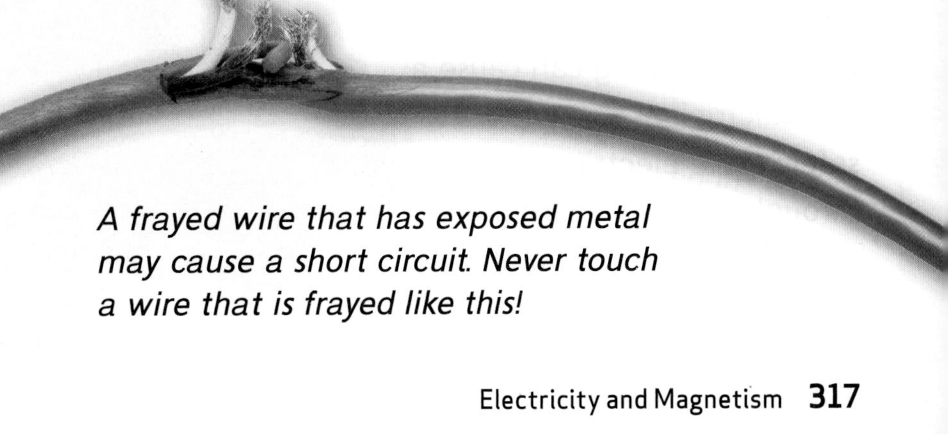

A frayed wire that has exposed metal may cause a short circuit. Never touch a wire that is frayed like this!

Electromagnets

Electricity and magnetism are related. When charged particles move, they form magnetic fields, which means we can use electric current to make magnets. An **electromagnet** is an electric circuit that produces a magnetic field.

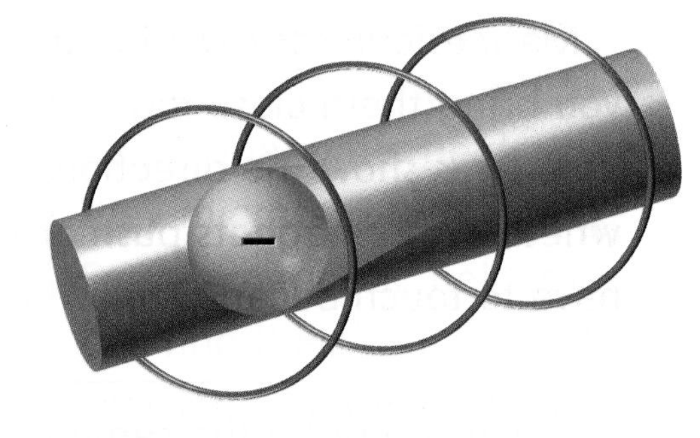

Electrons moving in a wire produce a magnetic field.

Effect of Current The simplest electromagnet is a straight wire. A magnetic field circles the wire when current is flowing through the wire. The more current, the stronger the magnetic field is. Turn the current off and the field disappears.

Effect of Coils When you wrap a wire into a loop, you increase the strength of the magnetic field. Many loops together can make a coil. The magnetism from each loop adds up to make the coil a stronger electromagnet. The coil's magnetic field is shaped like a bar magnet's.

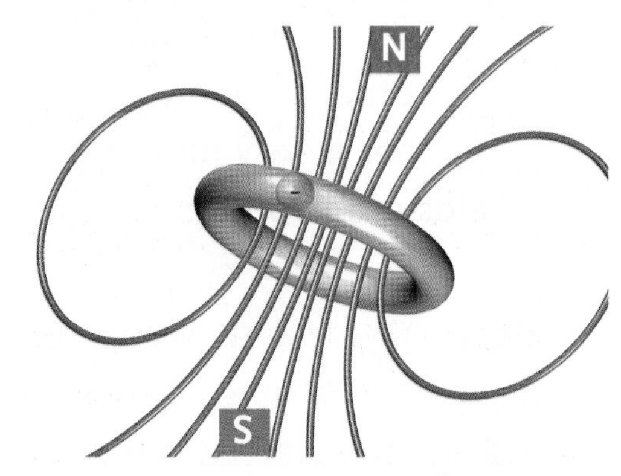

A loop of electric current will have north and south magnetic poles.

Effect of an Iron Core Placing a rod of iron inside the coil of an electromagnet will magnetize the iron. This rod of iron adds to the strength of the electromagnet's magnetic field.

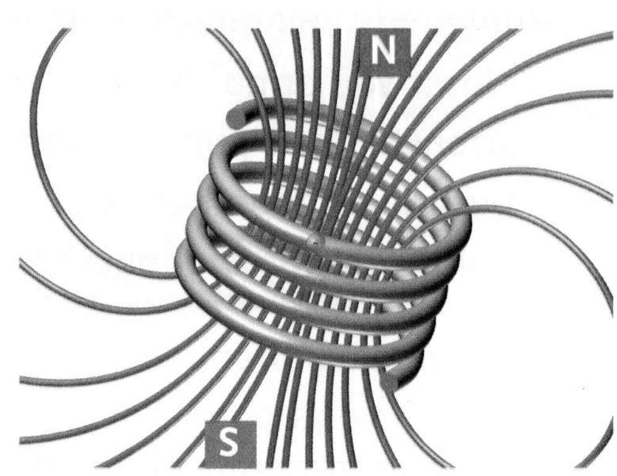

The magnetic field of a coil is like that of a bar magnet.

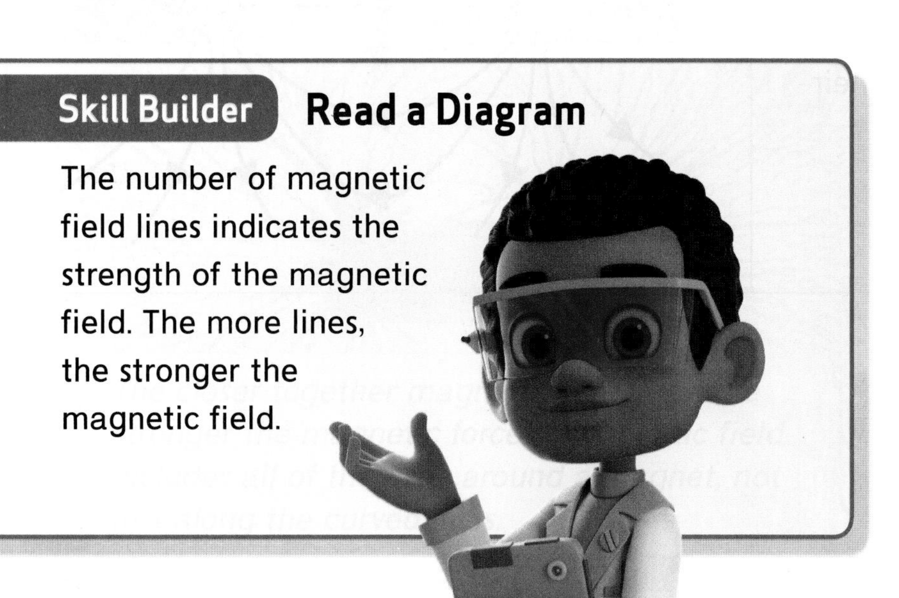

Skill Builder **Read a Diagram**

The number of magnetic field lines indicates the strength of the magnetic field. The more lines, the stronger the magnetic field.

Using Electromagnets

Electromagnets must be part of an electric circuit in order to work. Because circuits usually include a switch, an electromagnet can be turned on and off. The ability to turn the magnetic field on and off makes electromagnets useful in many devices.

Audio Speakers A *voice coil* is an electromagnet in an audio speaker. The voice coil sits in the magnetic field of a nearby permanent magnet. Electric current in the voice coil can change. That changes the coil's magnetic field. The coil's changing magnetic force interacts with the magnetic field of the permanent magnet. Pushes and pulls between the two magnets make the coil magnet move back and forth. The coil is connected to a cone of paper or metal. The coil's vibrations make the cone move back and forth, producing sound waves in the air.

Electric Motors A simple electric motor has three parts. It has a power source, a magnet, and a wire loop attached to a shaft. The *shaft* is a rod that can spin. The power source produces electric current. The current runs through the wire loop, making it an electromagnet. The magnet pushes and pulls on this electromagnet. The force then causes the loop and shaft to spin. The spinning shaft usually attaches to a wheel or a gear used to do work.

Skill Builder

Read a Diagram

Follow the direction of the arrows to understand what happens to the shaft of the motor when the loop of wire that surrounds it spins.

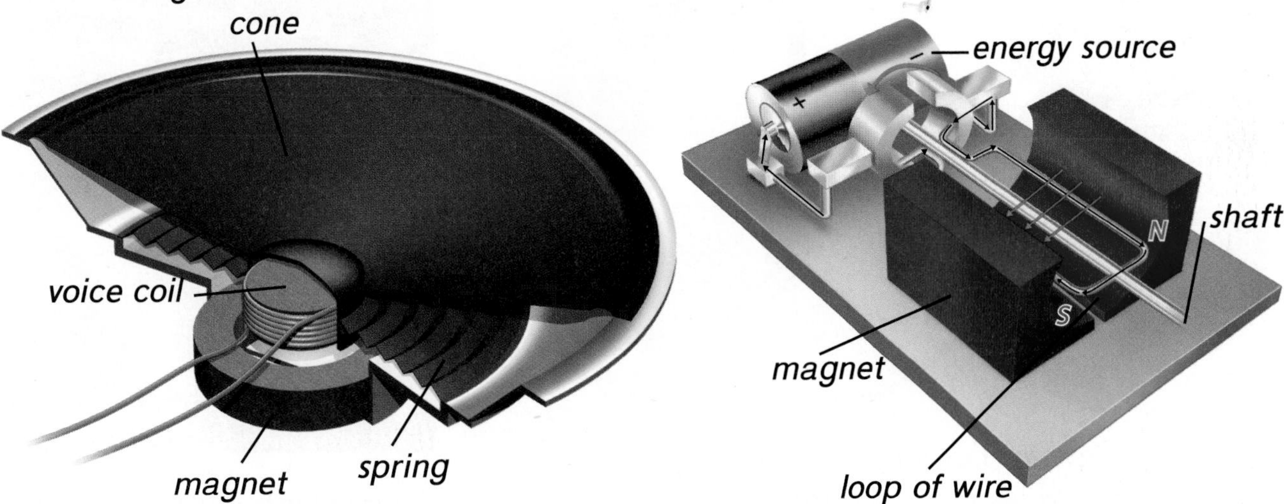

A voice coil vibrates next to a permanent magnet to produce sound in a speaker.

Magnets and Electricity

A motor uses electricity and magnetism to produce motion. Motion and magnetism can also be used to produce electricity. Suppose you turn the axle of an electric motor by hand. You would be rotating a wire coil in a magnetic field. This rotation produces electric current in the wire. Using the motor to produce electricity instead of motion makes it a generator. A **generator** is a device that creates electric current by spinning an electric coil between the poles of a magnet.

Energy of motion is used to turn the axle of the generator. As the coil moves through the magnetic field, forces push on its electrons. This movement produces an electric current. Wires attached to the loop allow the current to flow as the loop spins.

Power plants use very large generators to produce electricity. That electricity travels through power lines to the electrical circuits in homes and other buildings.

What Could I Be?

Power Plant Engineer

Power plant engineers design, build, and maintain generators that produce the electricity that powers homes, businesses, and modern society. Learn more about becoming a power plant engineer in the Careers section.

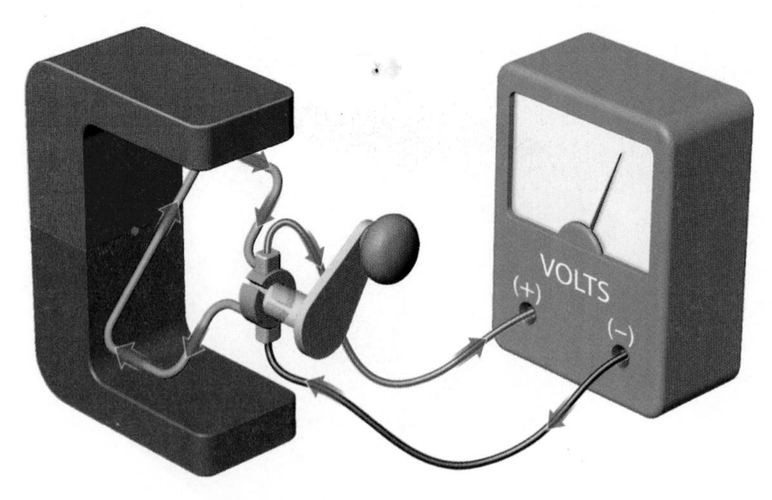

A simple generator has a metal coil in a magnetic field. As the coil rotates, an electric current is produced.

Power plants must use some other form of energy to cause the motion, or do the work, that spins the coil in a generator.

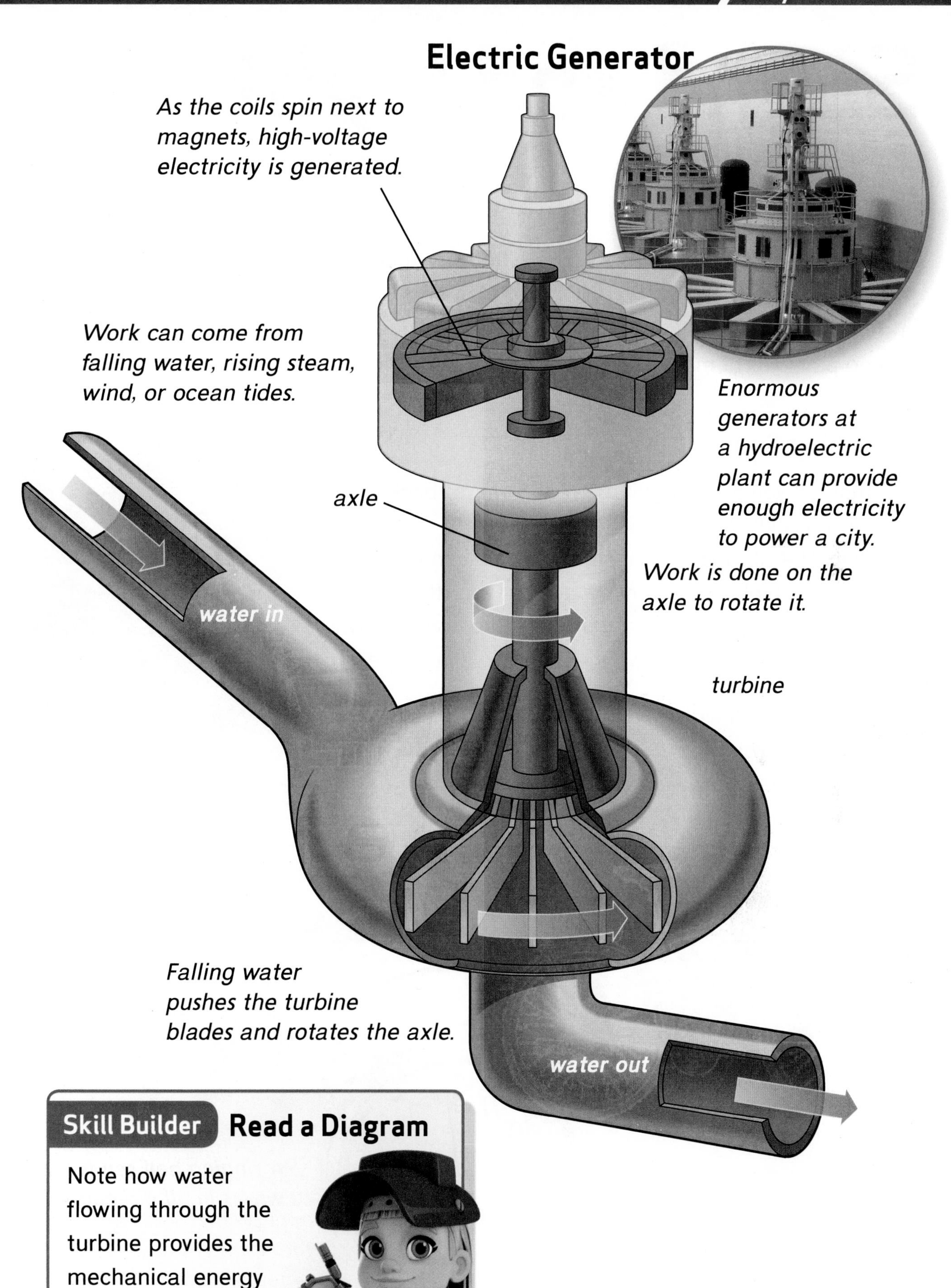

Courtesy of US Army Corps of Engineers

Skill Builder **Read a Diagram**

Note how water flowing through the turbine provides the mechanical energy to spin the turbine.

Sound and Light

Sound Energy

If you drop a book onto the floor, you hear a sound. Some of the falling book's energy of motion has changed to sound. Sound is a type of energy. Sound is produced by moving particles of matter, so it is a type of kinetic energy.

Producing Sound

Vibrations If you pluck a guitar string, it moves back and forth quickly. This back-and-forth motion is called a *vibration.* The vibration produces sound. You can feel a vibration if you place your fingers against your throat while you talk or hum. The vocal cords in your throat vibrate when air moves past them. These vibrations allow you to speak.

The sound of the drum is produced by vibrations of air particles.

All sounds begin with a vibration. The drum shown here is surrounded by air. The vibrating parts of the drum bump into air particles. Those particles bump into other air particles. The drum's vibration squeezes air particles, and then they alternately spread out.

A ringing bell sends sound waves in all directions.

The air particles spreading out produces regions of air that have many particles, called compressions, and regions of air that have fewer particles, called *rarefactions.* The compressions and rarefactions move through the air as a wave, carrying sound energy.

Sound Waves A series of rarefactions and compressions traveling through a substance is called a **sound wave.** The substance through which the wave travels is called the *medium* for the wave. Mediums can be solids, liquids, or gases. Examples of mediums include metals, water, and air. Like all waves, sound waves carry energy. When they pass through a medium, the medium is not permanently moved. Energy, however, is moved through the medium.

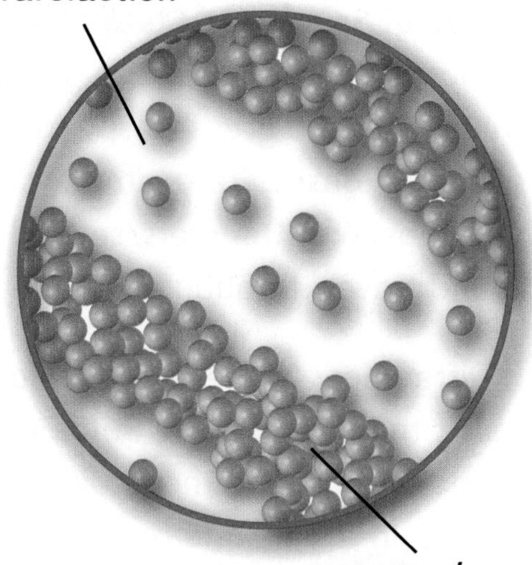

As a sound wave moves through air, the density of the air changes. The air does not move along with the wave.

Direction of Sound Waves Sound waves vibrate the medium in the same direction that the energy moves. They are called **longitudinal waves.** Think about how a coiled toy or spring moves when you push it together and pull it apart. Sound waves move in the same way.

We can also show sound waves as a series of peaks and dips. The peaks represent the high density of air in compressions. The dips represent the low density of air in rarefactions. However, remember that air does not move up and down like the peaks and dips.

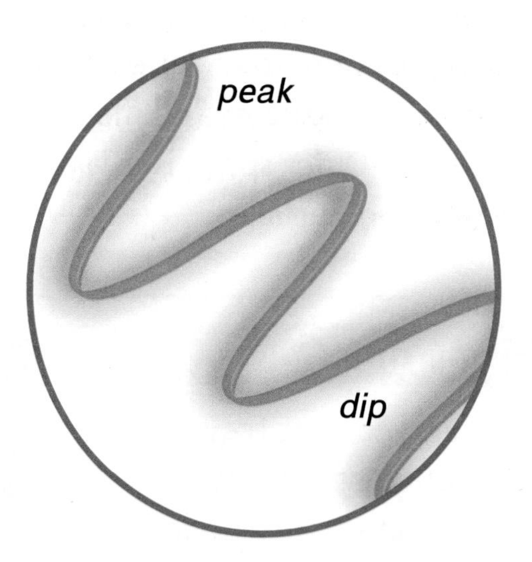

The alternating high and low density of the air can be shown as a series of peaks and dips.

When sound waves hit an object, the object starts vibrating. The object is moved by the energy of the wave. This movement is how sound from a loud airplane or stereo can rattle windows. You can feel the vibrations caused by such loud sounds.

Sound Energy - How Sound Moves

Speed of Sound and State of Matter Sound can travel through solids, liquids, and gases. In fact, sound tends to travel with the greatest speeds in solids and the slowest speeds in gases. For example, sound travels through steel at almost 6,000 meters/second (19,685 feet/second), but sound travels through air at only 343 m/s (1,125 ft/s).

Differences in the speed of sound result from how far apart the particles are in the medium. The particles carry sound energy, and their collisions are how sound energy travels. In a solid, the particles are close together, so they quickly collide. In gases, particles are far apart. Collisions are less frequent, so sound travels more slowly.

Speed of Sound and Temperature The temperature of the medium also affects the speed of sound. In warmer air, particles move faster. As a result, they collide more often and transmit sound faster.

Sound in Space Sound cannot travel in any area without particles of matter. For example, outer space has few particles, so there is no medium for sound to travel through. Outer space is a *vacuum,* a region that contains few or no particles.

From the far side of a ballpark, you can see the ball being hit before you hear the crack of the bat against the ball.

Water is a good medium for sounds such as dolphin songs.

Did You Know?

Sound cannot travel through space. Two astronauts on a spacewalk can talk using radios and can hear one another only because there is air inside of their helmets.

Absorption of Sound When a sound wave hits a wall, some of the energy of the wave is absorbed. *Absorption* is the transfer of energy when a wave disappears into a surface. The energy itself does not disappear. Absorbed sound energy transforms to thermal energy, heating the surface that absorbs it.

Reflection of Sound When sound waves hit a flat, firm surface, much of their energy bounces back. An **echo** is a specific, reflected sound. You can hear an echo if you make a loud noise in an empty room.

Echolocation Echoes can be useful. Bats make sounds that echo off their prey. The returning echoes tell the bat where the prey is located. This skill is known as **echolocation.** Whales and dolphins also use echolocation to learn about their surroundings and to find food.

Sonar Scientists have developed a system called *sonar* that works the way echolocation does for animals. Sonar stands for "**so**und **n**avigation **a**nd **r**anging." It is used underwater to find objects. The sonar system sends out sound waves that reflect off objects. The return time and direction of the sonar echoes are used to calculate the location of the object.

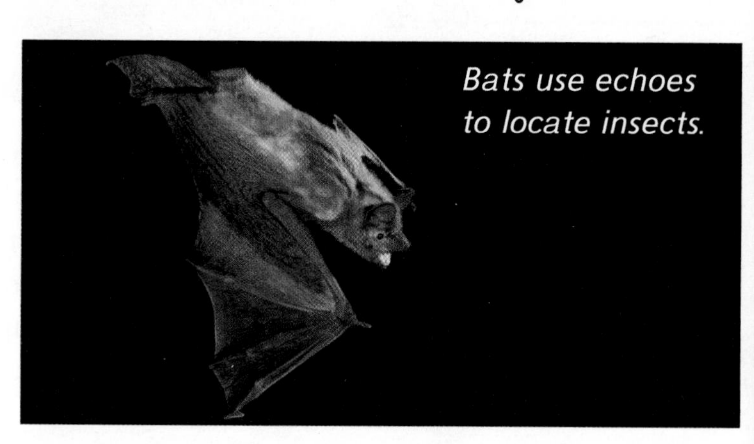
Bats use echoes to locate insects.

What Could I Be?

Sonar Technician

Interested in finding out what is beneath deep ocean waters? Sonar technicians use sonar equipment to locate all sorts of objects in the water. Sonar aboard ships can be used to find marine life, other vehicles, or even sunken treasure! Learn more about becoming a sonar technician in the Careers section.

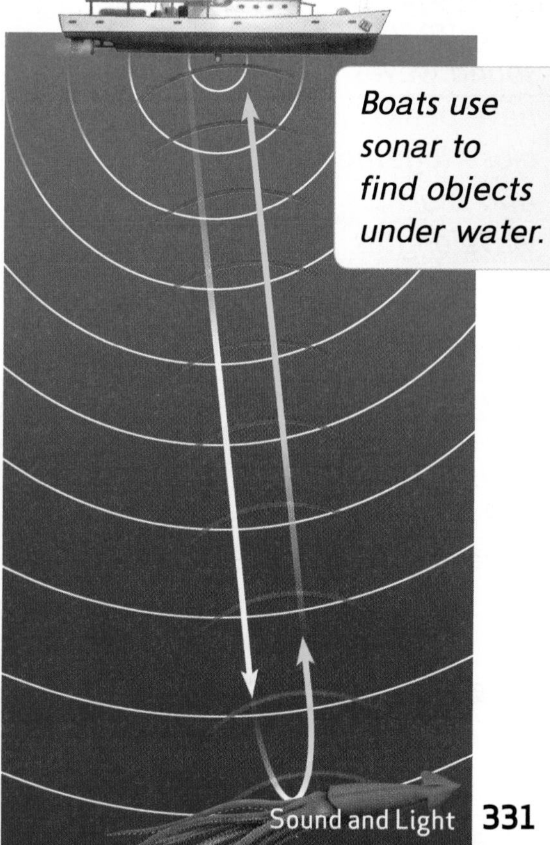
Boats use sonar to find objects under water.

Sound Energy - Pitch

All sounds are made by vibrations, but you hear them differently because of differences in their properties. Like all waves, each sound wave has a wavelength and a frequency. In sound waves, **wavelength** is the distance from one rarefaction to the next. If you represent a sound wave as a repeating pattern of peaks and dips, wavelength is the distance from one peak to the next peak.

Frequency is the number of vibrations a source of sound makes each second. The more vibrations there are, the higher the frequency of the sound. When you strike a bell, it vibrates quickly. The vibrations produce sound waves with a high frequency. When a drummer hits a base drum, it vibrates slowly. The vibrations produce sound waves with a low frequency.

Pitch is how we hear the frequency of a sound wave. If a sound wave has a high frequency, we hear the sound as having a high pitch. If a sound wave has a low frequency, we hear the sound as having a low pitch.

We hear the sound made by this bird as a high pitch because the sound waves produced by the bird have a high frequency.

High-frequency sound waves have peaks close together and short wavelengths.

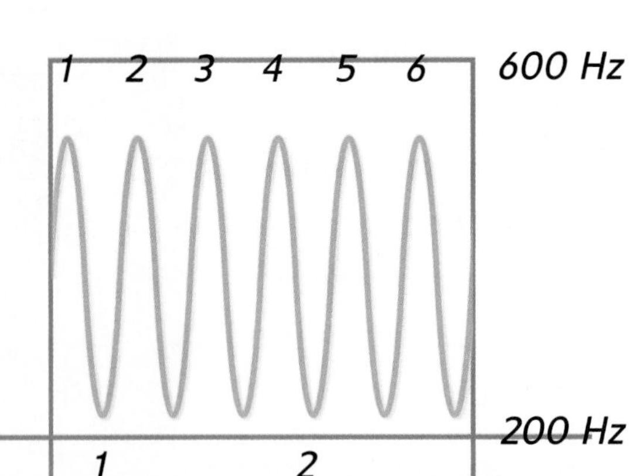

Low-frequency sound waves have peaks farther apart and longer wavelengths.

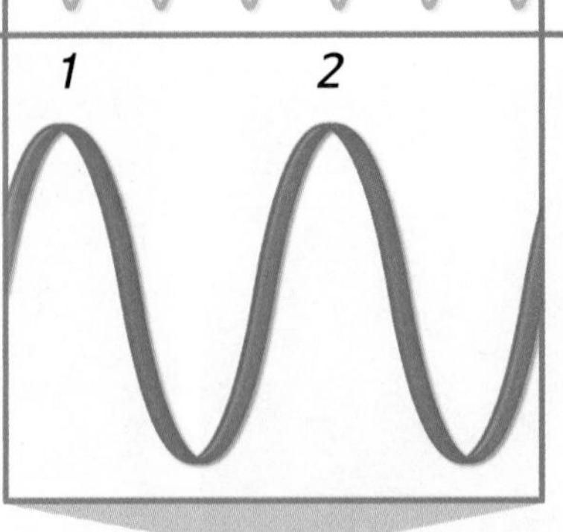

We hear the sound made by this big dog as a low pitch because the sound waves produced by the dog have a low frequency.

Changing Pitch

Changing the Pitch of Musical Instruments To make a sound higher in pitch, increase the number of times it vibrates each second. On a string instrument, shortening the string increases the pitch. On a wind instrument, shortening the tube increases pitch. A shorter tube produces a higher pitch because the air inside vibrates faster.

The pitch of a trombone changes with the length of its tubes.

Doppler Effect As you move toward the source of a sound, the pitch sounds higher. As you move away from the source of a sound, the pitch sounds lower. Recall that frequency is the number of peaks of a wave that pass by a location each second. If you move toward the source of a wave, the peaks move past your ear faster than if you were standing still. If you move away from the source of a wave, the peaks arrive at your ear more slowly and the pitch is lower.

A change in frequency produced by moving toward or away from a sound wave is called the *Doppler effect.* Any movement can cause a Doppler effect, but only fast speeds will change a pitch enough for you to notice it.

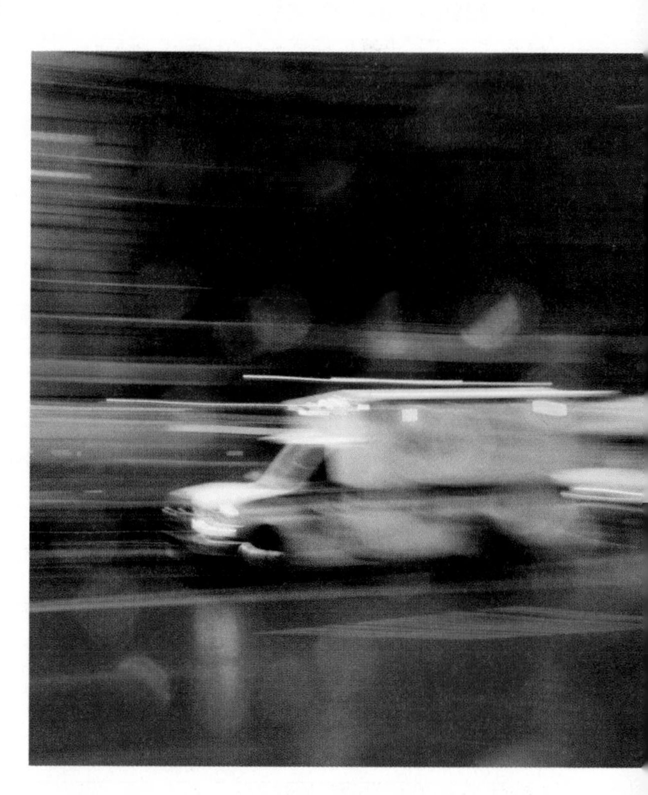

The sound of an ambulance siren changes as the ambulance gets nearer to or farther from you because of the Doppler effect.

Sound Energy - Volume

The height of a sound wave is its **amplitude.** The amplitude describes how dense the air is in the compressions or rarefactions compared with normal air. Amplitude is a measure of the amount of energy a sound wave has.

Amplitude affects the **volume**, or how loud a sound is perceived. Scientists measure the volume of sounds with units called *decibels (dB).* The table lists decibel levels for various sounds at certain distances.

Volume of Sounds

Decibel Level	Sound
130 dB	jet takeoff (at 100–200 feet)
110 dB	chainsaw
100 dB	garbage truck
85 dB	level at which hearing damage can begin
80 dB	average city traffic
60 dB	normal conversation
30 dB	whisper
10 dB	human breathing

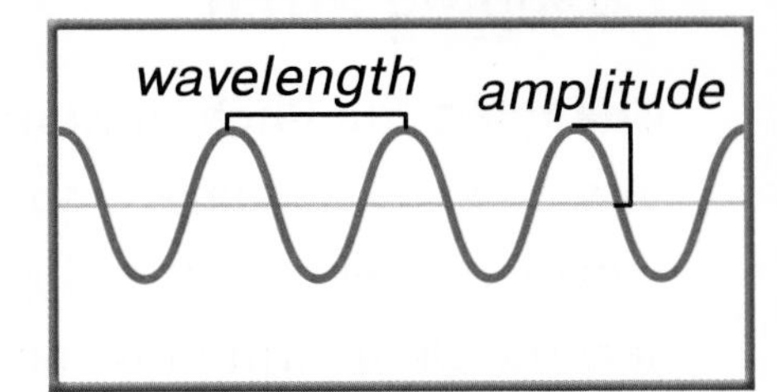

long wavelength low amplitude

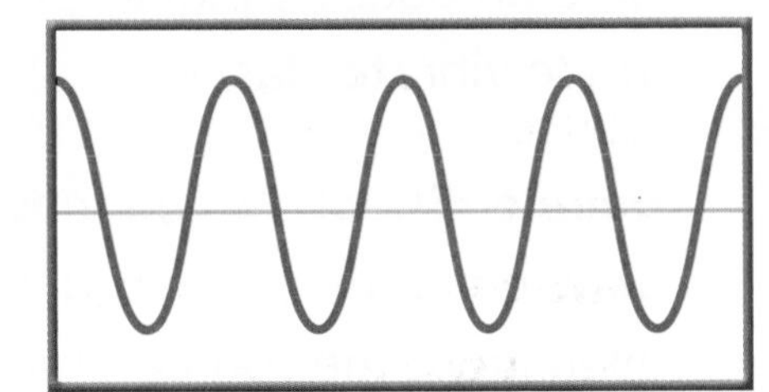

long wavelength high amplitude

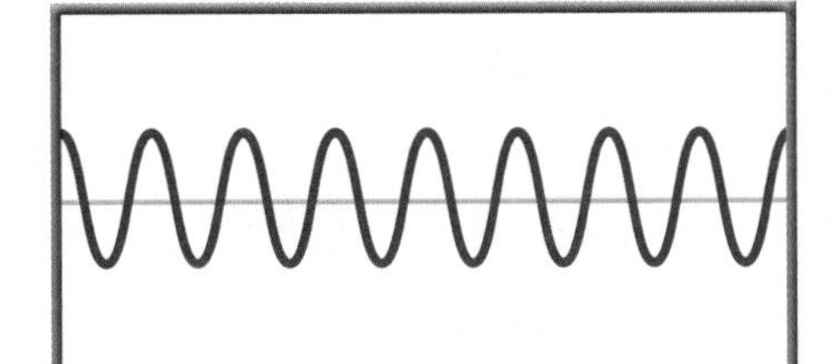

short wavelength low amplitude

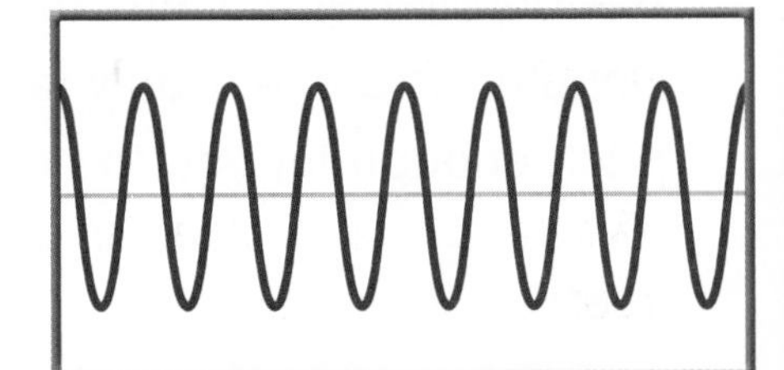

short wavelength high amplitude

Read a Diagram

Compare amplitudes and wavelengths to determine which sound has a loud volume and low pitch.

Changing Volume

Energy You can make sounds louder by using more energy. If you hit a drum harder, for example, you give the drum more energy. The extra energy increases the density of particles in the compressions of the sound wave, increasing its volume. The rarefactions will be less dense than before.

Medium A sound wave has a smaller amplitude when it travels through a dense material. The wave has the same amount of energy. Even though the amplitude is smaller, there are more particles moving in the medium.

Distance Volume is lower farther from the source of sound. The energy of the sound wave spreads out as it travels. Less energy at any one point means a lower volume.

Hearing Sound

We hear when sound waves travel into our ears. The diagram shows how the sound waves move from the ear to the brain.

1 *outer ear The outer ear collects sound waves. Like a funnel, it directs the waves into the ear.*

2 *eardrum Sound waves make the eardrum vibrate like the head of a drum.*

3 *middle ear The vibrations are picked up by three tiny bones in the middle ear. The bones are the hammer, anvil, and stirrup.*

4 *Inner ear The stirrup passes the vibrations to a coiled tube in the inner ear. The tube is filled with fluid and lined with tiny hair cells.*

5 *nerve to brain The moving hair cells signal a nerve in the ear. The nerve carries these signals to the brain. The brain interprets the signals as sound.*

What happens when a sound wave reaches your ear?

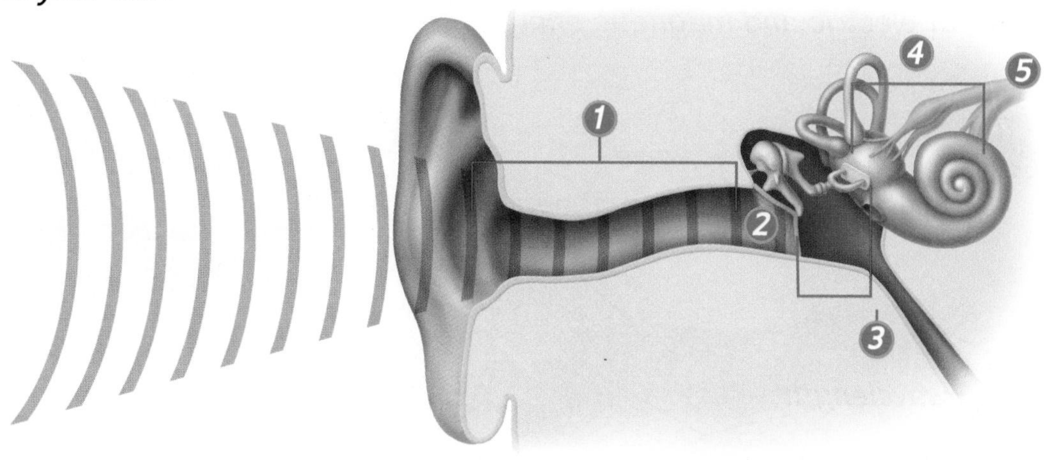

Light Energy

If you stand outside on a sunny day, you can feel the warmth of sunlight on your face. This warmth shows that light is a form of energy.

Make Connections

Jump to the Life Science section to learn more about sunlight as a source of renewable energy.

Light Waves

Vibrating Fields Light is made of vibrating electric and magnetic fields. These fields travel together as waves, but they do not vibrate in the same direction. In the diagram below, one of the fields vibrates left and right. The other field vibrates up and down. Both of the fields vibrate perpendicular to the direction that the wave travels. This type of wave is called a **transverse wave**.

Speed Unlike sound, light does not depend on compressions or rarefactions. Light waves can travel with or without a medium. Nothing travels faster than light in a vacuum. In a vacuum, light travels about 300,000 km/s (186,000 mi/s). Light travels more slowly through mediums like air or glass. In glass, light travels about 197,000 km/s (122,000 mi/s).

Wavelength and Frequency The wavelength of a transverse wave is the distance between one peak and the next. Frequency is the number of waves that pass a point each second. When you multiply the wavelength of a wave by its frequency, you get the speed of that wave.

Light is a wave of electric and magnetic energy.

electric wave
magnetic wave
direction of travel
wavelength

Malcolm Fife/age fotostock

Particles of Light

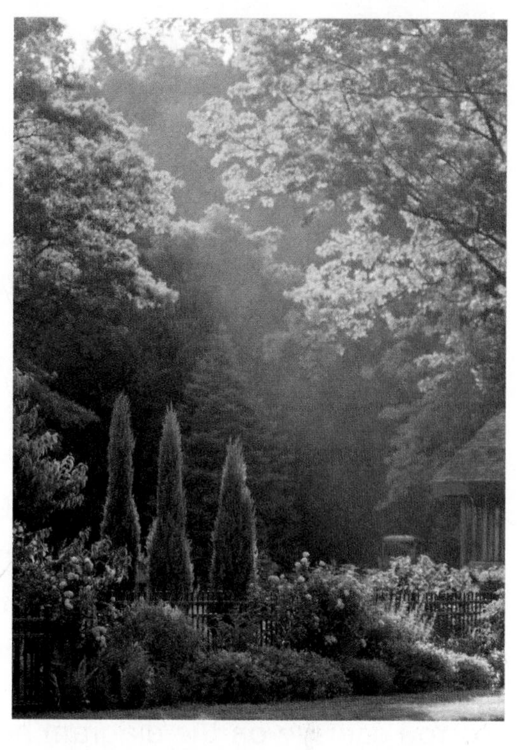

Light rays traveling in a straight line show that light is a type of particle.

Although light is a wave of energy, it is also a particle. Scientists were confused about the nature of light for a long time. They performed many experiments in an effort to understand light. They found that light has properties of both waves and particles, so they concluded that it is both.

Light is like a particle in several ways. It travels in straight lines called *light rays*. Light does not have mass like a particle, but it does have momentum like a particle. *Momentum* is the tendency of an object to move. When light hits an object, it acts like a tiny particle. It can change the direction of atoms and other small particles. Light looks like tiny dots when it strikes camera film. If light continues to strike the film, those dots eventually form an image.

Light travels as particles called photons. A **photon** is a tiny bundle of light energy. The energy of a photon is very small. A photon of red light, for example, has about $\frac{3}{1,000,000,000,000,000,000}$ J of energy! Each photon also acts like a wave with a frequency. If a photon has a higher frequency, it also has more energy.

Photons hit a piece of film individually. When enough of them have hit, the image taken by the camera appears.

The Electromagnetic Spectrum

The way in which electric and magnetic forces interact is called **electromagnetism**. Light is a form of electromagnetic radiation.

The different forms of radiation are together known as the **electromagnetic spectrum.** All of the forms of radiation travel at the same speed. They all can travel through either a vacuum or a medium. The different forms of electromagnetic radiation differ in wavelength, energy, and frequency.

You can see on the diagram how properties of radiation are related. Waves that have a long wavelength have low frequency. This low frequency means the waves do not vibrate back and forth as quickly. The energy of these waves is low. Waves that have a short wavelength have high frequency. These waves vibrate back and forth much faster. The energy of these waves is high.

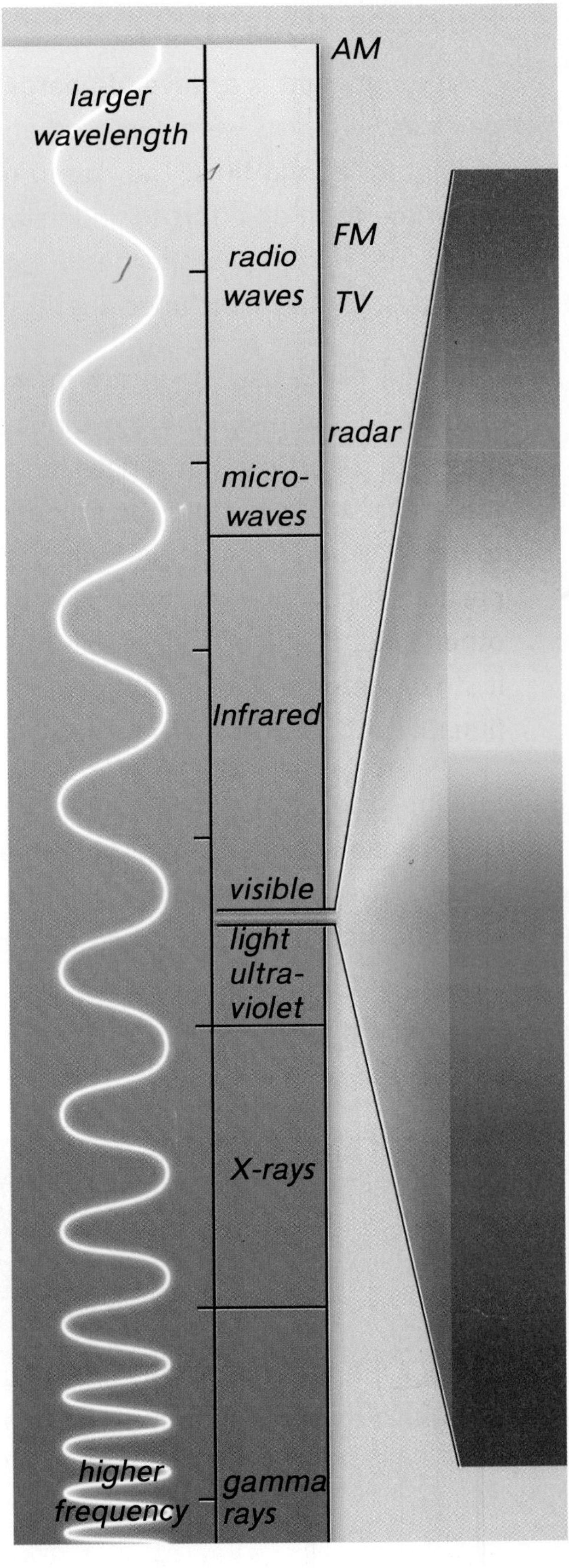

Skill Builder **Read a Diagram**

Higher-frequency photons have more energy.

Long-Wavelength Radiation Some forms of radiation have wavelengths longer than those of visible light. *Radio waves* have the longest wavelengths. Some radio stations send signals by radio waves. *Microwaves* are used in microwave ovens. They are also used to send television and cell phone signals. *Infrared waves* are used in remote controls. You also feel heat from a fire as infrared waves.

The cones at the top of cell towers send and receive microwaves used for cell phone signals.

Visible Light The light we see is just a tiny part of the electromagnetic spectrum. The colors of a rainbow show the visible spectrum of light. Red light has the longest wavelength of visible light. Violet light has the shortest wavelength.

Short-Wavelength Radiation Forms of electromagnetic radiation with wavelengths shorter than visible light can be dangerous because they carry so much energy. *Ultraviolet (UV) light* from the Sun, for example, can cause sunburn. UV light can also be useful. Hospitals use UV light to kill germs on surfaces. *X-rays* are used to form images of the inside of the body. This type of radiation has enough energy to pass through soft tissue but not through bone. *Gamma rays* carry the most energy. This type of radiation is sometimes used to treat diseases. A beam of gamma rays can kill cells of the body that have the disease.

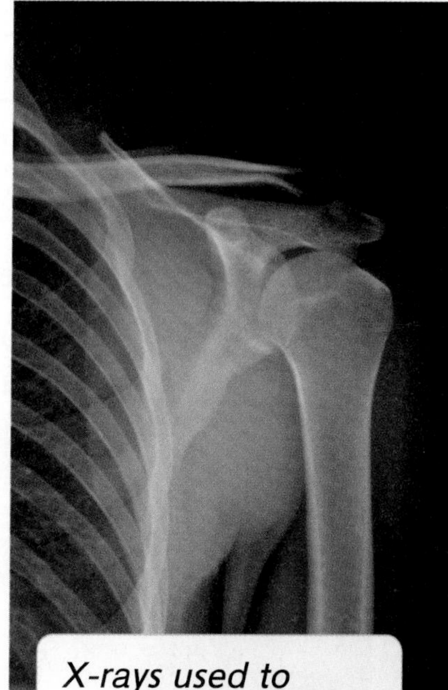

X-rays used to check for broken bones are a type of electromagnetic radiation.

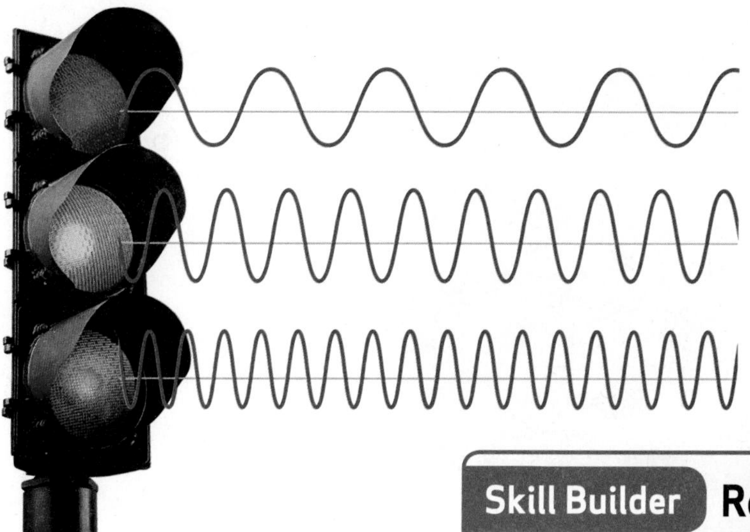

Skill Builder **Read a Diagram**

Look at the distances between the tops of each wave to determine which color has the longest wavelength.

Bouncing and Bending of Light

Reflection of Light

Reflection is the bouncing of a wave off a surface. Objects that reflect off a flat surface obey the *law of reflection:* the angle of an incoming light ray equals the angle of the reflected light ray.

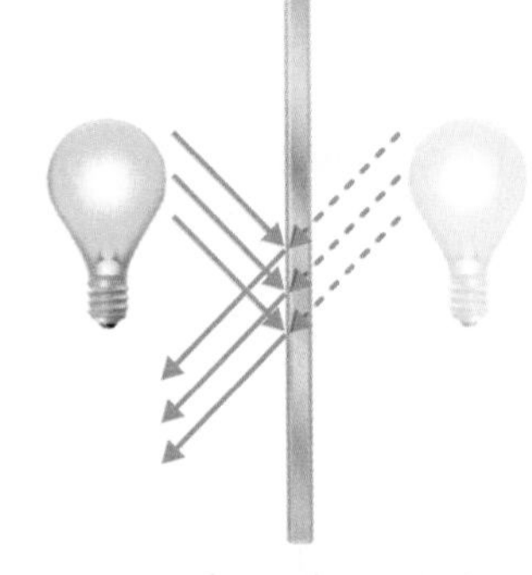

flat mirror

Flat Mirrors When you look in a flat mirror, the image you see appears to be behind the mirror. Of course, you are not really behind the mirror. It seems like light rays are coming out of the mirror, but in fact you are seeing rays that reflect off the mirror.

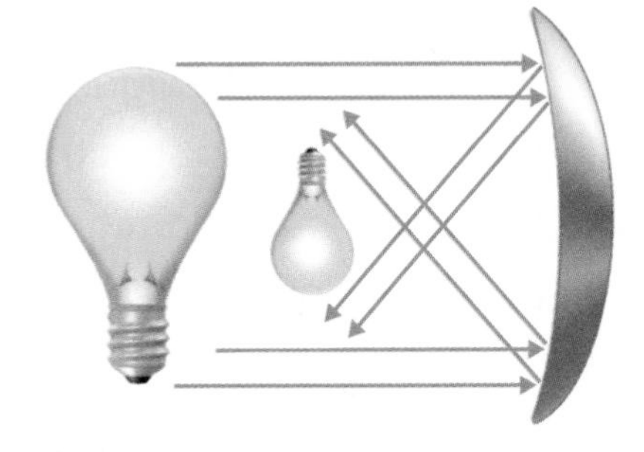

concave mirror

Concave Mirrors A mirror that is curved inward is a *concave mirror.* Light rays that strike this type of mirror bounce inward. The image you see depends on how close you stand to the mirror. If you stand very close, your image will appear larger than you are. If you stand far away, your image will be upside down and smaller than you are.

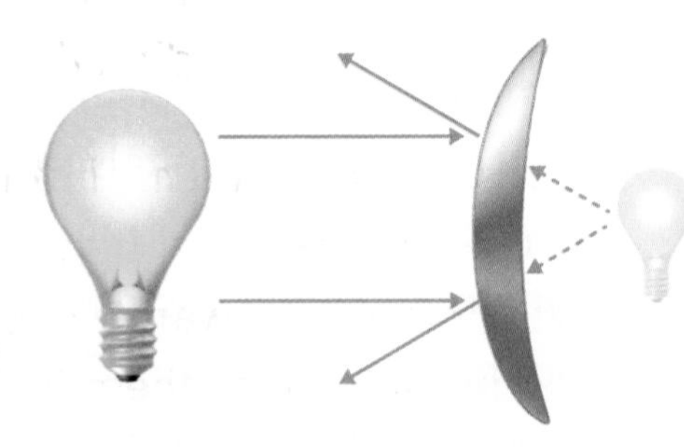

convex mirror

Reflected images may appear behind the mirror (faded bulbs) or in front of the mirror (bright bulbs).

Convex Mirrors A mirror that is curved outward is a *convex mirror.* Images in this type of mirror always appear smaller, and they always appear to be behind the mirror.

A convex mirror allows the driver to see a wide area behind the car.

Refraction of Light

Refraction is the bending of a wave as it changes angle passing from one substance into another. Light slows down when it moves from one material into a denser material. This decrease in speed causes the light's angle to change, or its direction to bend. Light speeds up when it moves into a less dense material. This increase in speed causes the light to bend in the other direction.

Flat Lens A clear piece of glass or plastic through which light travels is called a *lens.* Light slows down when it enters a lens and speeds up when it exists the lens. If the lens is flat, the light's path shifts a little, but its final direction does not change.

Concave Lens A lens that is thinner in the middle is a *concave lens.* Light that passes through a concave lens spreads outward. Objects always look smaller when you look at them through a concave lens.

Convex Lens A lens that is thicker in the middle is a *convex lens.* The image that you see when you look at an object through a convex lens depends on how far away the object is. Up close, the lens will cause the object to look bigger. Far away, the image will appear upside down and will be smaller.

Concave lenses are used in eyeglasses for people who are nearsighted, or have trouble seeing objects that are far away. Different lenses are needed for people who are farsighted, or have trouble seeing objects that are up close.

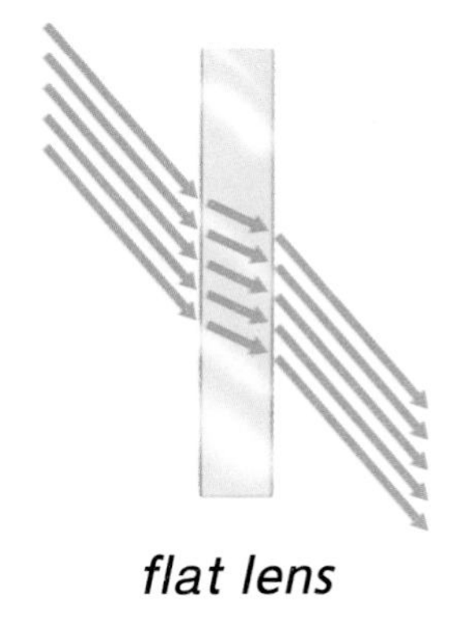

flat lens

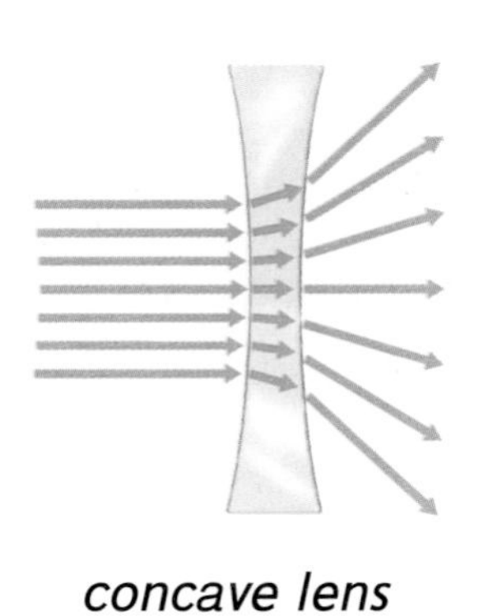

concave lens

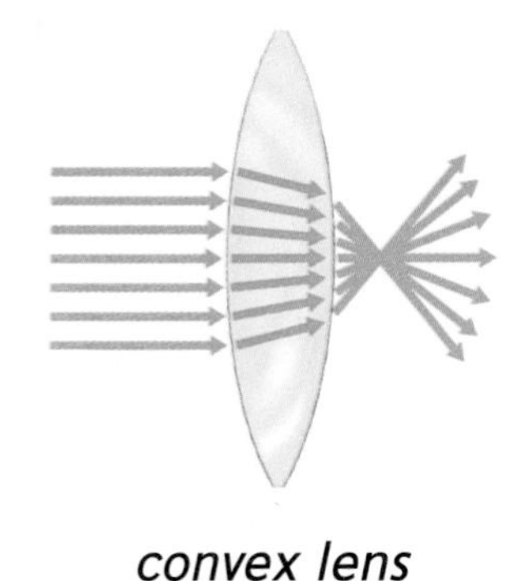

convex lens

Bouncing and Bending of Light - The Human Eye

You see images when light reflects off an object and enters your eye. The diagram shows the different parts of the eye that light passes through. Both the cornea and the lens refract the light so that it bends toward the retina on the back of the eye. The retina sends signals to the brain that you interpret as images.

Transparent, Translucent, and Opaque

Light that strikes a material may pass through it. Materials that allow most light to pass through are called *transparent*. Clear glass and plastic are transparent. You can clearly see objects when you look at them through a transparent material.

Materials that blur light as it passes through are called *translucent*. Wax paper and frosted glass are examples of translucent materials. If an object allows little or no light through, it is called *opaque*.

Objects look blurry through a translucent material.

Skill Builder **Read a Diagra**

Trace the path of the light as it enters the eye.

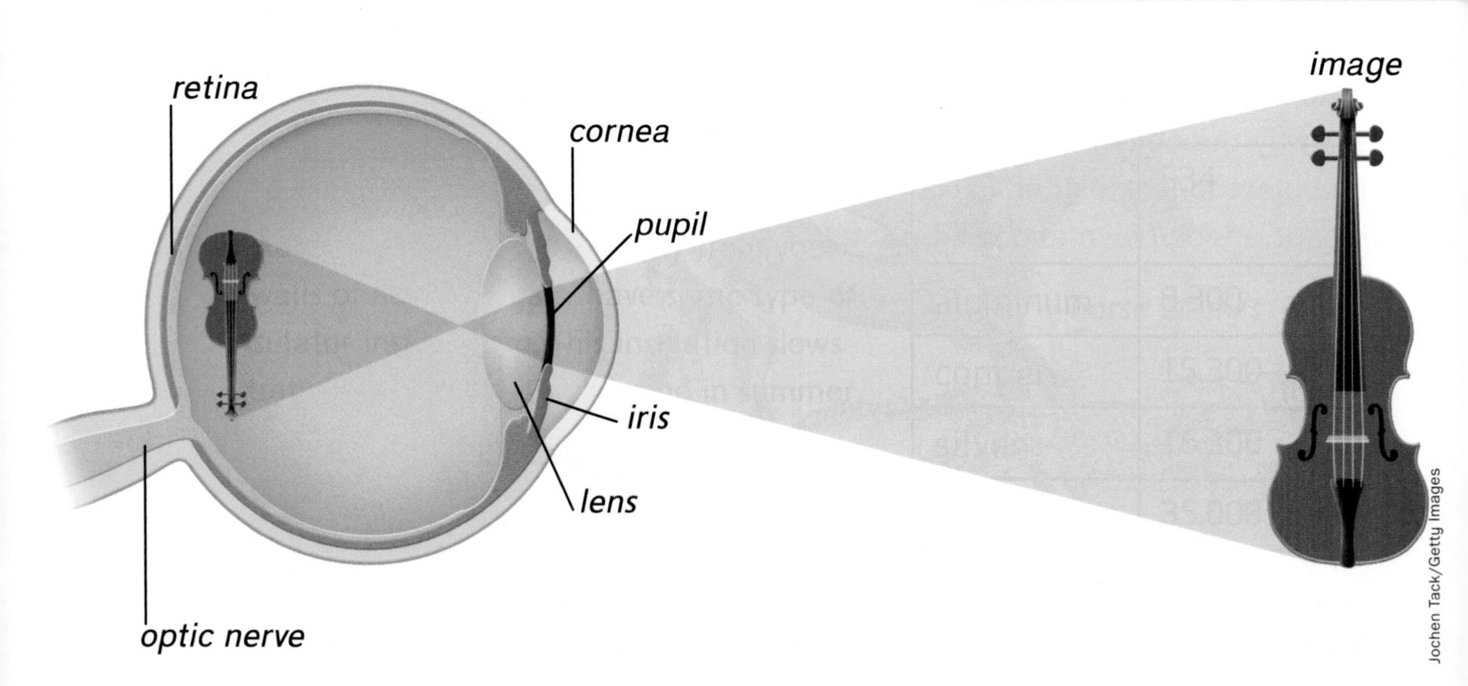

Jochen Tack/Getty Images

Factors That Affect Light Transmission Whether an object is transparent, translucent, or opaque depends on the type of material, its thickness, and the color of the light. Thicker objects have more particles to absorb photons, so they are more likely to be opaque. Some objects will be opaque, transparent, or translucent for only one color of light.

Shadows

Opaque and translucent objects block light. The area behind these objects is darker. The darker area is a shadow. *Shadows* are the absence of light. When an object is between a light source and another object, it casts a shadow on the other object.

Early in the morning, when the Sun is low in the sky, you cast a long shadow. Sunlight travels toward you at a small angle. At this angle, there is a long distance before the sunlight hits the ground behind you. As midday approaches, however, the Sun rises in the sky and the angle of the sunlight increases. Throughout the morning, your shadow becomes shorter and shorter.

Shadows depend on the angle and the distance between a light source and an object. They also depend on the distance between the object and the place where the shadow is cast. The closer a light source is to an object, the larger the shadow an object will cast.

You cannot see through opaque materials because light cannot pass through them.

Shadows are longer in the morning and short in the midday.

Light and Color

When sunlight hits raindrops in the sky, a rainbow appears. The colors are already in the sunlight that produces the rainbow.

Colors of Light Our eyes see light waves of different wavelengths as different colors. Visible light waves with longer wavelengths look red. Visible light waves with shorter wavelengths look violet. All the colors between red and violet have wavelengths that are between them too. White light, such as sunlight, is actually just a collection of many different wavelengths mixed together.

Color and Wavelength Different wavelengths of light will reflect and refract at different angles. This fact explains why white light that is refracted by raindrops in the sky is spread out into a rainbow. A *prism* is a cut piece of clear glass or plastic in the form of a triangle or other geometric shape. The band of color in a rainbow, or from light passing through a prism, is called a *spectrum*.

Opaque objects are the color of the light they reflect.

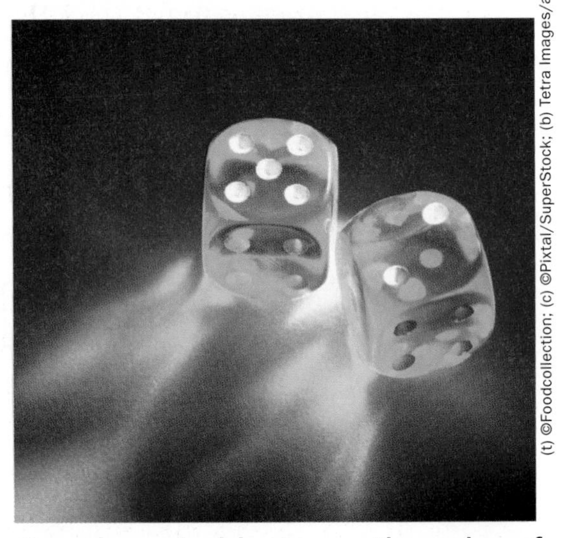

Translucent objects are the color of the light that passes through them.

Skill Builder

Read a Diagram

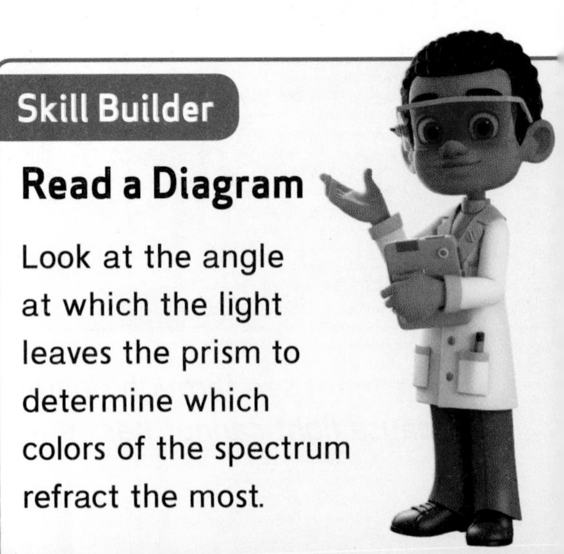

Look at the angle at which the light leaves the prism to determine which colors of the spectrum refract the most.

Mixing Colors

Colors of Light You see when light strikes the retina of your eye. The retina of your eye detects three ranges of wavelengths. These ranges of wavelengths are called the *primary colors of light*—red, green, and blue.

If you shine a red light and a green light on the same spot, the spot will appear yellow. If you shine a green light and a blue light on the same spot, the spot will appear light blue. You can see all the colors of the rainbow by different combinations of these primary colors of light. A combination of red light, green light, and blue light produces white light. This combinaton is why the light from a flashlight appears white. It is a mixture of all wavelengths of light.

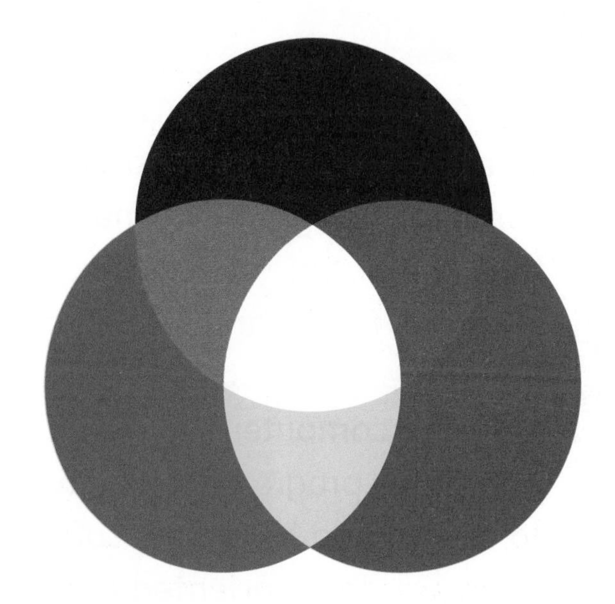

When equal parts of red, green, and blue light rays are mixed, they form white light.

Mixing Paints Colors mix differently when they are formed by the reflection of light. Remember that the color of an object is the color of light that it reflects. If you mix red paint and green paint, you will not get yellow paint. The primary colors used for paint are magenta (a type of pink color), cyan (a type of light blue color), and yellow. Mixing cyan and yellow paint makes green paint. Mixing magenta and yellow paint makes orange paint. Mixing all colors of paint makes black paint.

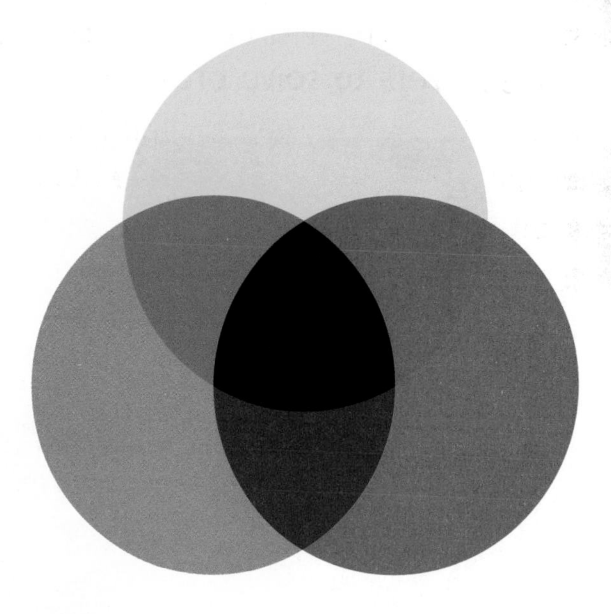

When equal parts of magenta, cyan, and yellow paints are mixed, they absorb all light and appear black.

Future Career

Computer Programmer How often do you use a computer? A computer programmer makes sure that computers work correctly. By creating, fixing, and testing computer language called code, these scientists help the computer run smoothly. Computer programmers also help to design and build new software. They are detail oriented and find patterns and answers quickly. New and improved technology is made possible by the work of computer programmers. They work closely with engineers to solve problems.

Engineering and Technology

(t-b) Nikada/Getty Images; Digital Vision/Photodisc/Getty Images; Comstock Images/Alamy; Pixtal/age fotostock

Introduction to Technology and Engineering

Many animals change their environment to help themselves meet basic needs. Humans do this in far more complex ways than other living things. We use tools and change the environment to help us survive and be comfortable. We also change things around us to make our jobs easier and to entertain ourselves.

Word Study

The word technology comes from the Greek *technē*, which means "art" or "skill," and -*logia*, which means "systematic treatment."

Technology and Engineering

People improve nearly every aspect of their lives with technology. *Technology* is the use of scientific information to solve problems and invent useful things. Some technology, such as a pencil and paper, is simple. Other technology, such as a smart phone with a camera, Internet access, and GPS, is more complex.

Technology is the result of a way of thinking about problems. *Engineering* is a planned design and testing process used to develop technology and solve problems. Engineering and technology are closely related concepts. The goal of engineering is to produce better technology. Engineers both develop technology and use it to conduct their work. They transform raw materials and ideas into all sorts of designed inventions and solutions.

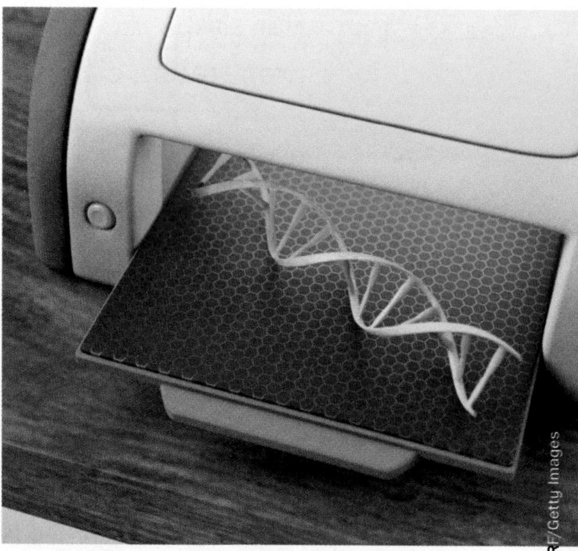

Maciej Frolow/Photographer's Choice RF/Getty Images

Engineers use technology such as 3-D printers to help them visualize the products they are working on. This model was made by programming a computer, connected to a printer, that builds up a plastic material layer by layer.

Even the earliest people had technology. Prehistoric people used technology when they chipped stones into sharp tips to make hunting weapons or when they made clothes from animal skins. Technology and engineering have advanced greatly through human history. People are now able to live in modern cities and are surrounded by useful gadgets. Today, a few people even work in outer space!

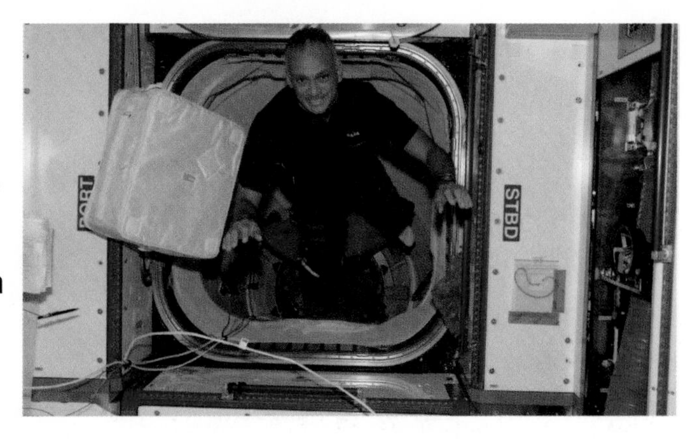

Engineers designed this astronaut's environment. Every detail of the space station is engineered to keep astronauts alive and enable them to work in orbit.

Food Technology

The earliest farmers used simple technology, such as tools made from sticks, to plant crops for food. Modern agriculture is an advanced industry. Large-scale farming relies on huge machines. Factories process and package food so it can be stored, shipped, and sold all over the world. In the past, people cooked food over simple fires. Now modern food preparation can take place in kitchens full of engineered appliances.

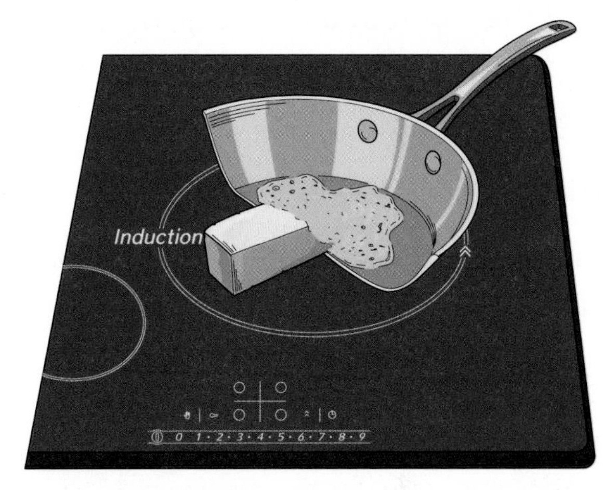

Induction cooktops use electromagnetism to heat food. Pans get hot, but the cooktop stays cool.

Shelter

Prehistoric people sought shelter in caves as protection from weather and predators. Today's homes are complex products of engineering. They are built from sturdy and long-lasting materials. In large cities, hundreds of homes may be built in a single skyscraper. Modern homes are equipped with plumbing to bring water in and take waste out. They have temperature control and electric lighting. Homes can be full of furnishings that make them comfortable. Many more technologies help people clean and care for their homes.

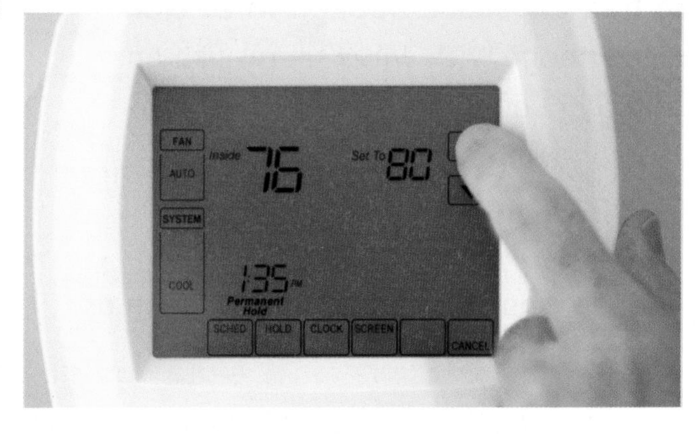

Smart thermostats learn patterns of when people come and go. They adjust temperature settings automatically to save energy.

Technology and Engineering - Transportation

The first person to build a cart with wheels was using transportation technology. We have come a long way from the days of horse-drawn wagons. Paved roads and motorized vehicles made it much easier for people to get from city to city. Powered flight gives us a fast and easy way to travel between countries and continents. Modern transportation technology moves people around the world and moves all of the goods that we consume, too.

Maglev trains use electromagnets that pull the train along above a track. Moving above the track eliminates friction and makes the trains very efficient.

Communication

Written language might be the most important technology in human history. The invention of the printing press enabled people to reproduce messages and share their ideas with many other people. The telegraph and the telephone provided another milestone in communication progress—the ability to transmit sound through wires. Now a telephone with wires seems like an antique! Engineers determined how to use radio and other types of waves to send and receive data through the air. Today we can carry convenient, wireless communication devices with us wherever we go.

Modern smart phones function not only as phones, but also as cameras, computers, and video screens.

Recreation

We rely on technology and engineering to stay fed, sheltered, employed, and healthy, but we also use technology to have fun. A game of marbles is an example of using technology for entertainment. So are playing in a soccer game, going to a movie, doing a crossword puzzle, and building a tower out of interlocking blocks. Toys, games, digital media, art supplies, bikes, boats, and sporting gear are all types of technology engineered to use for recreation.

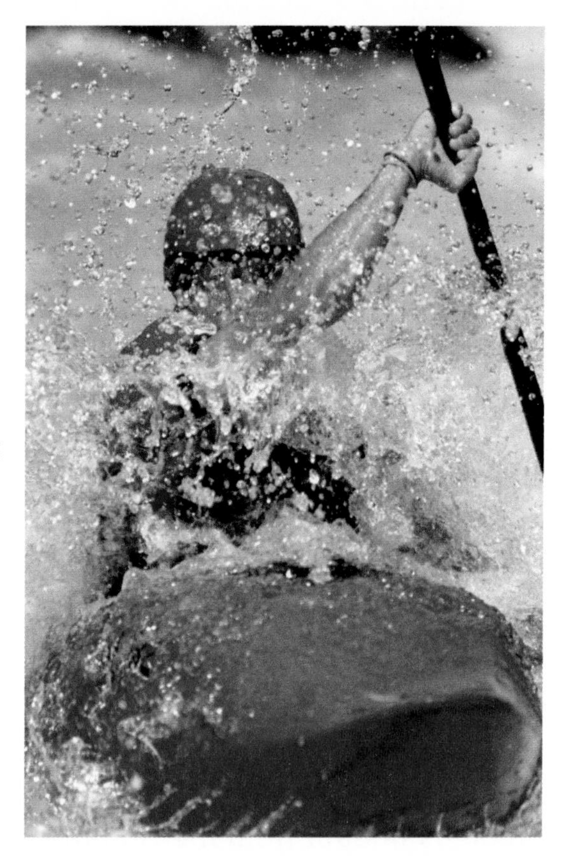

This kayak is an example of recreational technology.

Medicine

Technology such as medicines help people get well and stay healthy. Imaging devices such as ultrasound machines help doctors diagnose illnesses. The engineering of prosthetic limbs and other adaptive devices helps people with physical limitations perform tasks more easily. One of the most widely used adaptive devices ever engineered is eyeglasses!

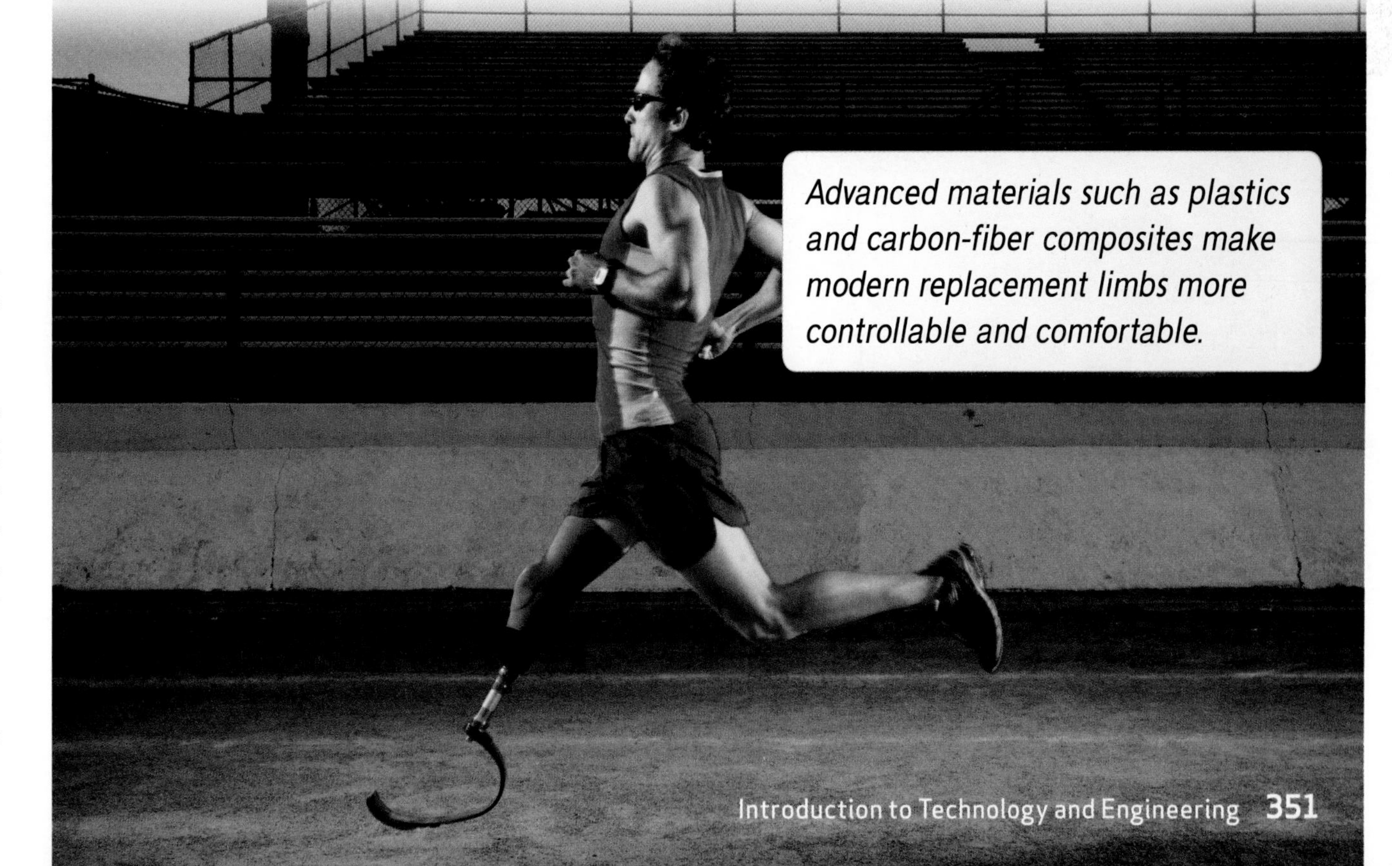

Advanced materials such as plastics and carbon-fiber composites make modern replacement limbs more controllable and comfortable.

The Math and Science Connection

Engineers rely on science when they develop technology and solve problems. All the devices that we use are made from materials that come from nature. Engineers use science to understand properties of the materials and how they behave under different conditions. For example, a bicycle wheel is usually made of an aluminum alloy. The metal is rigid and strong. It holds its shape under the load of a rider, but it is also lightweight. The wheel holds a rubber tire. The tire holds air to cushion the wheel against bumps on the ground. The rubber is treated so that it does not crack easily when it gets cold and does not melt or stretch too much when it gets hot.

Scientists and engineers rely on math in their work. Both science and engineering use processes of testing and retesting ideas and then comparing the results. Scientists and engineers compare numbers in data from measurements. Engineers also describe the shapes of the parts they design with detailed measurements. When an invention is reproduced or manufactured, specific measurements and instructions are followed to make sure all the parts of devices work the way they should.

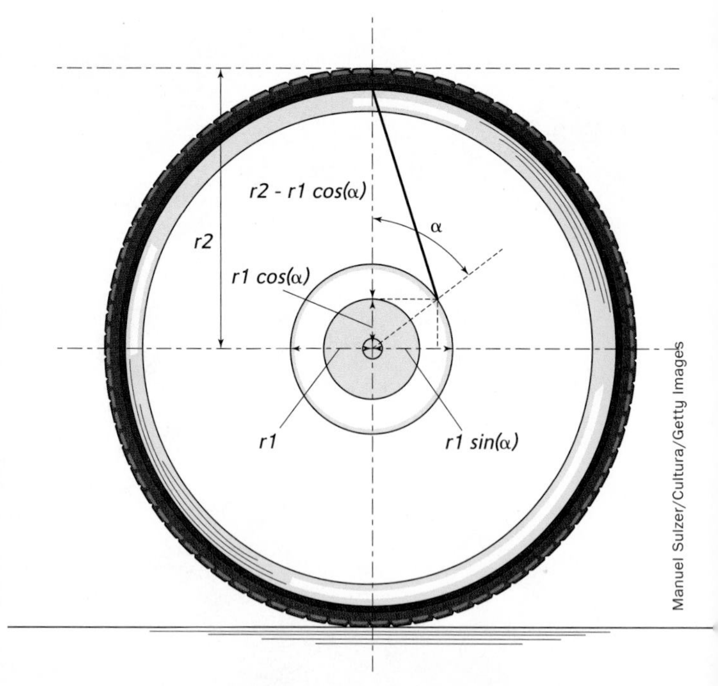

Manuel Sulzer/Cultura/Getty Images

Engineers use math to communicate their designs with precise measurements.

Engineering depends on science and math, but modern science relies on technology and engineering. The devices that scientists use to detect and measure properties, to collect data, and to control variables are all products of engineering design. A simple ruler is a piece of technology that scientists use. Complex devices, such as lasers and spectrometers, are also pieces of technology.

Perhaps one of the most impressive devices built by modern scientists and engineers is the Large Hadron Collider (LHC). Built at the CERN laboratory on the borders of France and Switzerland, the collider is engineered to study the smallest particles that make up matter, which are known as atoms. Accelerators boost beams of particles to speeds close to the speed of light. Then the beams are made to collide with each other or with stationary targets. Detectors observe and record the results of these collisions. Physicists hope to use the results to better understand physical laws. More than 10,000 scientists and engineers from over 100 countries worked together to design and build the LHC. The more advanced science becomes, the more it relies on devices that can make powerful detections and precise measurements.

The LHC runs for 27 kilometers (16.5 miles) in a circular tunnel 100 meters (328 feet) below Earth's surface.

Pascal Boegli/Getty Images

Engineering Design Process

Do you follow the same steps each time you solve a problem? Engineers try to approach a technology problem in a similar way each time. The steps are called the *engineering design process.* The process is sometimes called a loop because steps are repeated until the solution is successful in solving the problem.

Identify a Problem First, engineers think about the problem they want to solve. Making safer construction tools or buildings might be their goal. They pinpoint the exact detail they will focus on. They may even come up with a question to answer, such as: How can a building be made safer or stronger to withstand the effects of a tornado?

An engineer may discover a new opportunity they would like to explore. Suppose a new material was made to help astronauts live in space. An engineer might think about how that material could be used for everyday products. The engineer might start with the question: How else can this material be used?

Define the Project Limits Next, engineers think about how big their project will be. Is there a time limit for completing their work? How much money will they have for their project? Do they have experts in building and weather to help them? Will they need to come up with new materials, or do the materials already exist? These questions are important for an engineer to answer before starting a project.

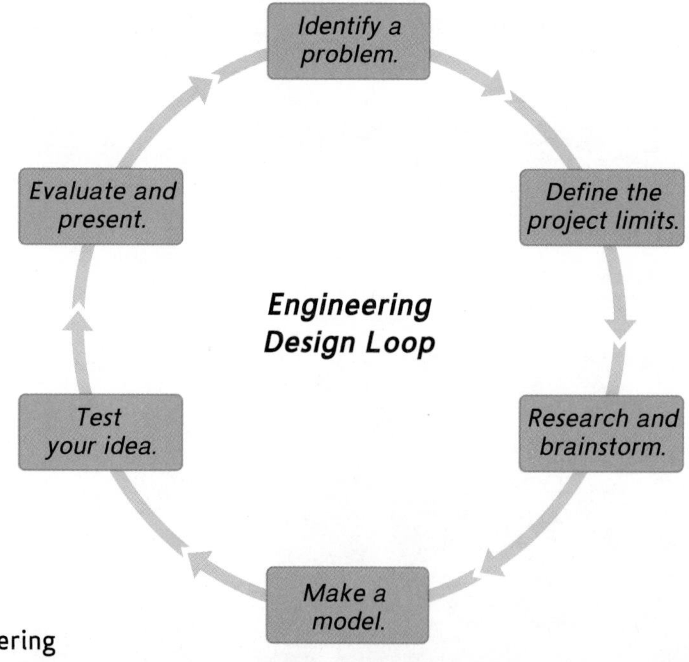

Research and Brainstorm Engineers rarely start from scratch. They begin by researching how others have tried to solve the problem before. The more information engineers have, the more questions they can answer. Then they brainstorm new ideas for solving the problem. Often this research step changes the project limits.

Make a Model Next, an engineer makes a model of his or her tool or technology. When building a model, an engineer may come up with new problems that need to be solved. They may even make several types of models to see which works best.

Test Your Idea Once a model is made, it must be tested. Testing a new product or process often shows that more work must be done. The test may raise new questions about the design problem. After the testing stage, an engineer usually returns to the model stage to make adjustments.

Evaluate and Present Once a working model is made, the design can be presented to others. This type of model is called a **prototype**. It may be tested with a wider group of people. At this stage, plans may develop for the technology to be sold. These plans do not mean the product is finished. There may be more improvements made to the product in the future.

If you were designing a paddle for a boat, you would model and test several shapes to see which one worked best.

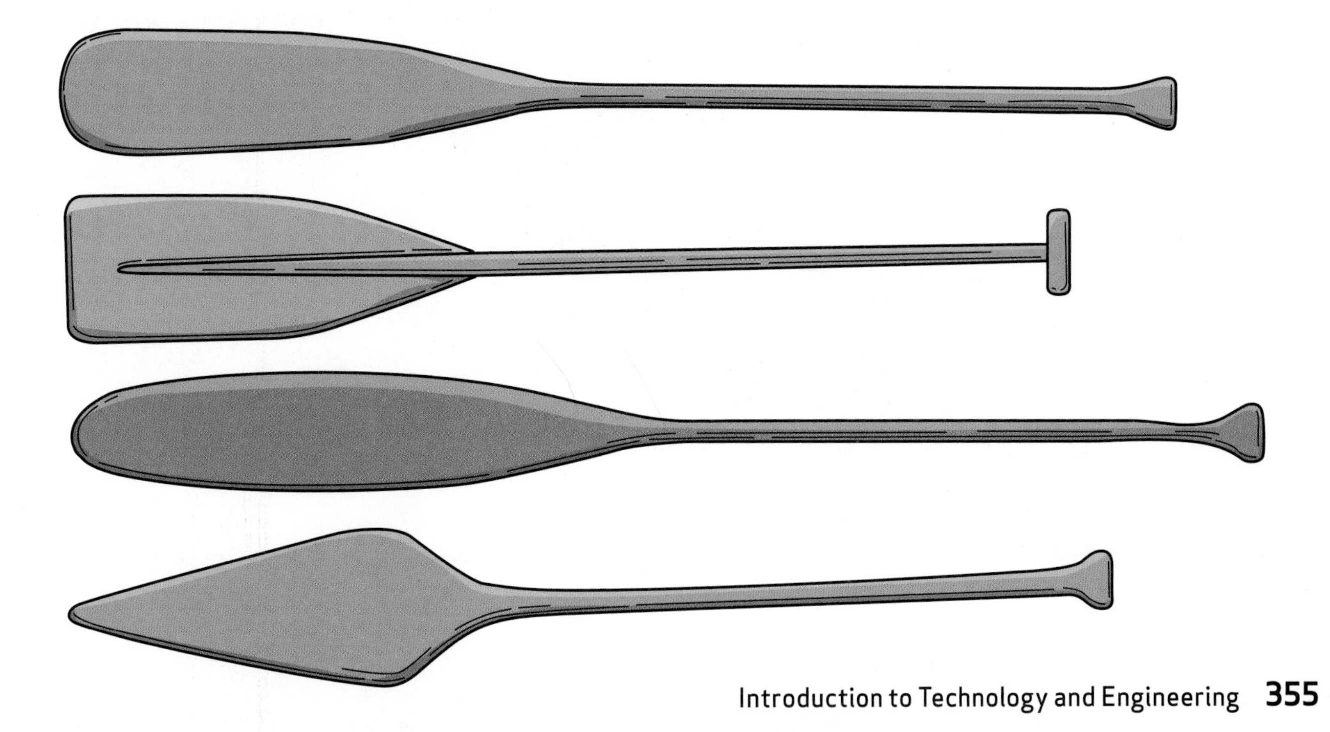

Design and Modeling

Sketches and models are important tools for communicating ideas. Sketches show what a model should look like. A sketch usually starts on paper. A sketch should show as many details as possible about the plan. It should include measurements and detailed labels of what the model will look like from all angles.

A sketch might label the materials used for each part. Arrows can show which parts of the model move and in which direction. The sketch may be used to tell how much material is needed and help engineers think about how different materials work together. Sketches help engineers think through the details of the project.

Many final sketches are made on computers. Scientists and engineers use computer programs to make a very detailed sketch of a product. These pictures can be made so that you can view all sides of the model by shifting the image on the screen. You can see what the back, sides, and top of the model will look like. This helps engineers see many details at once.

A computer sketch can also keep track of changes to measurements. If an engineer decides that a product must be larger or smaller, the computer can make the changes quickly.

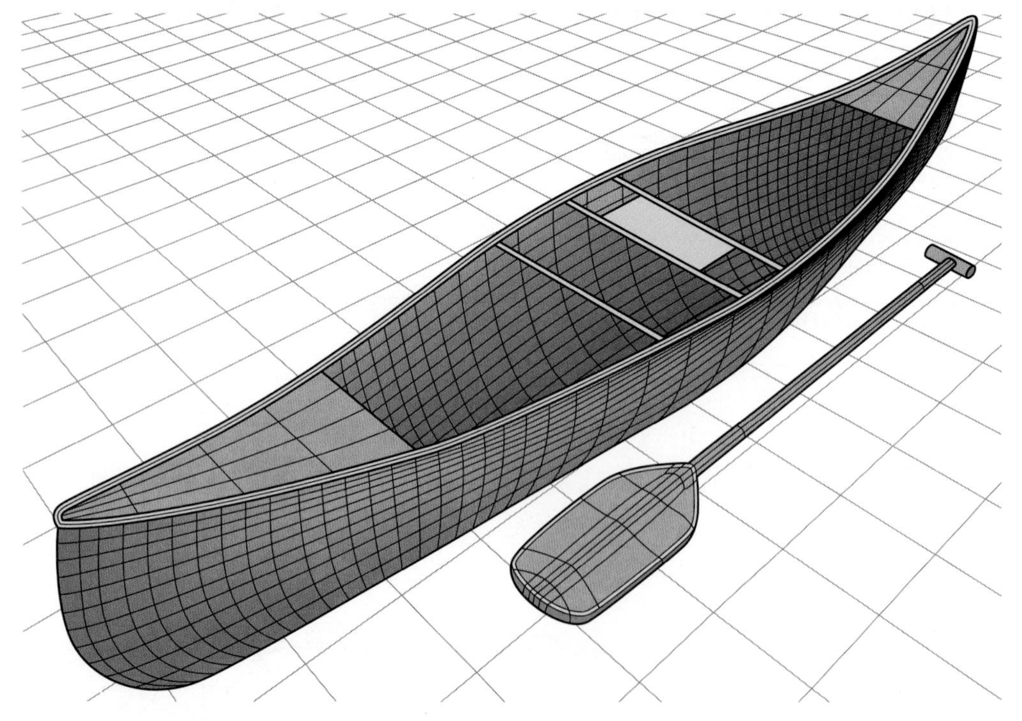

Computer sketches use complex math and can make fast calculations.

Once a detailed sketch is made, it's time to make the model. Think about trying to build an airplane before a model is built first. Many expensive or dangerous mistakes could be made. A model allows engineers to spot problems before the real airplane is made.

A *scale model* shows an accurate relationship between all the parts of the model. All the parts of the model have the same size ratio as the parts of the real product. A computer program can help engineers determine the measurements of all the parts in a scale model in proportion to the real thing.

Suppose a scale model of an airplane is 100 times smaller than the real plane. Everything else about the plane will be the same. This model helps engineers know how the plane will look and how it will work. Different engineers might work together to make the model of a plane. One engineer might be thinking about how the plane will fly. Another might be thinking about the safety of the design. Another might be concerned with adding the newest communication technologies into the design.

A scale model might be used to see how much cargo the plane can hold too. The model may be designed to hold 100 times less weight than the real plane. Models help engineers test their designs.

This model is a smaller version of a plane that may be built after the model is tested.

John Kaplan/NASA

Systems and System Thinking

Engineers think about how their work fits into a system. A *system* is a collection of parts that work together. The way the parts of a system work affects how the whole system works. Suppose one wheel on your bike is low on air. The wheels are part of a system. The problem with the wheel affects the whole bike. With low air in your bike tire, the bike becomes harder to pedal. It moves more slowly with the same amount of effort. It might not move as straight as it did with the right amount of air in the tire.

When you are riding the bike, you are part of a system that includes you, the bike, the road, and other factors. When working on an engineering design, it is important to consider how each part fits into a whole system.

Many of the products we buy also work as a system. The parts of a smart phone or a computer tablet are part of a complex system. Even something as simple as a pair of eyeglasses is part of a system.

Natural Systems

Systems are all around us. Some systems are part of nature. The solar system is made of the Sun, planets, moons, and other objects. The human body is also made of systems. A system of muscles and bones work together to help us move. Natural systems give engineers many of the ideas they use to build human-made systems.

An ecosystem is a natural system. Its parts include living things along with the nonliving parts of their environment.

PHOVOIR/FCM Graphic/Alamy

Artificial Systems

Your school is a system of grade levels and classrooms. Your bus route to school is a system of roads. The Internet is a complex system of connected information. An engineer thinks about systems when designing new tools or technologies. A factory must work as a system. Each part of a factory machine has parts that work together. Factory workers also work in a special order. Consumers who use the technology are also part of the system.

Thinking About Systems

Engineers know that the environment affects how their technology will work. They know the technology affects the environment too. For technology to work the way it is supposed to, engineers must think about how several systems affect one another. Good technology must solve one problem without creating other problems.

Engineers also think about the relationship between *inputs* and *outputs*. They ask questions, such as, what do we have to put in to this solution to make it work? Will the outcome be worth the effort and material we put in? If the answer to the second question is yes, then the technology will probably be put to use. Engineers do not just guess at the answer, though. They gather *feedback*. Feedback is information about how a solution is working. Ideas from feedback provide some of the input for the next version of a design.

Traffic lights are part of a system designed to keep drivers safe. People and cars also affect how well the system works.

Mmphotos/Photodisc/Getty Images

Materials and Fabrication

Metals, fibers, plastics, and ceramics are common types of building materials. Before making a product, engineers must find out as much information about the materials as possible. Each product has materials that give it its traits, or characteristics. The materials used to make a football are durable while still being somewhat flexible. The materials used to make a baseball bat are strong but also lightweight.

Materials for new products must be tested well. Testing materials often takes place in laboratories. Engineers put the materials under strain and stress. Machines pull or bend them over and over. Engineers will not use materials that do not stand up well in their tests. For example, a new material for firefighter safety suits must be fireproof. It also must be lightweight and flexible enough to move around in.

Making New Materials

Chemicals may be added to or taken away from a material engineers already know of. This action will change the properties of the material. It may make the material stronger or more flexible. It may make it resistant to fire or rust. The temperature at which a product melts or freezes will also give engineers an idea about its properties. Heating a material is another way engineers can change how it performs.

Tempered glass is an example of a material that engineers change with heat and chemicals to make it safer. If tempered glass breaks, it forms small rounded pieces instead of long, sharp shards.

After engineers have completed the design of a technology, the final step is to *manufacture* it. Manufacturing means making something from raw materials by hand or with a machine. *Fabrication* is the making of a material or product for many people to use. Anything that was made in a factory was manufactured. It went through a fabrication process.

Engineers work to find the best way to make many products quickly. The process of manufacturing a toy car might start with a mold in the shape of the car. Liquid solutions are then poured into the mold. The liquid will dry or cool and harden into a solid part. Many products are formed from materials at high temperatures. The materials may be stretched or pushed into shape by machines. Then they cool and harden.

Many products are very complex to manufacture. They may have moving parts, or pieces that are joined together. There are hundreds of pieces to assemble on a car. When making a real car, some materials must be treated with special chemicals before being assembled. The car seat fabrics may be treated with chemicals so they will resist fire. Some materials are treated with a special finish. Paint or weather treating are kinds of finishes that protect the outside of a car.

Even the parts in machines used on assembly lines are manufactured in other factories.

Digital Vision/Photodic/Getty Images

Types of Engineering

Civil Engineering

Think of any product, service, or process, and there is likely an engineer behind it. There are so many different kinds of technologies that engineers have to specialize in different areas. A *civil engineer* works on projects that are used by the public. Civil engineering projects are usually large structures, such as airports, highways, bridges, and tunnels. They may include skyscrapers, roads, canals, or waste disposal systems. Some of the most amazing structures around the world were designed, planned, and built by civil engineers. Civil engineering projects are designed for many people to use at once, such as a highway system. They usually need to last a very long time. Due to growing populations today, more civil engineering projects are needed all around the world.

Civil engineers must consider many things when building a large public project. For example, when building a hydroelectric power plant, what should be considered? The right place along a river must be chosen for the location of the plant. The safest, newest procedures must be used when building the plant. The strongest building materials must be used as well. Researching other power plants is a good place to start. Addressing any problems those plants have could help engineers make a better and safer plant. Public safety is the most important thing for civil engineering to think about, while also considering how their new structures will affect the environment.

A hydroelectric power plant turns the motion of water into electricity.

Electrical and Electronics Engineering

An *electrical engineer* works on projects that involve electrical currents. Electrical engineers plan the systems that bring power to homes and neighborhoods. They design and plan the systems that power amusement park rides, factories, and schools. An electrical engineer might make sure the right amount of electricity is generated at a power plant and then sent through wires to your home. Or an electrical engineer might work on the electrical systems within an office building or factory. Safety is a big concern for electrical engineers. They may create backup systems, called generators. That way if the power goes out in an area, people can still have the electricity they need.

An *electronics engineer* designs and develops products that use electronic signals. These products include smart phones, video games, digital recording devices, and even cable television. Electronics engineering is a growing field because many products are developed today to use electronic signals.

Electrical and electronics engineers understand how electricity moves. They think of new ways to use electricity to help us solve problems and make our lives easier. They think about the tools and materials that might interfere with electric signals and try to overcome these problems. They design safety features for the devices they develop. Electrical and electronics engineering are used on many other types of engineering projects as well.

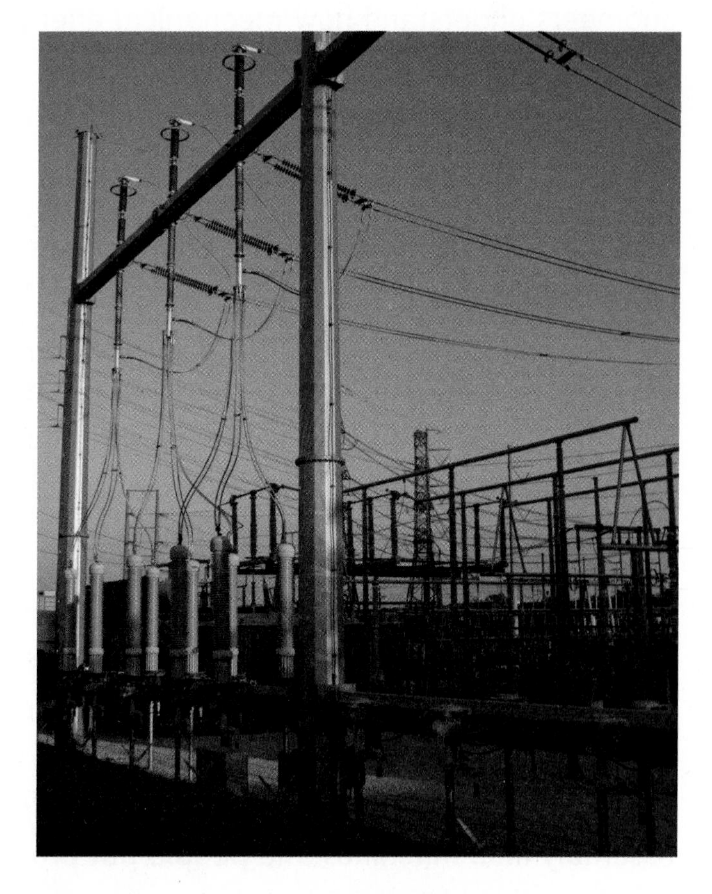

We often depend on reliable electricity. It reaches us from power plants through complex systems designed by electrical engineers.

Types of Engineering - Mechanical Engineering

Mechanical engineers design projects that have moving parts. They specialize in forces and motion. For example, mechanical engineers plan many details that go into the movement of a wind turbine. They might calculate the number of spins per minute needed for the wind turbine to produce energy or test how much spinning could make the system overheat or stop working. Mechanical engineers must be concerned with the mass of objects and how they start and stop moving.

Wind turbines are designed using mechanical engineering as well as electrical engineering.

Moving parts are used in many industrial machines. Simple machines also play a big part in many mechanical design plans. Mechanical engineers understand how mechanical energy is changed to other kinds of energy.

Manufacturing Engineering

Manufacturing engineers design processes to produce large quantities of an item. Today, humans manufacture most of the items we use. Every part of every airplane, microwave oven, refrigerator, and computer goes through a manufacturing process. Even many of the foods we eat are made and packaged in a factory process.

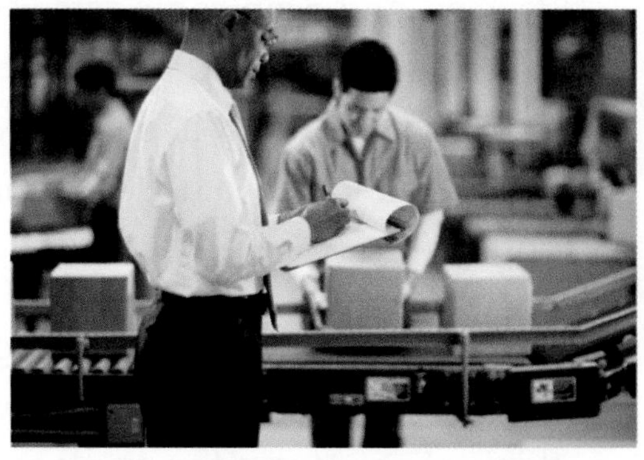

Manufacturing engineers look for new and better ways to produce goods with less waste.

The manufacturing engineer finds the best process for making thousands or millions of an exact item. They help decide how to assemble small parts quickly. A team of manufacturing engineers might come up with a new kind of assembly line. In an assembly line, machines and people put products together piece by piece.

Biomedical Engineering

Biomedical engineers solve problems related to medicine and human health. Their work can help people live healthier lives. There are many different specialties within biomedical engineering. Some biomedical engineers design products that are used inside the human body, such as artificial hearts. Others work on designing more helpful tools to be used in surgery. After patients have operations, they often need physical therapy to get stronger again. Biomedical engineers design the equipment used in physical therapy.

Biomedical engineers try to find ways to detect diseases earlier, and they also work to find cures to diseases. Biomedical engineers must understand a lot about how the human body works. They must know what artificial materials can be used inside the body. If the right materials are not used, a person can become sick or have a bad reaction to the materials. They also understand a lot about the chemical makeup of the body. Understanding how the body functions helps biomedical engineers come up with methods for solving problems in the human body.

Michael Svoboda/iStock/Getty Images

Engineers design devices that help people adapt to injuries or limitations of body parts.

Types of Engineering - Chemical Engineering

Every material involves chemicals, so *chemical engineers* help in some way with the development of almost all technology. That ranges from the dyes in our clothes to the additives in our foods. It includes the types of plastics that are used in packaging nearly everything sold in stores. Chemical engineers are involved in many areas of developing new products.

Some types of chemical engineers specialize in products that will be used inside the human body. From aspirin and cough medicine to a new snack food, the products must be safe for humans to consume.

Chemical engineers design the processes that are involved in making materials. Chemical engineers worked on the entire process of making the paper that you write on. They developed the chemicals used in the process of making the trees into pulp. There are also chemicals in the dyes used in the paper and in its packaging.

Chemical engineers also help develop products that can be recycled. Changing the formula of how some plastics or other products are made will make it easier to recycle them. Over the years, even the recycling process has been designed and improved by chemical engineers.

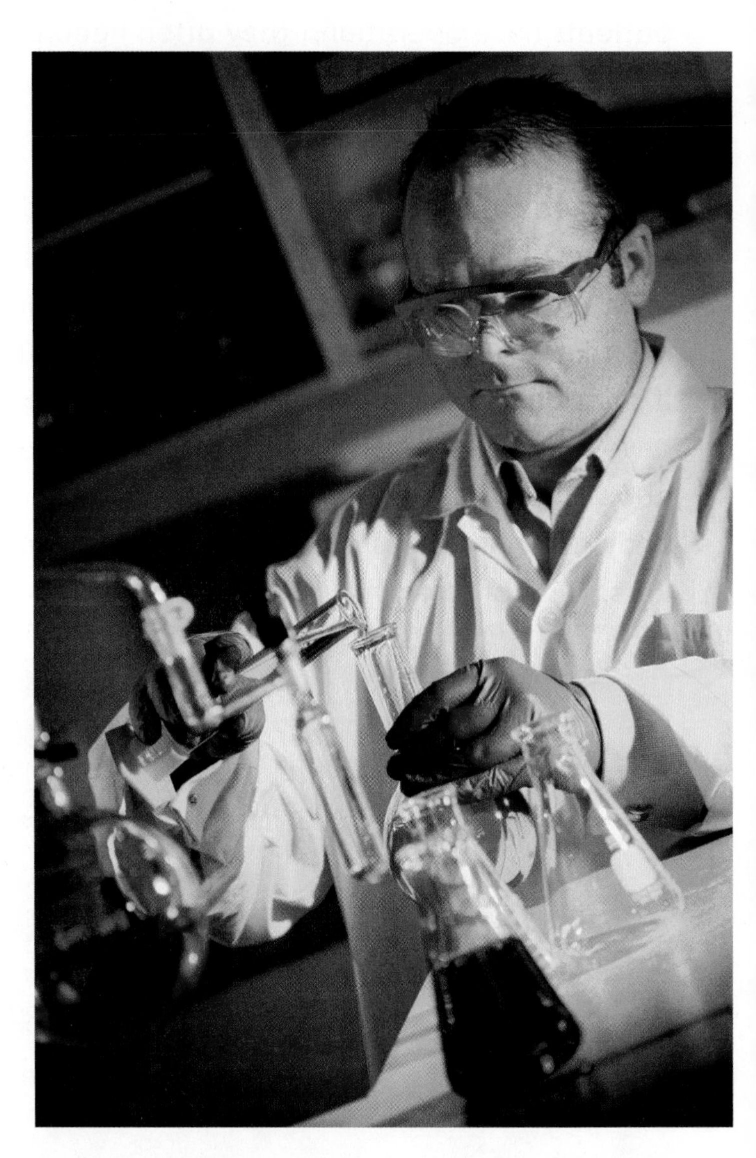

Chemical engineers are involved in many parts of the manufacturing process.

Design Pics/Kristy-Anne Glubish

Environmental and Agricultural Engineering

Engineering does not only refer to careers in which people invent technology and objects from nothing. All of the raw materials we use originally come from the natural world. The ways people use natural resources affects the environment. *Environmental engineers* develop processes to improve the natural environment so that humans and other living things will continue to have air, water, and land they need to stay healthy. Environmental engineers work on keeping the environment clean and on repairing the damage done by pollution.

Agricultural engineers also work on projects that include natural processes. Agricultural engineers work to maximize the production of food. Our food comes from animal livestock and plants grown through farming. Agricultural engineers have to know a lot about climate, soil, and water resources. For animal farming they also must know a lot about waste management and veterinary health.

Large-scale agriculture uses fertilizers and pesticides to increase food production. Often these can harm other parts of the natural environment. Large farms affect the habitats of wildlife. Agricultural engineers and environmental engineers both try to solve problems. Sometimes the easiest solution for one type of engineering presents another problem to be solved by another type of engineering. Both of these types of engineers must be very aware of ways their technology affects other systems.

Agricultural engineers try to get the most food out of a system. Environmental engineers try to reduce the impact of human activity on natural systems.

Engineering Teams and Skills

When engineers work together, amazing technologies can be created. The bigger the project is, the more engineers from different specialties are likely to be part of the process. Not all of the workers on large public projects are engineers. There are many skilled workers, technicians, and builders who are part of engineering teams. It takes a wide variety of skills and knowledge to make a successful engineering project.

Engineering Teams

The engineering process is often completed by teams. When an engineer has a good idea that can help solve a problem, he or she will make a plan. Different people might help the engineer in different parts of the design process. An engineer might consult with other engineers. Other people on the team can come from different backgrounds. Large building projects need many construction workers. They need electricians. They need people who know where to place pipes and how to mix concrete.

Another person on an engineer's team might be a drafter. A drafter is someone who makes technical drawings. They may use computer programs to make the drawings. The drawings are used to communicate details to the whole team about what the finished project will look like and details about measurements and materials.

Several types of engineers may work on the same project together as a team.

Pixtal/age fotostock

Some people who work on engineering projects are researchers. The very beginning stages of the design process involve research. Researchers find out as much as they can about the design problem. They might ask: How have other engineers tried to solve this problem? How many people does this design problem affect? Does a solution to this problem already exist? If so, can that solution be improved? What kinds of technologies would be needed to fix this problem? This information is useful for the planning stages of an engineering team.

The development stage of an engineering project involves ongoing research. Some scientists might test how materials react to different conditions. They might research the cost of materials and where they might be bought. Usually designers compare different materials to see which would work best.

Other people on engineering teams might work with the public. They can use a research method called focus groups. Engineering teams might ask people to test a product in its early stages. The focus group gives opinions about how the product worked. They give the researchers ideas about how to improve the product. Then they can work on making those changes.

Engineering teams may work with the public in focus groups to test products.

Savas Kesiner/iStock/Getty Images

Engineering Teams and Skills - Engineering Skills

An engineer uses many skills in his or her work, especially math and science skills. And each specialized type of engineering requires its own additional skills. For example, a biomedical engineer must study biology and know a lot about the human body. A civil engineer must know a lot about architecture and how buildings are made.

To become an engineer, a person typically needs to go to college for about four years to get a degree. However, some engineering technician jobs only require two years of classes and training. Many math and science classes are part of earning an engineering degree. Special classes make a degree in civil engineering different from a degree in mechanical or electrical engineering.

To make a new kind of medical tool, an engineer must know about how different chemicals work together to make products. If the tool has moving parts, he or she must understand mechanics. If the tool uses electricity, then knowledge of electricity is needed. Even when specialized teams work together, an engineer needs to understand the whole system.

Engineers must also understand the risks involved in their work. Engineers often lead teams of workers. For example, an electrical or chemical engineer must not ask a team member to do something dangerous.

What Could I Be?

Automotive Engineer

Automotive engineers try to make better, safer, and longer-lasting cars. Today's automotive engineers face new challenges in trying to develop cars that run on renewable energy and do not produce as much pollution. More automakers are now designing electric cars and hybrids, which run on a combination of electricity and gasoline. Learn more about automotive engineering in the Careers section.

Engineering teams use many science skills.

Teamwork Teamwork is a big part of working on an engineering project. Team members must communicate well with one another. On any single civil engineering project, there could be hundreds of people working on different tasks. Any change in the process must be communicated to everyone affected. Directions to each part of a team must be clear for people to follow. Teamwork helps engineering projects work smoothly and stay on schedule.

Communication From the start of a project, engineers must communicate in many ways. They must present their ideas to team members. They must describe their plans to people who will help them research. They must explain every step of how the project will work. Speaking and writing are important skills for an engineer. Even when a project is finished, engineers must explain how the new product will help people. They will need to explain how new technologies work.

Decision Making Making good decisions is also a part of the engineering process. There are many possible paths an engineer can take with a project. Remembering the main design problem can help engineers make the right decisions as they work. An engineer must decide when and how a design should be improved. They must decide when more research is needed. Teams of engineers may work together to make decisions that will make the best product or solution.

Engineers need communication and decision-making skills.

Pixtal/age fotostock

Engineering Challenges of the Future

The need for engineers is increasing every day. In our changing world, technologies are making life easier in many ways. But at the same time, some of our world's problems are getting bigger. Engineers are in an exciting position to solve some of these problems. Today engineers can solve some of the world's biggest challenges with some of the most advanced tools and technologies.

Energy

Many engineers today are focused on finding new ways to harvest energy. Our busy world is using more energy every day. The old ways of obtaining energy are harming the environment. Burning fossil fuels to transform their stored energy into electricity and heat will cause natural resources to run out. Engineers try to find ways to use renewable resources, instead of coal or oil, to obtain energy. That means finding ways to use energy from wind, water, or the Sun. However, these new technologies can be expensive to produce. Engineers try to find ways to make these technologies in less expensive ways. They also try to make them easier for people to use. Engineering systems are needed to bring new energy sources to the public.

When the car was first invented, many gas stations needed to be built. Methods were needed for drilling and shipping oil for gasoline. With today's renewable resources such as solar and wind energy, systems are also needed to bring the technologies to the public.

Engineers face the challenge of making renewable energy, such as solar power, available to everyone.

Infrastructure

The way a city or system is organized is called its *infrastructure.* The roads, bridges, and tunnels that help people move from city to city are infrastructure. The electrical and telephone lines in a city or town are also infrastructure. Many of the nation's infrastructures are old. Many bridges built a hundred years ago have never been replaced. The engineers did not expect as much traffic to pass over the bridges as we have today. Civil engineers are often involved in improving infrastructure and keeping citizens safe.

Some engineers replace parts of old roads or bridges. They replace old underground cables with new technologies. Improving old infrastructures can help us keep up with technologies that are always changing. Other engineers design brand new infrastructures. Large engineering projects might take place right next to older ones. A new bridge is often built right next to an old one that is still in use. When the new bridge is finished, the old one can be removed.

Infrastructure can come in a digital form as well. Computers are faster and more powerful than ever. People need ways to send information quickly, which requires modern infrastructure of high-speed cable. Improvements to the Internet are important so people can communicate quickly and securely. Infrastructure for Internet security is one of the biggest technology challenges of the future.

Updating infrastructures is an important challenge for engineers.

Krakozawr/iStock/Getty Images

Information Technology

Information technology is the use of systems that send, store, or receive information. The telephone was one of the first electronic information technologies. The telephone was invented in the late 1800s. It took time to set up new telephone systems. First, wires needed to connect from place to place. It took decades for the systems that allow phone communication to be set up around the country. Cities were wired for phones much sooner than rural areas. Phone calls were connected through large switchboards by operators.

Over time, technologies were improved. Soon, telephone cables went into people's homes. Calls could be made directly from phone to phone. Sounds became clearer over the wires. Telephone lines today connect cities and towns all around the world. One home may have telephone wires in many different rooms.

Information technology changes faster than most other technologies.

Cellular Communication

Technologies continue to change constantly. Today many people do not even use the phone wires in their homes because of cellular phone technology. Instead of sending messages through wires, cellular phones send signals through the air. Wireless signals are sent from tower to tower. Like earlier telephone technology, cell phone technology was slower at first. Then more and more cell phone towers were built. More people could be connected. Sound technology was improved on the phones. Many newer cell phones work like computers as well. These smart phones make people's lives easier. Information technology continues to change our lives.

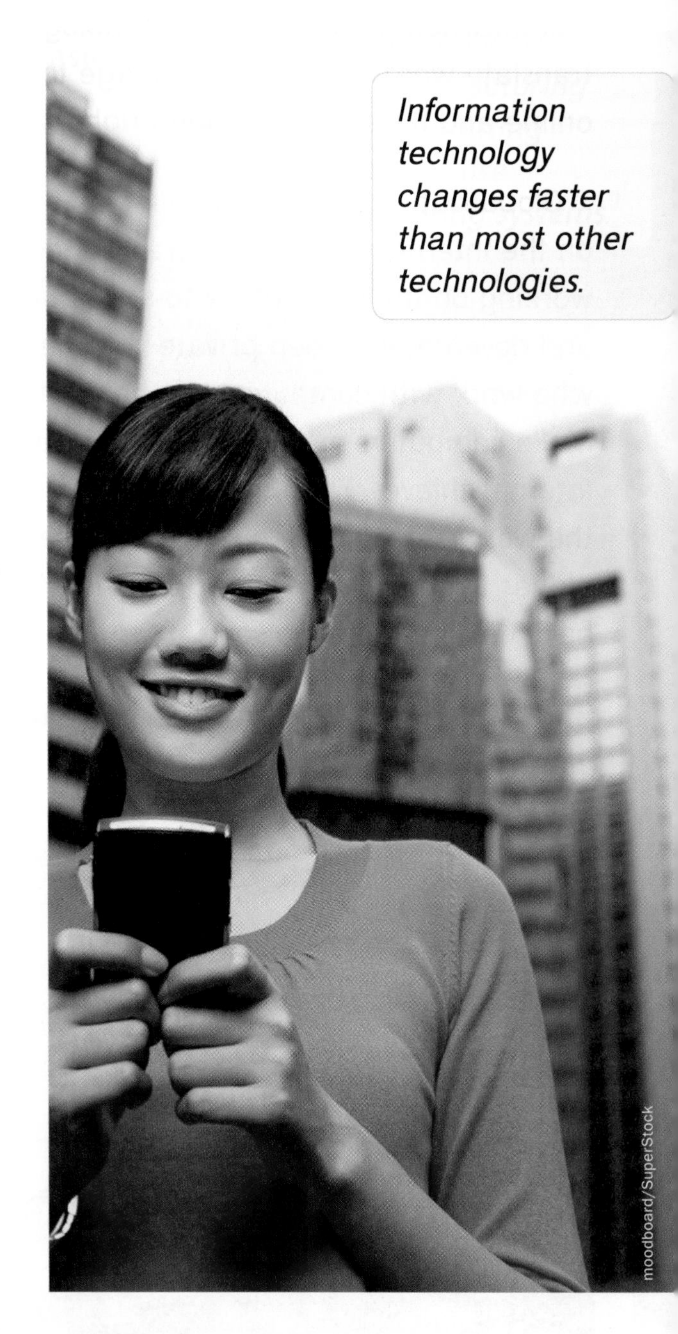
moodboard/SuperStock

Computer Engineering

Computer technology is also changing very quickly. These changes make information technology a growing field. Some people who work on computer technologies are called *software engineers*. They develop new computer programs. These new programs help people work in faster and easier ways. Software engineers may help companies come up with programs that meet their special needs. For example, a doctor has different computer needs than a banker does.

Computer technologies are growing quickly for students, too. One lightweight computer tablet can hold an entire library of textbooks. Students can also study for and take tests on computers. Teachers can keep track of students' progress with computer programs and online programs.

Software engineers also develop technologies that people can use at home for fun. The applications, or apps, you may use on a computer tablet or cell phone are made by software engineers. Some people use apps to play games; others use them to organize information, read books, or learn new things.

Software engineers use the design process to develop apps and other software programs. Then they test the programs to make sure they work with different types of computers. Like other technologies, computer technologies change often. A software engineer working today is likely to use different systems than one who worked the same job just a few years ago.

Software engineers design the computer programs you use.

Information Technology - Data Storage

Not long ago, storing digital information was a big challenge for engineers. A computer the size of an entire room was needed to hold large amounts of digital data. The amount of information such a computer held is small compared with the amount computers we use today can hold. Today a smart phone can hold hundreds of files of books and music.

The Internet is a global network linking millions of computers. Data that is shared on the Internet is stored and controlled by computer systems called servers. These servers can hold large amounts of digital data. When the data are needed, they are retrieved and sent to the computer that sent the request. Information can be sent from server to server in a matter of seconds. Internet services that send and receive information are faster than ever. The technology is also becoming less expensive for people to buy. That means more people can have access to information.

Like all technologies, Internet technologies are changing too. Data from many servers around the world can now be stored in an Internet system called the cloud. The cloud makes it possible for even more information to be stored and used again. In the future, data storage methods will likely change more. Engineers will play a large part in designing and planning new computer systems.

Computer servers hold and transfer data among millions of smaller personal devices.

Peshkova/iStock/Getty Images

Space Technology

Technology and science have come so far that people have even reached outer space. Humans have used technology to blast off into space and explore what is beyond Earth's atmosphere. Even the telescopes and cameras that help us view space from Earth have become more advanced than ever. Space technology is another growing field of science.

A satellite is an object in space that moves in an orbit around a moon or planet. Engineers have placed over a thousand satellites into orbit around Earth. These satellites send and receive signals from the ground. Satellites help us communicate around the globe. Some television signals are sent around the world using satellites. These signals allow one event to be viewed by people all around the planet at the same time as the event is happening.

Along with finding ways to communicate with Earth from space, engineers design robots to explore other planets. Mechanical and electrical engineers have designed probes and rovers to investigate other planets. Their work is helping us examine new places that are difficult for humans to reach safely.

Humans continue to learn to explore space. The International Space Station is an orbiting lab where astronauts can work in space. Today astronauts live at the space station for months at a time.

The engineering triumph of the International Space Station combines thousands of technologies.

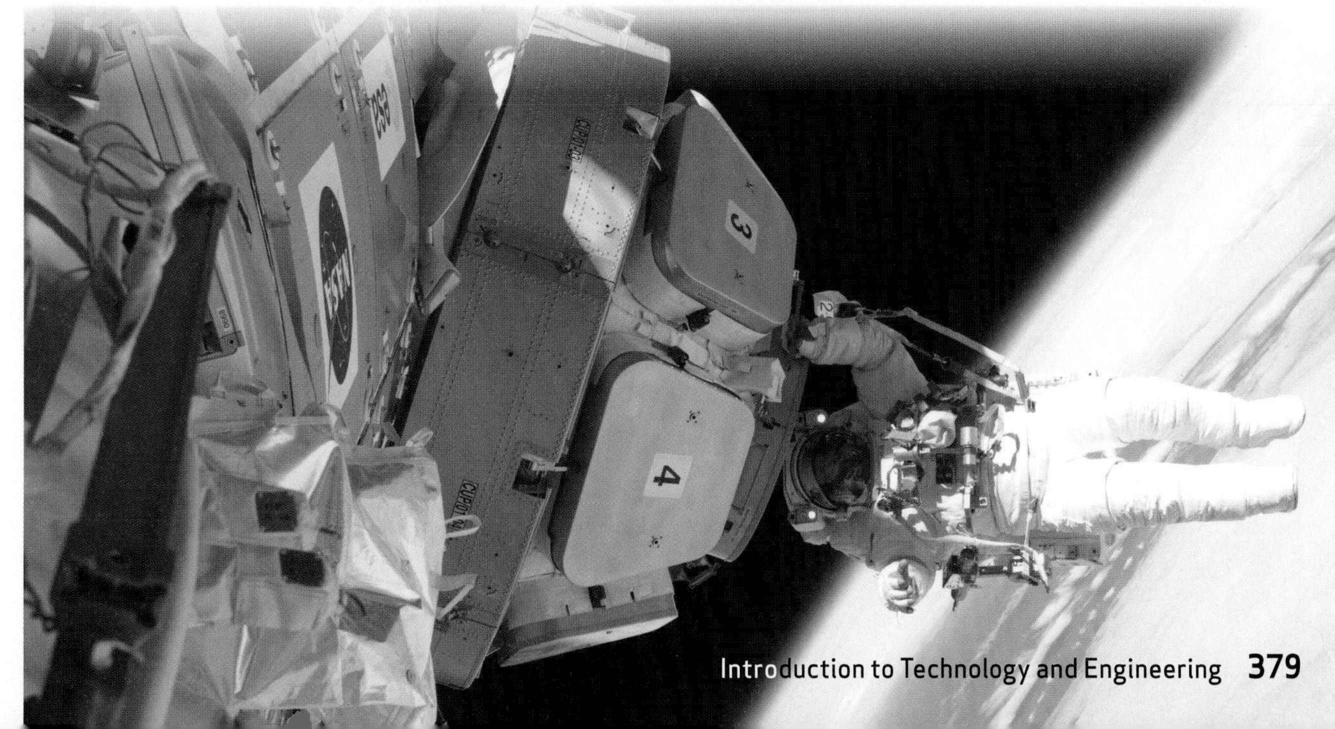

NASA

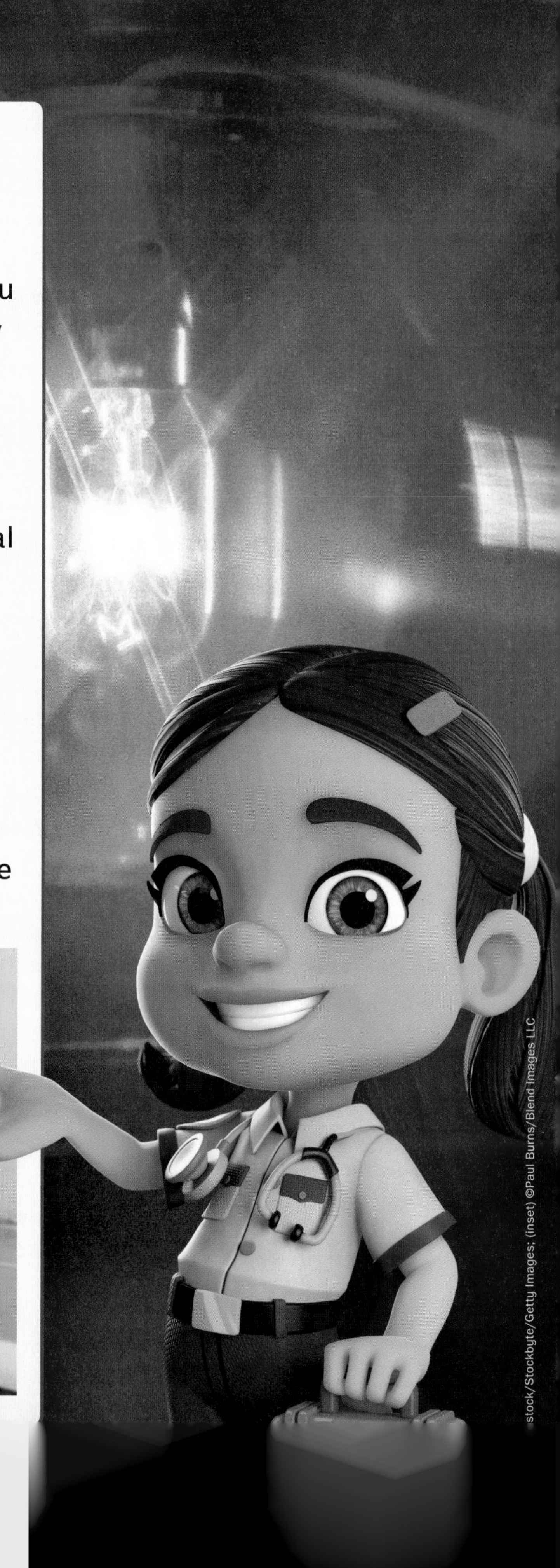

Future Career

Emergency Medical Technician (EMT) Do you like your days to be full of action? Do you enjoy helping others? An emergency medical technician has a fast-paced and very important role in society. They make sure that people who have been in an accident or have a health emergency get to the hospital as swiftly and safely as possible. EMTs need to know advanced first aid, CPR, and how to think quickly and critically about how to best help the person in need. These experts study throughout their career to make sure they are up-to-date on how to care for others when they are in need of help.

Human Body

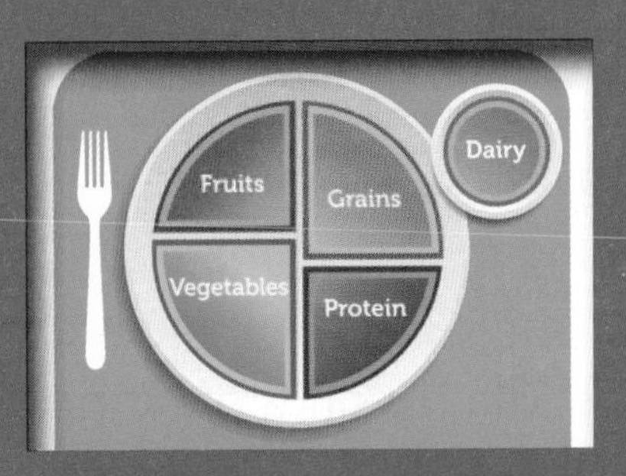

Nutrition and Nutrients

Body Systems

Nutrition and Nutrients

Nutrients are substances in food that provide the body with energy. Nutrients also provide building materials the body needs for growth, repair, and daily activities. Eating healthful foods provides the body with the nutrients it needs. The six kinds of nutrients are carbohydrates, fats, proteins, vitamins, minerals, and water.

Carbohydrates, Fats, and Proteins

Carbohydrates, proteins, and fats provide all of the body's energy. Carbohydrates are the body's main source of energy. Starches and sugars are two types of carbohydrates. Some foods, such as hard candy, are nothing but sugar. Many other foods, including fruits, some vegetables, and milk, contain sugars along with other nutrients. Starches are made of many sugars linked together. Beans, breads, and pasta are all rich in starches. The body breaks starches down into simple sugars. Fiber is a type of carbohydrate contained in whole fruits and vegetables, whole grains, and beans. Your body needs fiber to help move other foods through the digestive system.

The nutrients that provide the most energy to the body are fats. Fats help the body store vitamins and energy. They also help body cells work properly. The body uses the energy in fat when it does not get energy from other nutrients. The body needs only a small amount of fats. Fats are found in meats, eggs, butter, cheeses, oil, and nuts.

Proteins help build and repair cells and give the body energy. They help bones and muscles grow. They help the immune system fight diseases and heal wounds. Unlike fats, extra protein cannot be stored by the body. It needs a new supply every day. Foods high in proteins are milk, eggs, meat, beans, fish, nuts, and cheese.

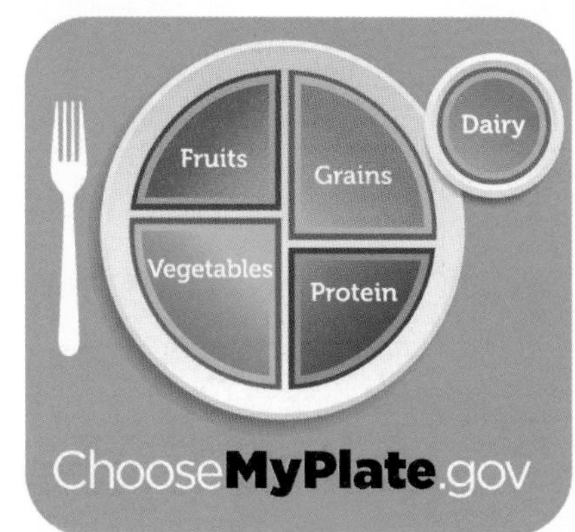

USDA's Center for Nutrition Policy and Promotion

MyPlate is a reminder to help people make healthful food choices. Eating foods in the proportions shown can help ensure the body gets all the nutrients it needs.

Vitamins and Minerals

Vitamins and minerals are nutrients that are needed by the body in small amounts. A balanced diet should provide the body with all the vitamins and minerals it needs.

Vitamins are nutrients that help the body perform specific functions. The body cannot make most vitamins. It has to get them from the foods a person eats.

Minerals are another type of nutrient. Like vitamins, different minerals have different functions in the body. Minerals help keep the bones and teeth strong, help the body release energy from food, and keep the body's cells working well.

Water

About two thirds of the body is made up of water. Water is needed to digest food, to transport nutrients to cells, and to build new cells. Water helps keep body temperature stable. It also helps remove carbon dioxide, salts, and other wastes from the body.

Vitamin A keeps your skin and eyes healthy. It is found in yellow and orange vegetables, tomatoes, and leafy green vegetables.

Vitamin B_1 is needed to break down carbohydrates. It is found in cereals, whole grains, meat, nuts, beans, and peas.

Calcium is a mineral that builds strong bones and teeth. It also helps muscles work and helps blood clot. It is found in milk, milk products, and broccoli.

Phosphorus also builds strong bones and teeth and helps cells function. It is found in meat, poultry, dried beans, nuts, milk, and milk products.

Organization of the Human Body

Like all organisms, humans are made up of a basic unit of life called a cell. In fact, the human body is made up of trillions of cells. Cells are organized into *tissues*, a group of similar cells that perform a specific function. There are four types of tissues in the body: muscle, nervous, connective, and epithelial tissue. The cardiac muscle in your heart is an example of a tissue.

A group of tissues that work together to perform a specific job make up an *organ*. The heart, lungs, and skin are examples of organs. The tissues in an organ all perform specific functions. For example, epithelial tissue lines the stomach. Muscle tissue helps the stomach contract as it digests food.

Organs work together as part of *organ systems*. For example, the heart and blood vessels are organs that make up the circulatory system. The organ systems in the human body work together to help the body function properly. The diagram shows the 11 organ systems of the human body. Organ systems work together to perform their functions. For example, the circulatory system and the respiratory system work together to bring oxygen to the cells of the body. The muscular and skeletal systems work together to give the body support and movement.

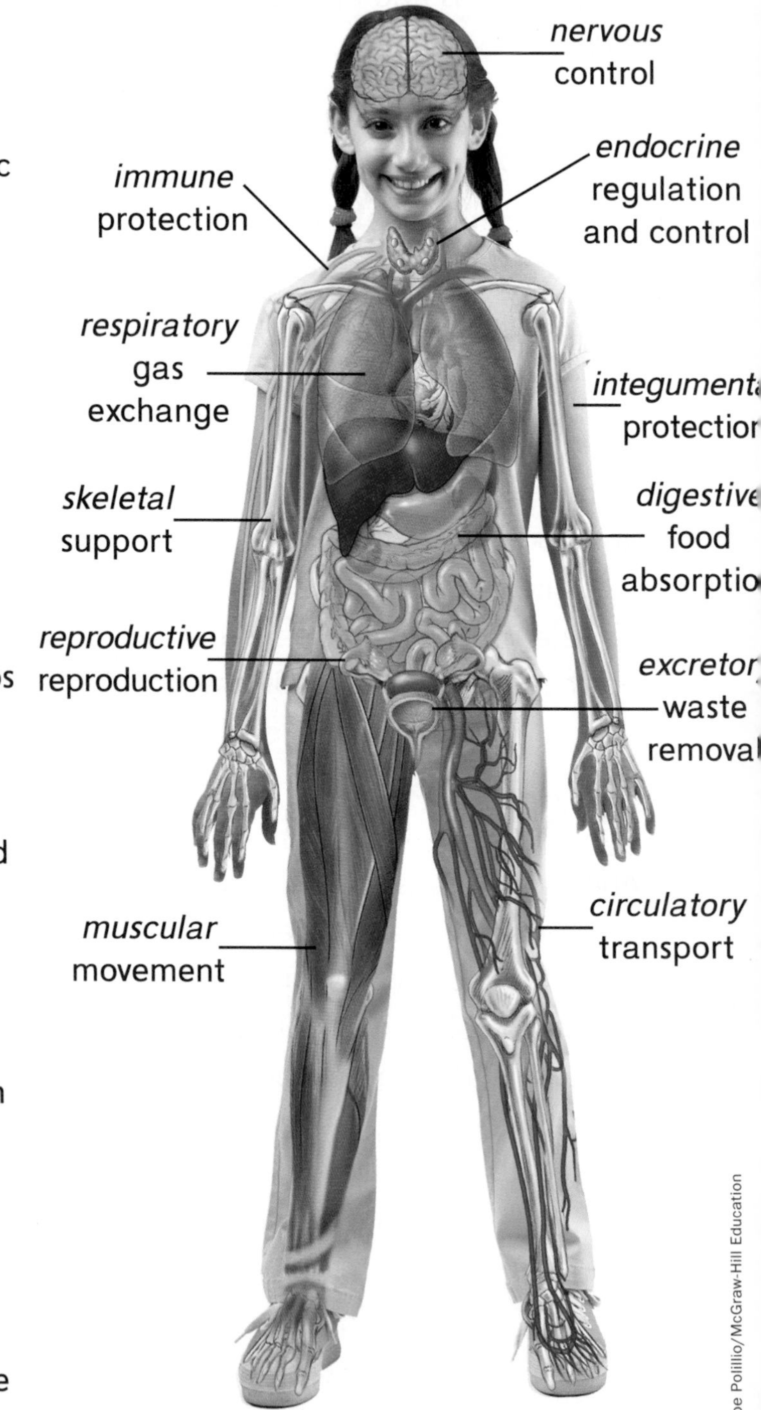

Joe Polillio/McGraw-Hill Education

The Skeletal and Muscular Systems

The supporting frame of the body is the skeleton, which is made up of 206 bones. The *skeletal system* gives the body its shape and works with muscles to move the body. Each bone in the body is the size and shape best fitted for a specific job. For example, the rib cage protects organs such as the heart and lungs. The bones of the skull protect the brain. The long bones of the legs support the body's weight. Bones also store minerals such as calcium and produce blood for the body in a tissue called bone marrow.

A joint is a place where the bones meet. Some joints are moveable. They allow movement at places such as the knees, hips, shoulders, and elbows. Other bones, such as the skull bones, cannot move. Partly moveable joints are found where the ribs are connected to the breastbone.

The *muscular system* is made up of three types of muscles: skeletal muscle, cardiac muscle, and smooth muscle. Tough, cordlike tissues called tendons attach skeletal muscles to the bones. Tendons and muscles work together with the skeletal system to move the body. Most muscles work in pairs to move the bones. Cardiac muscles are found only in the heart. The muscles contract, pumping blood through the body. Smooth muscles make up internal organs, such as the stomach, intestines, and blood vessels. These muscles perform functions automatically without us noticing.

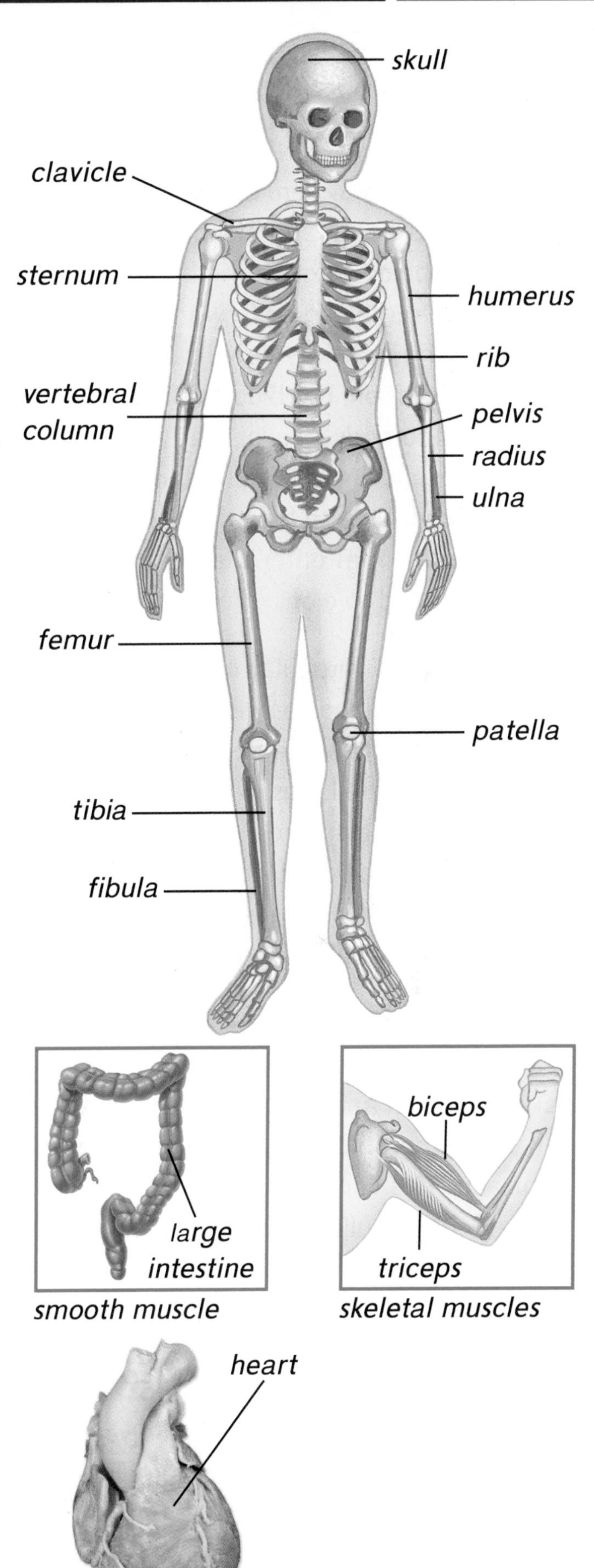

smooth muscle

skeletal muscles

cardiac muscle

The Circulatory and Respiratory Systems

The *circulatory system* consists of the heart, blood vessels, and blood. Circulation is the flow of blood through the body. Blood is made up of red blood cells, white blood cells, platelets, and a watery liquid called plasma. During circulation, red blood cells deliver oxygen to the body's cells. They also pick up carbon dioxide from the cells and transport it to the lungs. White blood cells fight and destroy germs that enter the body. Platelets allow the blood to clot.

The heart is a muscular organ that pumps blood through blood vessels. Arteries carry blood away from the heart. Some arteries carry blood to the lungs, where they pick up oxygen. Other arteries carry blood rich with oxygen to the body's cells. Veins carry blood back to the heart. Blood is exchanged from arteries to veins through tiny vessels called capillaries.

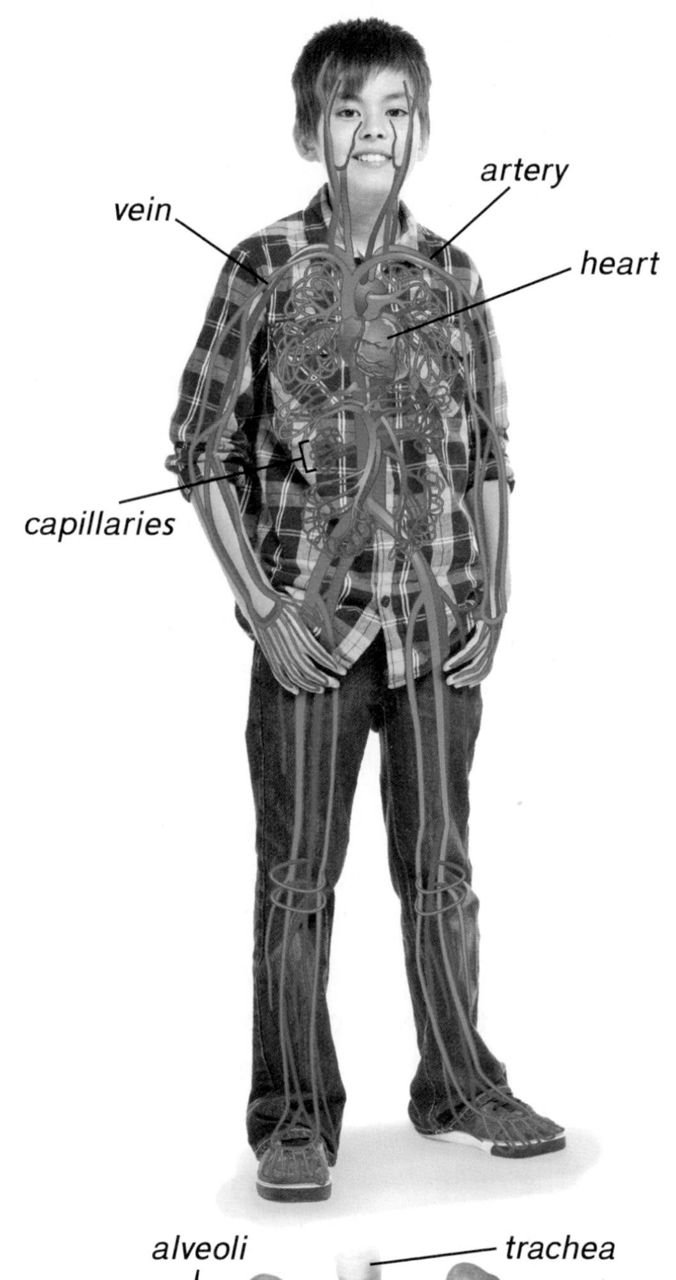

The *respiratory system* allows the body to take in oxygen and release carbon dioxide. A muscle in the rib cage called the diaphragm contracts. This contraction causes the lungs to expand. Air is pulled into the throat or nose and then travels down the trachea. In the chest the trachea divides into two bronchial tubes. These tubes further branch into smaller tubes called bronchioles. Finally, oxygen-rich air flows into tiny air sacs called alveoli. There, oxygen is exchanged with carbon dioxide.

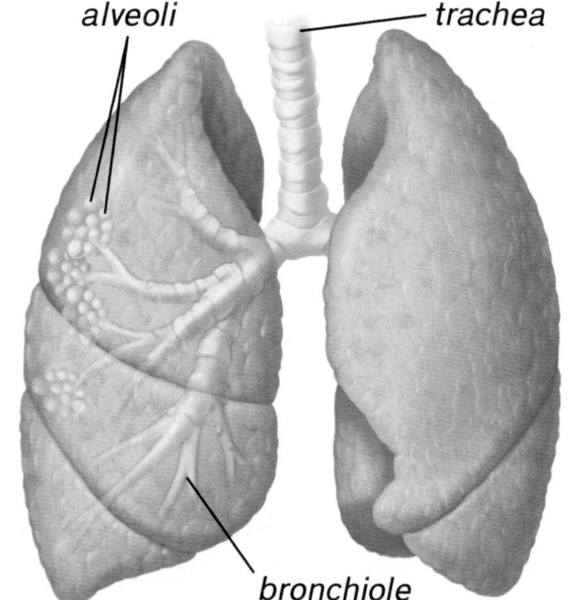

ranplett/E+/Getty Images

The Digestive and Excretory Systems

The *digestive system* breaks down food into nutrients that the body can use. Digestion begins in the mouth, where food is chewed and moistened with saliva. Food travels down the esophagus and into the stomach. Muscles in the stomach churn and mix the food. Acids present in the stomach help break down the food into simpler substances. Soon the food turns into a thick liquid.

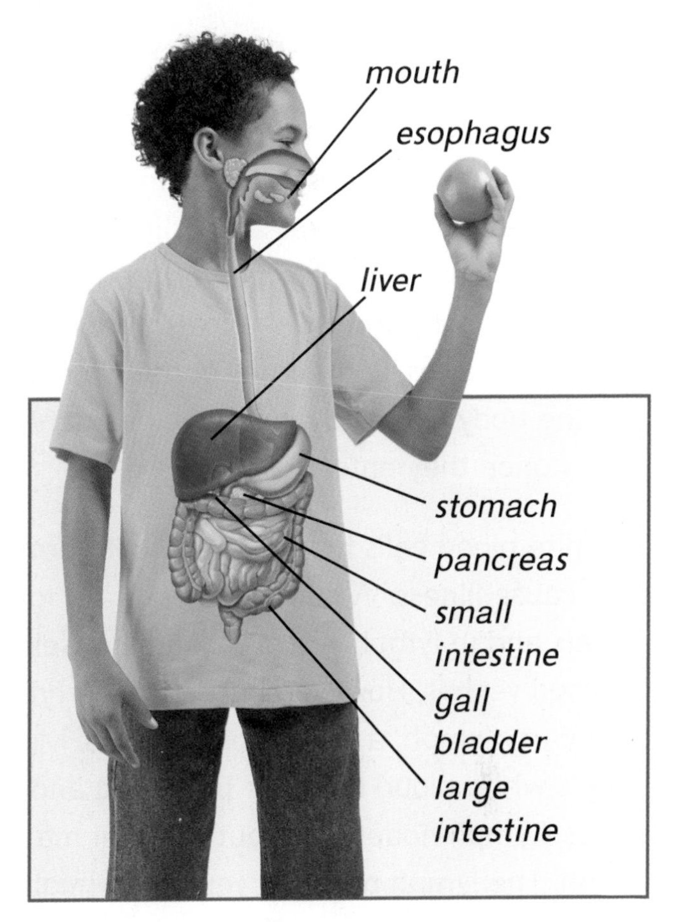

The liquid passes into the small intestine. The walls of the small intestine are lined with fingerlike projections called villi. Food is absorbed into the villi where capillaries pick up the digested nutrients. Blood vessels transport the nutrients to all parts of the body. Water is absorbed from undigested food in the large intestine. The undigested food passes out of the body as waste.

The *excretory system* removes wastes from the body. In the liver, nitrogen wastes are filtered from the blood and converted into urea. Urea is carried by blood to the kidneys. Structures in the kidneys called nephrons filter wastes from the blood. The purified blood returns to the circulatory system, and the wastes are converted to urine. From the kidneys, urine passes into the bladder, and then out of the body through the urethra. Wastes are also excreted through sweat glands in the skin. Sweat is mostly water with a tiny amount of urea and mineral salts.

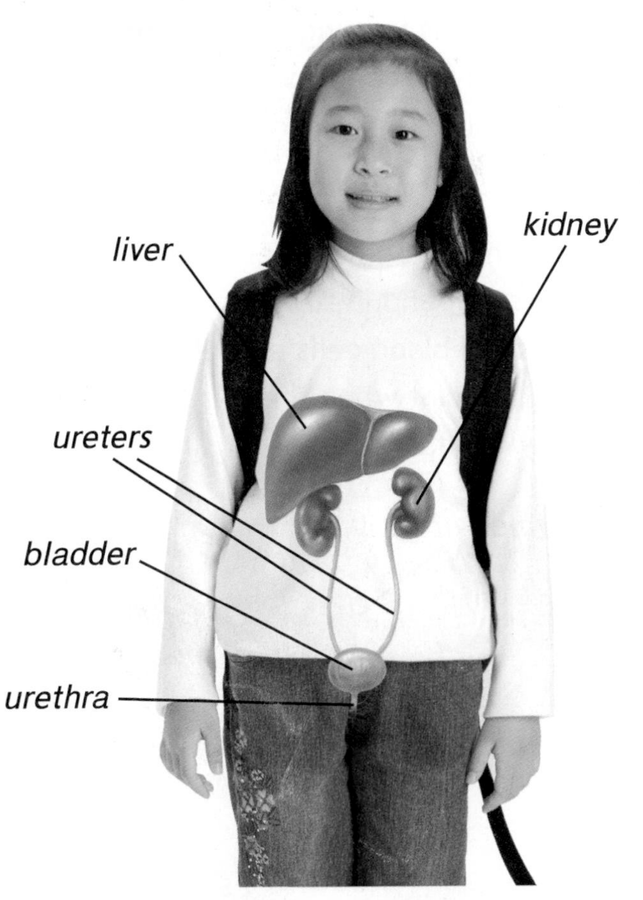

The Immune System

The *immune system* helps the body fight diseases and infections that are caused by germs, such as bacteria and viruses. The body has a first line of defense to help it prevent germs from entering. The skin acts as a barrier against germs. Chemicals in saliva, tears, and mucus kill germs. Sometimes germs do make their way into the body. The body, however, has defenses to kill germs once they enter.

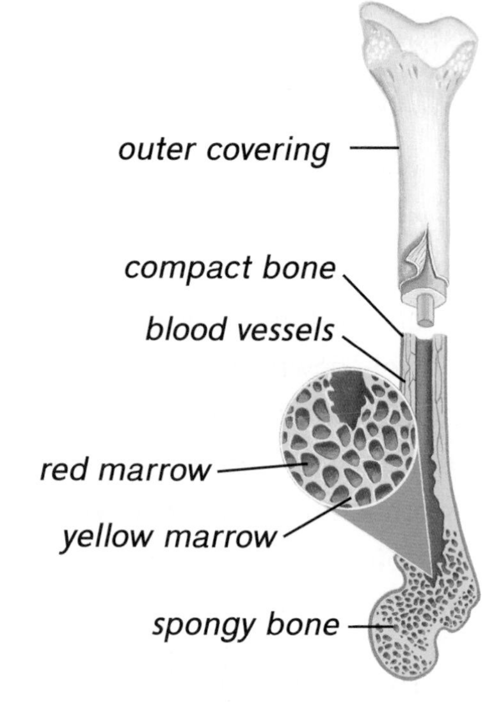

White blood cells help find and kill germs before they cause illness. White blood cells are found in blood vessels and in lymph vessels. Lymph vessels are similar to blood vessels. However, instead of carrying blood, they carry lymph, a straw-colored fluid. Many of the body's white blood cells are produced and live in lymph nodes. Lymph nodes filter out harmful materials in lymph. The lymph nodes in your neck might become swollen when you are ill. The swollen lymph nodes show that the body is fighting invading germs. As germs reproduce in your body, white blood cells move into the infected tissue. They surround and fight off the invading germs that may enter your body.

Some white blood cells are made in red bone marrow, a soft tissue found in some bones. Red marrow also produces red blood cells and platelets. Platelets cause the blood to clot when there is a break in the skin, allowing wounds to heal over time.

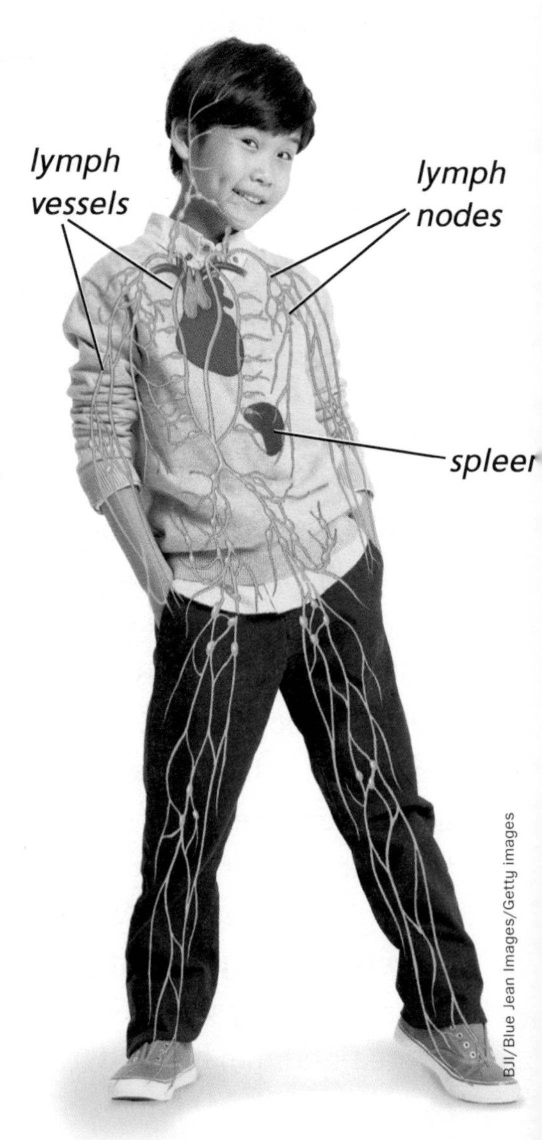

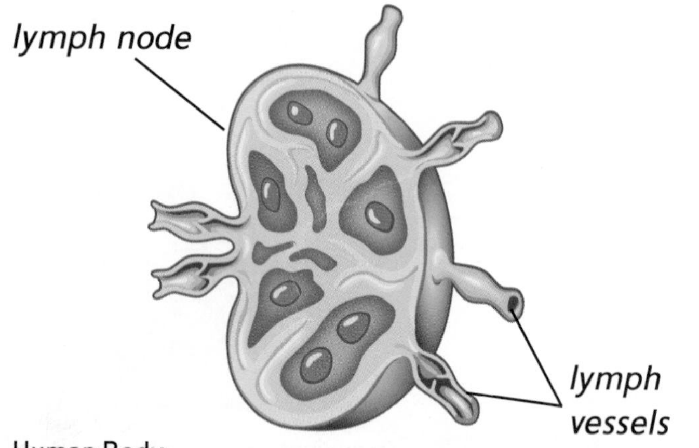

Communicable Diseases

A *disease* is anything that interferes with the normal functions of the body. A disease that can be passed from person to person is called a communicable disease. Disease-causing agents called pathogens can infect the body with disease. Pathogens include some types of bacteria and viruses.

A *virus* needs to be inside a living cell to reproduce. As the virus reproduces, it takes away nutrients and energy from the cell. A virus may cause a fever when it is present in the body. The high temperature of the body kills off the virus. For some viral diseases, such as the flu, measles, or chicken pox, vaccinations protect against the disease.

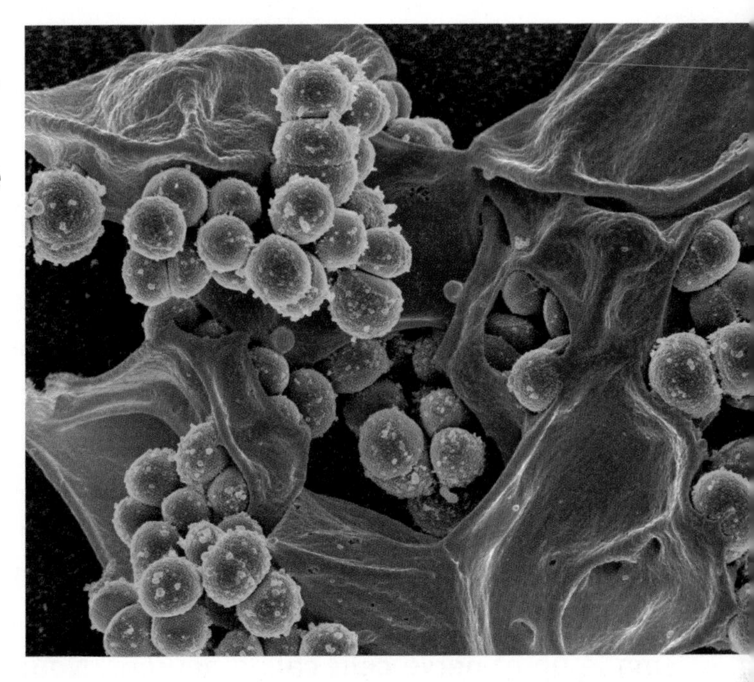

Bacteria are one-celled organisms that can also cause illness or disease. Bacteria can live outside the body on many surfaces. When a person comes into contact with the bacteria, it may enter the body and cause illness. That is why it is important to wash your hands frequently and not share cups or utensils.

Human Communicable Diseases

Disease	Pathogen	Organ System Affected
common cold	virus	respiratory system
chicken pox	virus	skin
smallpox	virus	skin
polio	virus	nervous system
rabies	virus	nervous system
influenza	virus	respiratory system
measles	virus	skin
mumps	virus	digestive system/skin
tuberculosis	bacteria	respiratory system
tetanus	bacteria	muscular system
meningitis	bacteria or virus	nervous system
gastroenteritis	bacteria or virus	digestive/excretory system

The Nervous System

The **nervous system** is responsible for receiving and responding to information. The nervous system is composed of two parts. The brain and spinal cord form the central nervous system. All other nerves make up the outer, or **peripheral**, part of the nervous system.

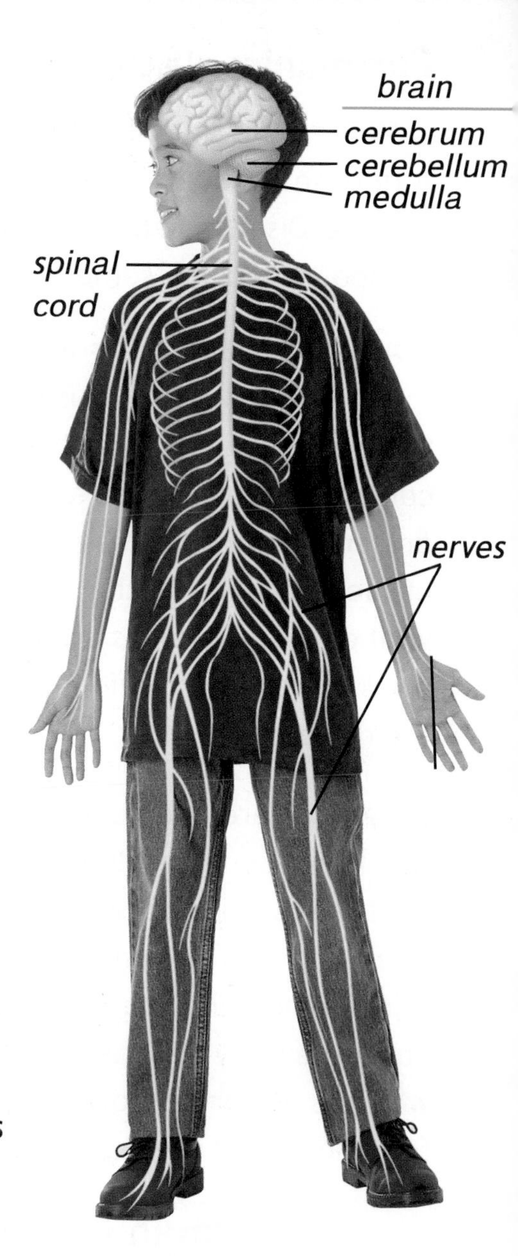

The largest part of the human brain is the cerebrum. The cerebrum stores memories and is the control center for the senses. Thoughts occur in the cerebrum. The cerebellum lies below the cerebrum. It controls balance and directs the skeletal muscles. The brain stem connects to the spinal cord. The lowest part of the brain stem is the medulla. It controls automatic functions such as heartbeat, breathing, blood pressure, and the muscles in the digestive system.

The spinal cord is a thick band of nerves that carries messages to and from the brain. Nerves branch off from the spinal cord to all parts of the body. The nerves are made up of nerve cells called neurons. Neurons receive information from cells of the body. Each neuron has three main parts—a cell body, dendrites, and an axon. Dendrites carry impulses, or electrical signals, toward the cell body. An axon is a nerve fiber that carries impulses away from the cell body. The gap between neurons is called a synapse. Impulses travel from one neuron to the next, passing information to the brain through the spinal cord.

Parts of a Neuron

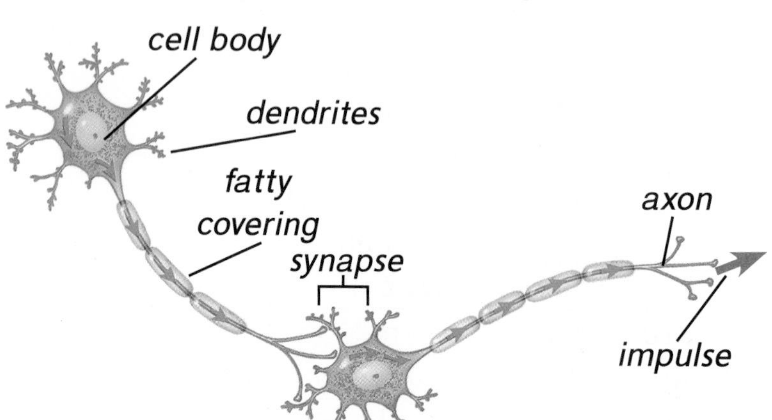

Stimulus and Response

The nervous system, the skeletal system, and the muscular system work together to react to or adjust to the surroundings. A stimulus is anything in the environment that causes the nervous system to adjust. When the body reacts to a stimulus, it is called a response.

The body responds in different ways to stimuli. A **reflex** is a quick reaction that occurs without waiting for a message to be sent from neurons to the brain. For example, touching something hot causes the hand to quickly pull away. No conscious thought is involved in this response. Instead, this reflex is an action controlled by the spinal cord.

The body responds to external stimuli with neurons. There are three kinds of neurons: sensory, associative, and motor. Each type of neuron does a different job in response to stimuli. Sensory neurons receive stimuli from your body and the environment. Associate neurons connect the sensory neurons to the motor neurons. Motor neurons carry signals from the central nervous system to different parts of the body.

Nerve Response

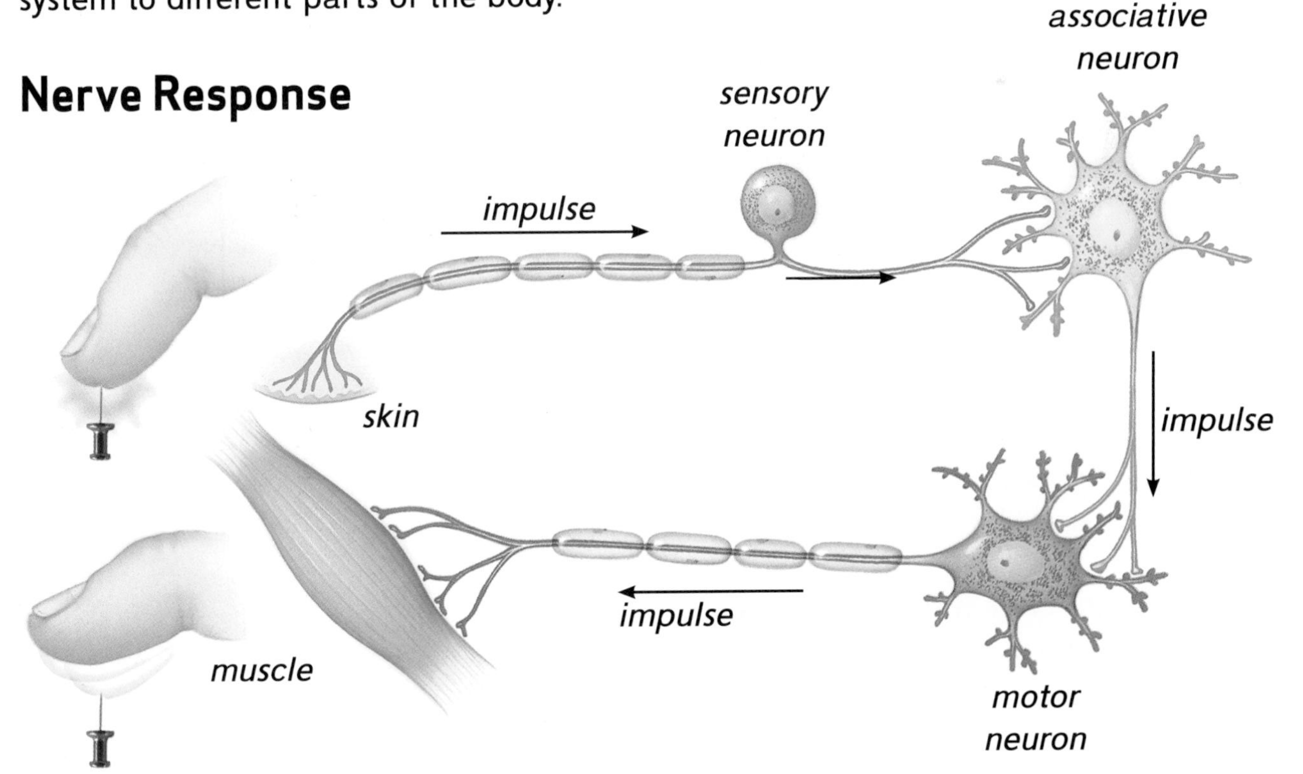

The Nervous System - Senses

The senses are part of the nervous system. They detect stimuli in the environment and send impulses to the brain. The brain interprets the stimuli as images you see, sounds you hear, odors you smell, flavors you taste, and textures of objects you touch.

Sight Light is reflected off objects and into the eye. Light passes through the pupil, a small opening in the eye, and falls on the retina. Here, receptor cells on the retina change the light into electrical signals, or impulses. The impulses travel along neurons in the optic nerve to the brain.

Hearing Sound waves are collected by the outer ear. The sound waves travel down the ear canal and reach the eardrum, causing it to vibrate. Receptor cells in the ear change the sound waves into impulses. The impulses travel along the auditory nerve to the brain.

Smell The sense of smell is the ability to detect chemicals in the air. Chemicals enter the nose and dissolve in the mucus in the nasal cavity. Receptor cells send impulses along the olfactory nerve to the brain.

Taste Chemicals in food dissolve in saliva and are carried to the taste buds. Inside each taste bud are receptors that sense five main tastes—sweet, sour, salty, savory, and bitter. Receptors send the impulses along a nerve to the brain.

Touch Receptor cells in the skin can sense hot, cold, wet, dry, light, and heavy touches. The receptor cells send impulses along sensory nerves to the spinal cord. From there, the impulses go to the brain.

Sense of Smell

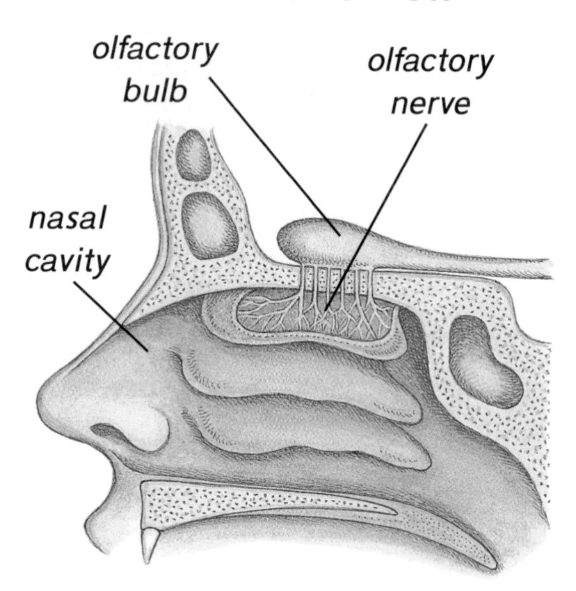

Sense of Touch

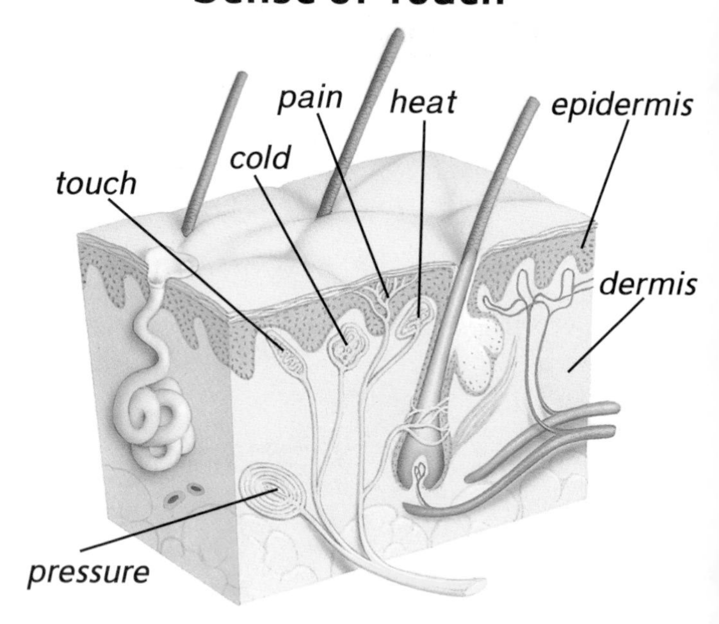

The Endocrine System

The *endocrine system* is made up of your body's glands. A gland releases chemicals into the bloodstream. These chemicals, called hormones, control many of the body's functions. An organ that produces hormones is called an endocrine gland.

Endocrine glands are found in different places in the body. Each gland releases one or more hormones. Each hormone has a specific function. The hormones seek out a target organ or organ system, the place in the body where the hormone acts. For example, the pancreas produces insulin, which regulates blood sugar level. The parathyroid gland secretes a hormone that controls calcium levels in the blood.

Endocrine glands allow the body to maintain a constant, healthy balance. Hormones control many of the body's daily activities as well as overall development. For example, the pituitary gland controls the body's growth rate. Endocrine glands can turn the production of hormones on and off whenever the body produces too much or too little of a certain hormone.

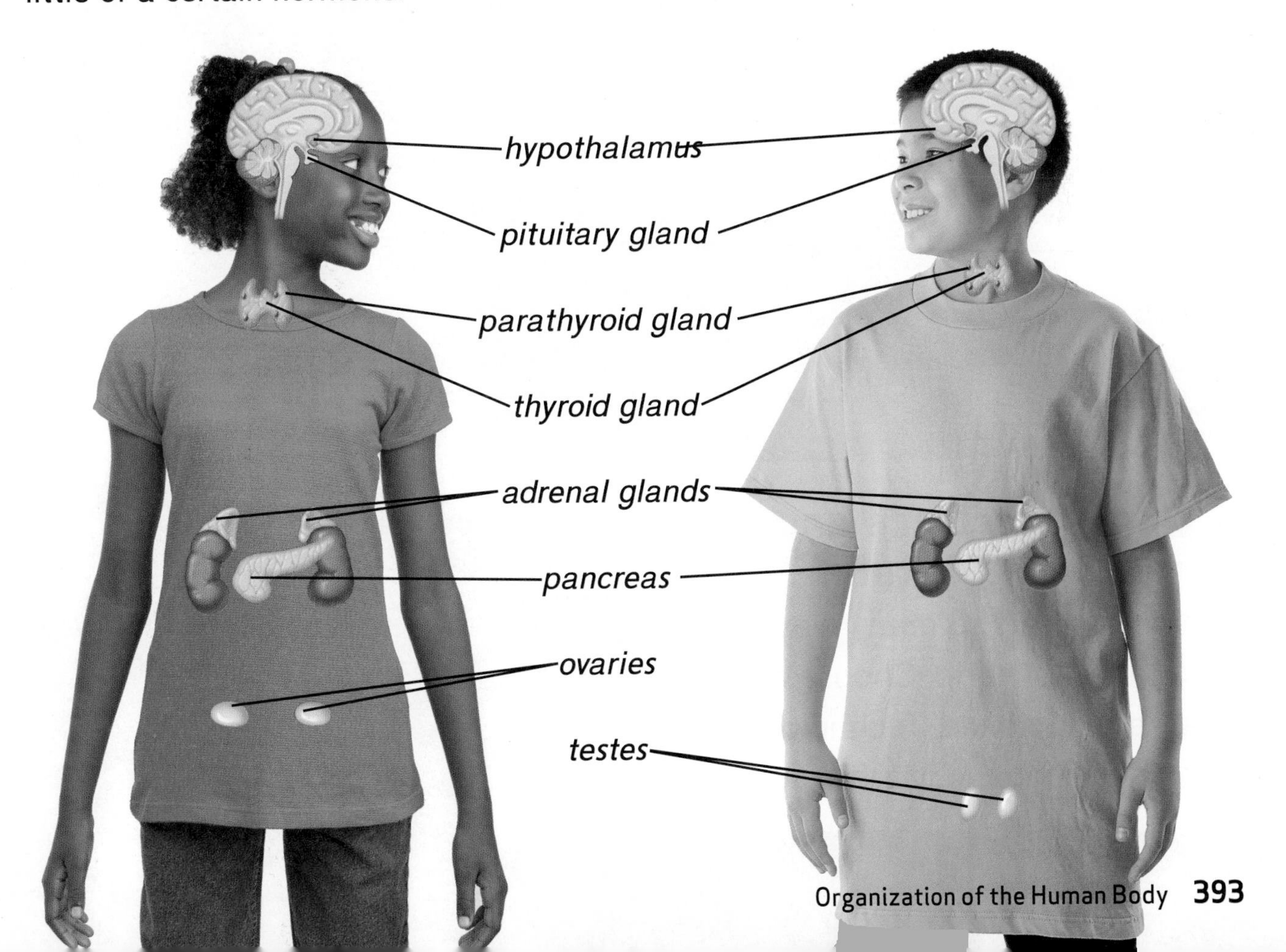

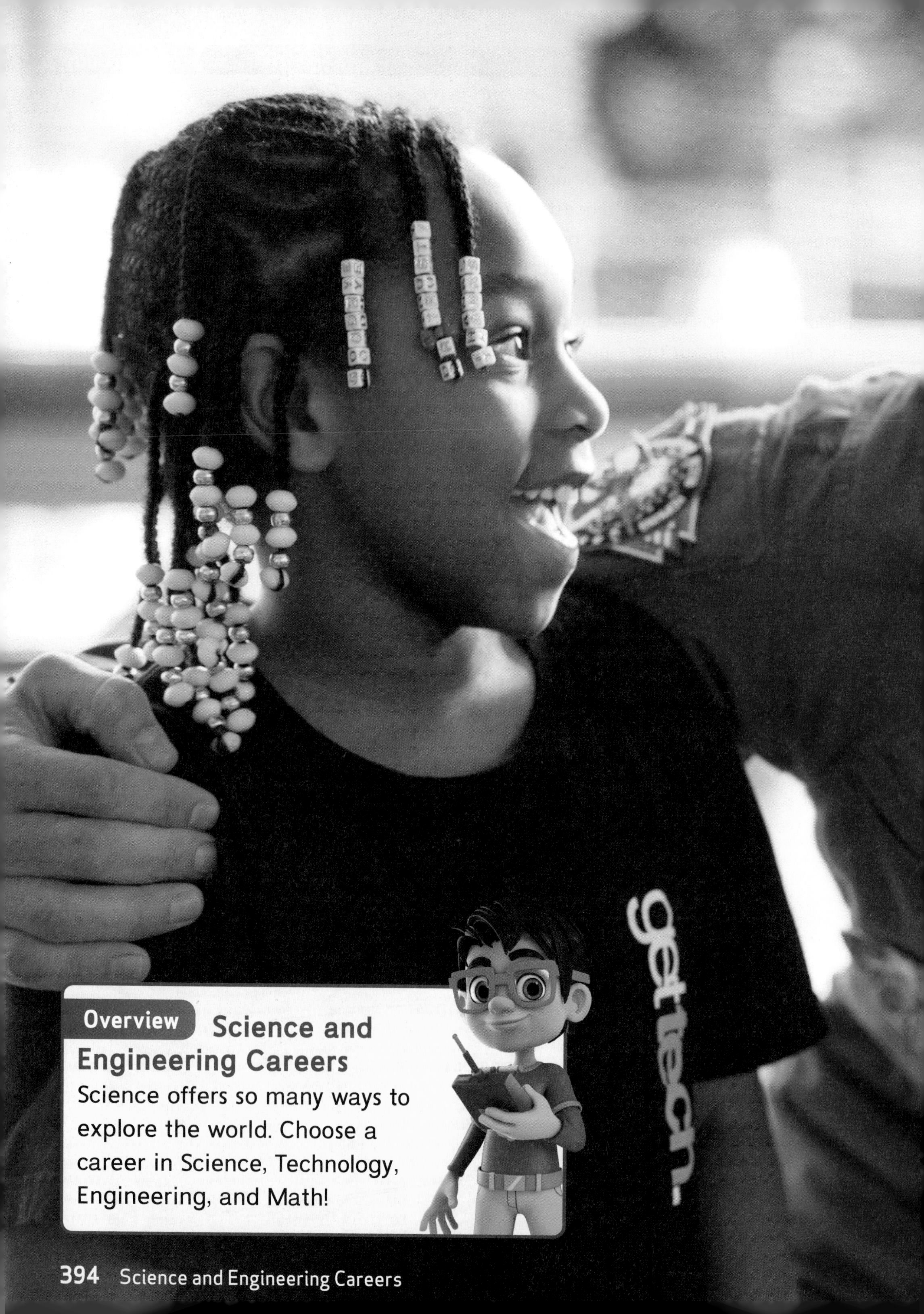

Overview Science and Engineering Careers

Science offers so many ways to explore the world. Choose a career in Science, Technology, Engineering, and Math!

Science and Engineering Careers

Engineering Careers

Civil Engineer

Do you have an interest in construction, architecture, and design? If so, you might want to consider a career as a civil engineer. Civil engineers design and maintain roads, bridges, water systems, and energy systems. They also design and maintain public facilities such as ports, mass transit systems, and airports. They even design and construct structures in our national parks. In addition to having a good understanding of forces and motion, civil engineers must have knowledge of the advantages of different construction materials and which ones would be best to use for a specific purpose. Civil engineers usually work in an office or at a construction site. Students interested in becoming civil engineers need a solid background in math and the physical sciences. After earning an engineering degree, civil engineers must pass a test to be licensed.

Updating structures such as bridges is an important challenge for civil engineers.

Electrical and Electronics Engineers

Do you have an interest in the science behind electricity, or in the technology that powers electronics such as computers and global positioning systems? If so, you may want to consider electrical or electronics engineering. These engineers plan, design, produce, and test electrical equipment. The equipment could be inside the power plants that generate electricity around the world. They also design communication systems, certain components of automobile engines, and radar and sonar equipment. While earning their degrees, students study math, physics, computer-aided drafting, circuitry, and computer sciences.

Engineers develop, design, test, and manage the manufacture of electrical and electronic components used in solar systems.

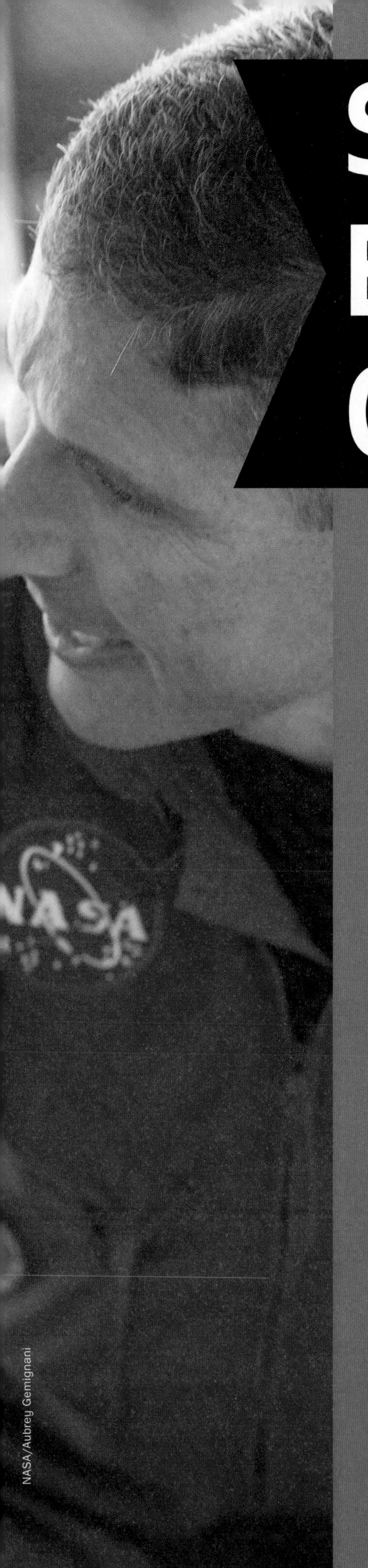

Science and Engineering Careers

Life Science Careers

Cell Biologist

Are you interested in studying cells up close? Do you want to learn about how they function and support life? If so, you may want to consider a career as a cell biologist. Cell biologists study the parts, structure, and function of cells. They may run tests to find out how cells respond to medications. They may also study how cells respond when attacked by viruses or bacteria. As a cell biologist, you could work for a university, a hospital, or even for the government. Courses in biology, zoology, chemistry, and physiology are just some of the classes required to become a cell biologist.

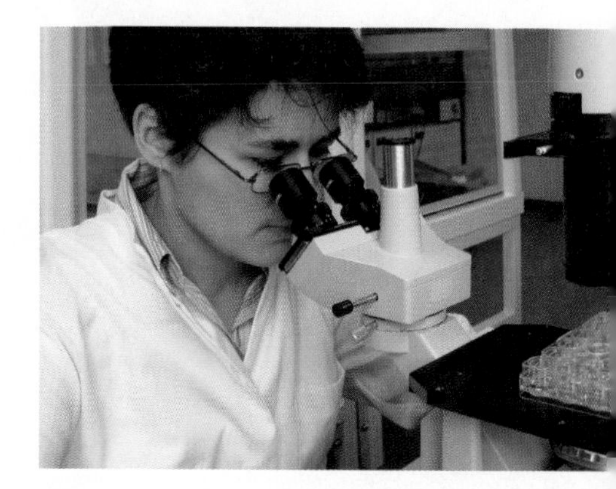

Some cell biologists study how cells reproduce in the hope of finding treatments or cures for diseases that affect cells, such as cancer.

Biomedical Engineer

Do you enjoy designing and building objects and also have an interest in medicine? Biomedical engineering may be a career option for you. Biomedical engineers design, develop, and maintain medical devices. Some of these scientists perform tests on new or proposed products, and help develop safety standards for medical devices. One exciting area of biomedical engineering is tissue engineering. Through tissue engineering, a biomedical engineer can use cells to grow new tissue, such as skin, blood vessels, muscle, bone, and cartilage, right in his or her own laboratory. Biomedical engineering students study both engineering and medicine. They also may take courses in cell biology, anatomy, chemistry, and math.

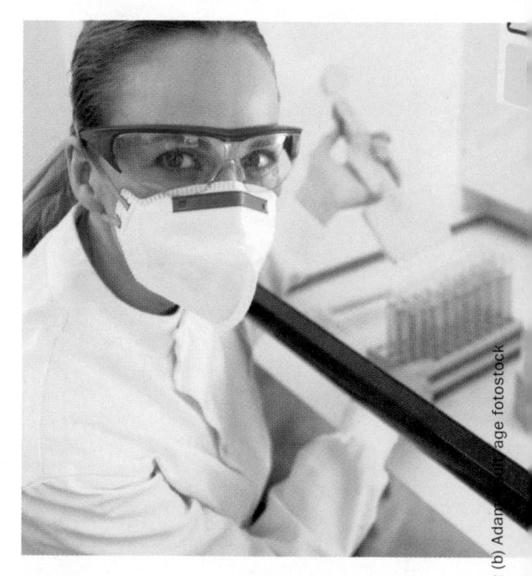

In the near future, biomedical engineers may be able to produce new hearts and other organs using cells and a machine called a bioprinter.

Landscape Architect

If you love plants and have a knack for art and design, then a career in landscape architecture may be a good choice for you. Landscape architects plan and design areas of land in places such as waterfronts, corporate and college campuses, golf courses, and other open spaces. They also work to restore areas disturbed by humans, such as wetlands and mined areas. They spend much of their time building models of their plans, estimating budgets, and interacting with clients. Landscape architects must know whether the plants they choose can tolerate shade, sunlight, drought, and other conditions in a particular climate. They must also have a good understanding of soil types and how well they drain and absorb water. Some necessary college courses for landscape architecture students include planning and design, botany, and geology.

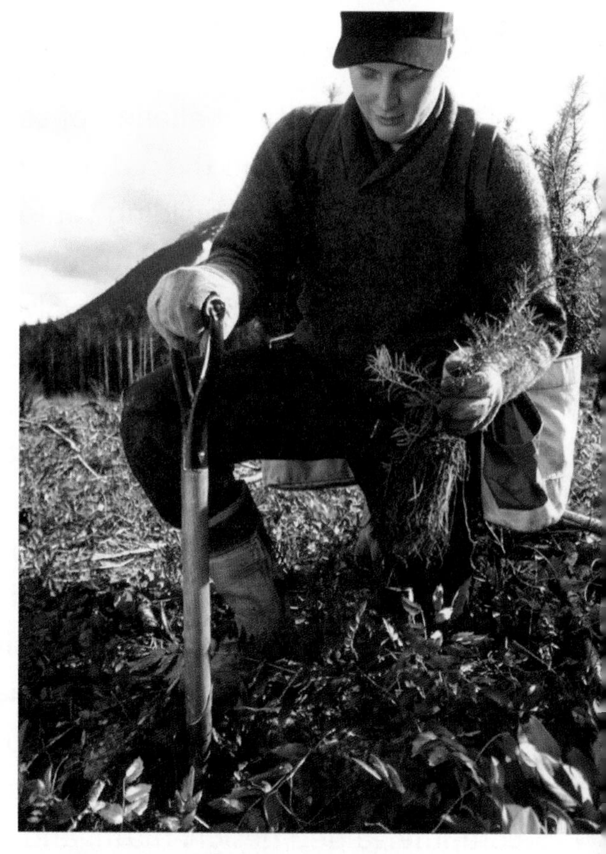

After plans are approved, landscape architects often oversee the completion of their projects.

Ornithologist

Many people enjoy observing birds and learning to identify them by sight and sound. If you enjoy this activity, then you may want to consider becoming an ornithologist. Ornithologists study all areas of bird life, including their habitats, life cycle, body systems, songs, flight patterns, and migration patterns. As an ornithologist, you may work as a researcher for a college or university, in a museum, in a zoo or wildlife area, or for a parks service. You may even work in the field, observing birds in their natural habitats and collecting information about nesting, mating, and rearing of young. Ornithologists usually have a strong background in many areas of science, including ecology, biology, and zoology.

Ornithologists rely on bird sightings made by nonprofessionals to help them understand changes in bird populations

Biomimicry Engineer

Do you enjoy the challenge of solving a puzzle? Are you interested in the natural world? If you also enjoy designing and building your own creations, then biomimicry engineering could be a great fit for you. Biomimicry engineering involves using nature to find solutions to engineering problems. For example, the hook-and-loop tape on your jacket was modeled after small, prickly plant seeds that stick to clothing and animal fur. Engineers have developed a coating based on the patterns in shark skin for use on boats and other watercraft. This coating helps the boats move through the water more efficiently and use less energy. Biomimicry engineers need a strong background in life science. They study biology, botany, and zoology. They also take courses such as computer-aided design, math, and materials science.

Biomimicry engineers imitated what geckos do to design a robot that could climb walls.

Animal Behaviorist

How does a rabbit avoid predators? Why do dolphins travel in pods? What does a red-winged blackbird do to attract a mate? If you want to find the answers to these questions and more, then you might want to become an animal behaviorist. Animal behaviorists study why animals do what they do. They are interested in how an animal responds to its environment and also how it responds to other living things. Some animal behaviorists work behind the scenes in a zoo or aquarium. Others conduct research in a laboratory or teach at a college or university. Animal behaviorists take courses in ecology, psychology, and ethology, which is the study of animal behavior.

An animal behaviorist studies animals in all stages of their life cycles to see how behavior changes over an animal's lifetime.

Environmental Engineer

Do you enjoy finding solutions to problems? Are you interested in finding better ways to help care for our planet? You may want to consider a career in environmental engineering. Environmental engineers develop new ways to treat water to make it safe for drinking. They work to improve recycling facilities, and devise ways of preventing landfills from contaminating soil and groundwater. Some help manufacturing plants reduce air pollution or deal with harmful wastes. They also inspect facilities to make sure they are following environmental regulations. Environmental engineers must earn a college degree and take many courses in math, physics, chemistry, biology, and geology. Many engineers enter cooperative engineering programs that provide college credit for job experiences.

Environmental engineers use their knowledge of engineering, geology, biology, and chemistry to craft solutions to environmental problems.

Park Ranger

Do you like to work outdoors and care for nature? Then a career as a park ranger may be for you. Some park rangers collect environmental data on wildlife and plant populations. Others teach the public about how to enjoy and protect nature. They explain how pollution, litter, and climate change affect the parks. Some even provide law enforcement and firefighting services. Park rangers have a college degree in park and recreation management, conservation, botany, geology, wildlife management, or forestry. Many park rangers serve as interns in National Parks before being hired as full-time rangers. Rangers who manage sites often have additional degrees.

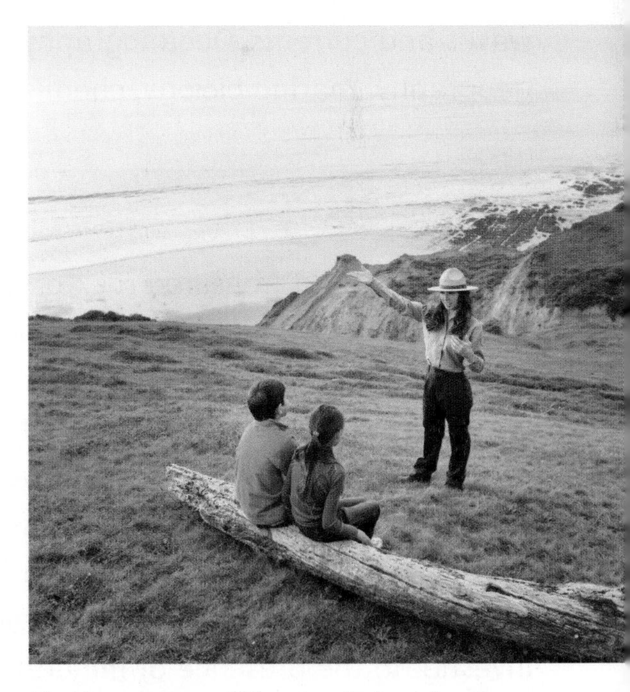

Park rangers fill a variety of roles in our National Parks.

Earth Science Careers

Oceanographer

A career in oceanography is perfect for anyone who is interested in Earth's oceans. There are different branches of oceanography. Biological oceanographers study the vast number of organisms that live in oceans. They want to know how the organisms grow, meet their needs, and adapt to the ocean environment. Chemical oceanographers study the makeup of salt water and how it changes and interacts with land and air. Geological oceanographers study the ocean floor and the development of underwater mountains, valleys, and canyons. Physical oceanographers study ocean waves and currents. Oceanography students study geology, geography, marine biology, chemistry, and physics.

Some oceanographers use submersibles to study life and landforms below the ocean's surface.

Mineralogist

Do you have a treasured rock or mineral collection? If so, then you may want to consider a career as a mineralogist. Mineralogists and gemologists explore the chemical and physical properties of minerals, which are naturally occurring solid substances. They examine the structure of minerals and study where minerals form in nature. They use this information to help companies locate likely sites for mines. Some mineralogists, called soil mineralogists, investigate the presence of minerals in soil. Others find ways to grow artificial mineral crystals for use in industry. Becoming a mineralogist requires a college degree. Students who wish to become mineralogists may study Earth's landforms, geology, chemistry, and math.

Mineralogists work both in the field and in labs.

Hydrologist

Water is one of Earth's most valuable resources—and there is a lot of it! If you appreciate the importance of water, you may want to be a hydrologist. Hydrologists study the movement and occurrence of water on and under Earth's surface. They apply scientific knowledge and mathematical principles to solve problems of water quality, quantity, and availability. Hydrologists also explore the properties of water and its role in the water cycle. Hydrologists have a number of different responsibilities. Some work with communities to help find water supplies for drinking and irrigation. Others work with government agencies to help clean up polluted or contaminated water. Much of a hydrologist's work depends on computers for organizing, summarizing, and analyzing data collected in the field. Students considering a career as a hydrologist need a strong background in math, statistics, geology, physics, computer science, chemistry, and biology.

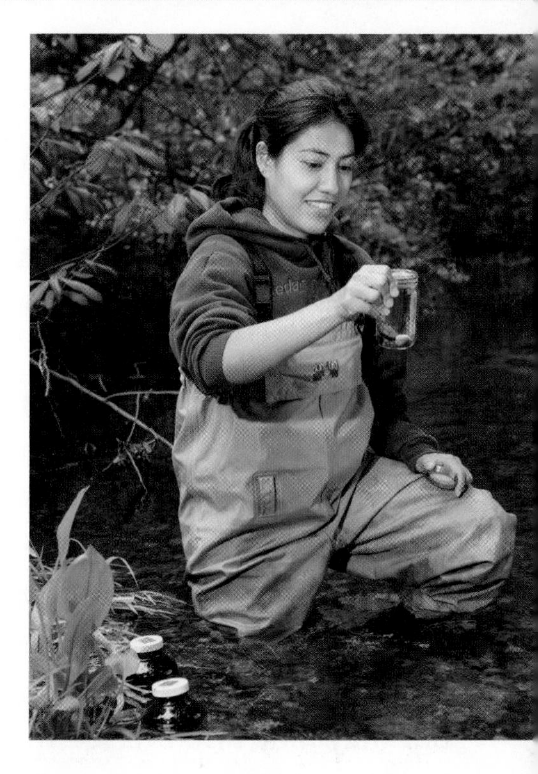

Testing water for pollutants is one of the jobs of a hydrologist.

Soil Scientist

Soil is another valuable resource. Soil scientists explore nothing but soil. Soil scientists perform a variety of different tasks. If a farmer wants to know which crops would grow best in a particular area, a soil scientist can help the farmer figure it out. When data is collected about soil samples on Mars, it is the job of a soil scientist to analyze them. Soil scientists even advise builders on the best areas for development or safe areas in which to build landfills or store harmful wastes. Soil scientists need a strong background in environmental science, geology, mineralogy, chemistry, and agricultural science.

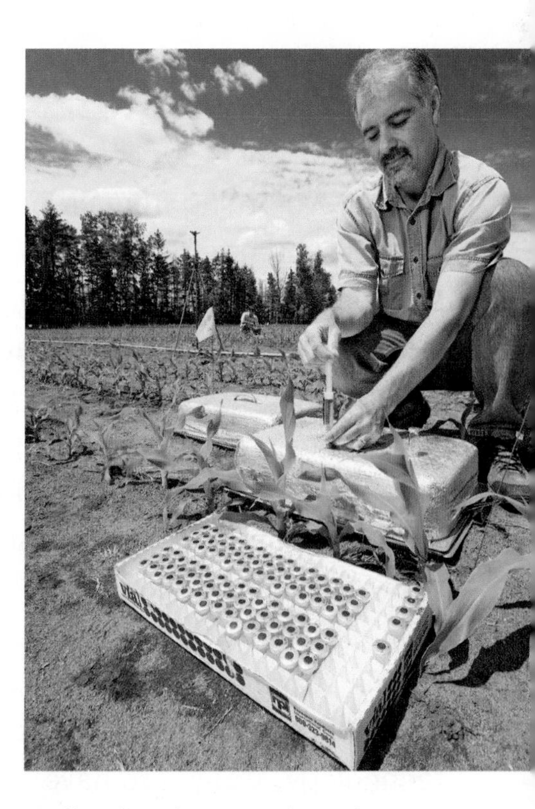

Soil scientists analyze how soils affect environmental quality and human health.

Structural Engineer

Earthquakes and other natural disasters can cause millions of dollars in damage. If you have an interest in planning and designing homes, skyscrapers, dams, and bridges that can withstand even the strongest earthquake, then a job as a structural engineer could be for you. Structural engineers have many duties, but one of them is applying their knowledge of materials, forces, soil, and Earth science to the design of buildings that can keep people safe and reduce damage in earthquake-prone areas. Not only do structural engineers design these structures, but they may also oversee their construction and help keep them on budget. Students who wish to become structural engineers take courses in physics, statistics, math, and geology.

Structural engineers work to design buildings that can withstand the forces of an earthquake.

Climatologist

Climatologists are concerned with how climate affects everything from crops to construction. They research temperature, wind, and precipitation patterns over many years. They use radar, satellite systems, and airplanes to collect data. They use these data to develop computer models that help predict weather patterns and dangerous weather conditions including hurricanes, heat waves, and snowstorms. Some study the composition of the atmosphere and monitor variations in air pollution. Climatologists have a strong background in environmental science, meteorology, physics, and computer technology. Many also have advanced degrees in statistics or math.

Climatologists collect and analyze data from ice cores, soil, water, air, and plant life to find patterns in weather.

Petroleum Geologist

Oil is an important resource in today's world, but, unlike water, it can be difficult to find. Petroleum geologists search for places to drill for oil. These scientists use their knowledge of Earth's rocks and soil to help oil companies locate areas where oil can be extracted. They work closely with petroleum engineers, who design the equipment used to extract the oil from below the ground. Although a petroleum geologist spends some time in the field, he or she works mostly in an office or laboratory using computer programs to analyze data. Most petroleum geologists earn a degree in geology and then go on to earn a second degree in petroleum geology.

Petroleum geologists use their knowledge of Earth and its structures to find and extract oil and natural gas.

Aerospace Engineer

If you have an interest in the technology behind aircraft, then you should consider a career in aerospace engineering. There are several types of aerospace engineers. Aeronautical engineers design aircraft, such as airplanes, helicopters, and missiles. Others, called astronautical engineers, design spacecraft such as space shuttles, launch vehicles, and satellites. After a prototype of a design is built, an aerospace engineer will perform tests and analyze the results to make sure that it operates properly. He or she may also help oversee the manufacturing process. Aerospace engineers take classes in math, physics, computer-aided design, drafting, and aerodynamics, which is the study of how moving air affects objects.

Aerospace engineers design all kinds of aircraft and spacecraft, from drones to satellites.

Physical Science Careers

Biochemist

Chemical reactions happen in nature. They also happen inside your body. If you are interested in studying these chemical reactions and want to help develop new products that improve people's health and wellness, then a career in biochemistry may be a good fit for you. Biochemists help research and develop new vaccines and medications. They investigate how heating and processing affect the nutritional value of different foods. Some help companies develop new products such as drinks that help refuel an athlete's body. Students who wish to become biochemists must have a strong background in both biology and chemistry. They may also take courses in anatomy, physiology, physics, and computer science.

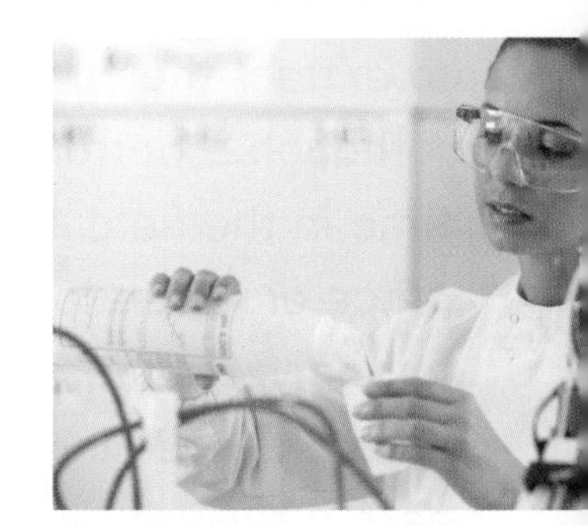

Some biochemists work in the cosmetic industry developing products that do not irritate skin.

Barge Pilot

Would you enjoy the challenge of navigating a large ship through narrow waters? That is exactly what a barge pilot must do. Barge pilots steer large cargo ships, called barges, up and down rivers and into and out of a port. They must have knowledge of water currents and the depth of the water at certain points. They must also be able to maneuver the barges around tight bends and curves. To do this maneuvering, understanding forces and motion and knowing when to accelerate and change direction is a must. Barge pilots earn a degree from a special academy and then undergo a year of additional training. During their training, they take courses in math, physics, computer science, and engineering.

Understanding how the forces acting on a boat affect motion is the job of a barge pilot.

Physical Therapist

Are you interested in how muscles, bones, and other tissues work together to help the body move? Do you want to help people regain strength and manage pain? If your answer is yes, then you may want to consider a job as a physical therapist. After a serious injury or illness, some patients need help strengthening and increasing mobility of certain parts of the body. As a physical therapist, you would devise a plan for a patient that includes stretching, mobility exercises, and strength training. Using a variety of tools and techniques and applying concepts of forces and torque, or rotational movement, your goal would be to help the patient restore function and reduce pain.

Physical therapists work in a variety of places, including hospitals, clinics, doctor's offices, and nursing homes. Some may specialize in sports-related injuries. Physical therapists also use numerous tools, such as balls, weights, crutches, balance beams, and resistance bands. Becoming a physical therapist can take up to seven years and requires a special license. As they earn their degree, students take courses in the fields of anatomy, physiology, biology, chemistry, and physics. They also receive hands-on experience through on-site training.

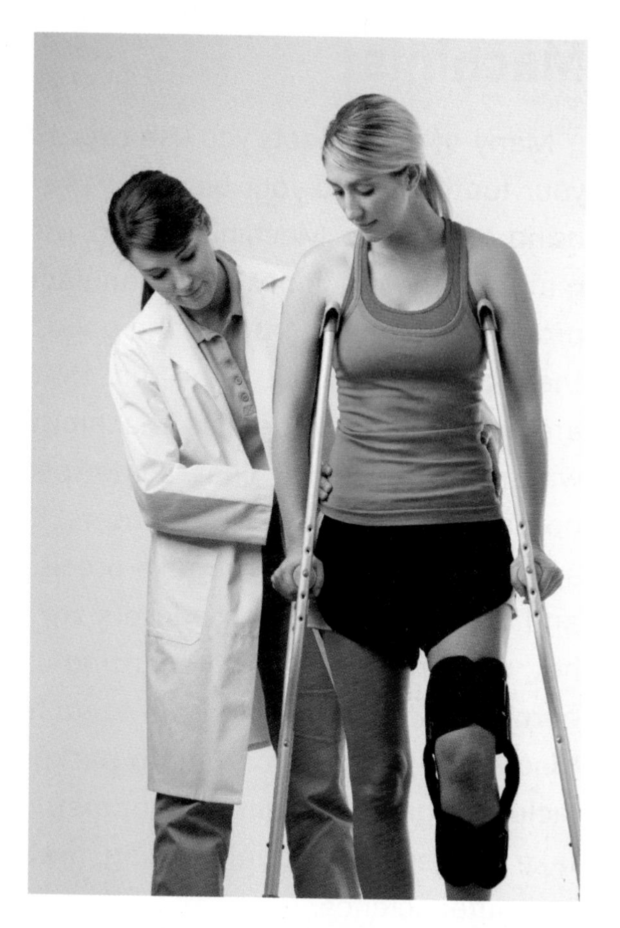

Physical therapists must understand how forces affect how the body moves.

Some physical therapists focus on working with children.

Machinist

Many of the objects you use every day, from your toothbrush to your bicycle, were made using a machine. Machinists build, use, and maintain the machines that manufacture these products. They also inspect the goods that the machines produce to be sure that there are no problems or defects. Machinists work with a variety of machines. Some specialize in operating machines that use lasers. These lasers are used to cut metal or make precise holes and other shapes. Because lasers are extremely hot, uneven edges from the cutting usually burns or melts, requiring little to no finishing. Machinists have a strong math background, including geometry and trigonometry. They also take classes in blueprint reading, drafting, and computer science.

Many of the tools machinists use are controlled by computer. The machinist programs the computer using information from blueprints in order to produce the needed parts.

Electrician

The next time you flip on a light switch or turn on your computer, thank an electrician! Electricians are responsible for safely wiring buildings, allowing access to electricity that provides light, heat, and power to appliances. Electricians also inspect electrical systems and repair or replace faulty wiring, transformers, circuits, fixtures, and breakers. They must be able to read and interpret blueprints and are responsible for making sure all electrical components meet safety codes. Electricians attend technical school and have on-the-job training. They have a strong background in math, blueprint reading, and electrical theory.

Wiring circuits to allow electricity to safely and efficiently flow to all areas of a home is one job of an electrician.

Power Plant Engineer

Are you fascinated with how electricity is generated and transmitted? Power plant engineers deal with the generation, transmission, distribution, and utilization of electric power. These engineers not only build the generators that convert energy from wind, water, or steam to electricity, but also maintain them. They often lead teams of engineers who must quickly and efficiently troubleshoot problems and devise solutions in order to keep the electricity flowing. To become a team leader, a power plant engineer must have a degree in engineering plus two or more years experience working with other engineers in a power plant.

Power plant engineers work to make sure the flow of electricity is maintained from power plant to homes and businesses.

Sonar Technician

Are you interested in working on a ship or submarine, using special equipment to detect objects in the deepest parts of the ocean? If so, then a career as a sonar technician might be for you. The word *sonar* is an acronym that stands for "*so*und *n*avigation *a*nd *r*anging." Sonar technicians operate equipment that sends out sound waves or pulses that bounce off objects. They analyze and interpret the data that are returned, estimating where and how far away an object may be. In addition to operating sonar equipment, these technicians also identify and fix malfunctions. In order to become a sonar technician, a person must join the United States Navy. The Navy is a branch of the military that conducts operations at sea. While in the Navy, sonar technicians take classes and gain hands-on training with sonar equipment.

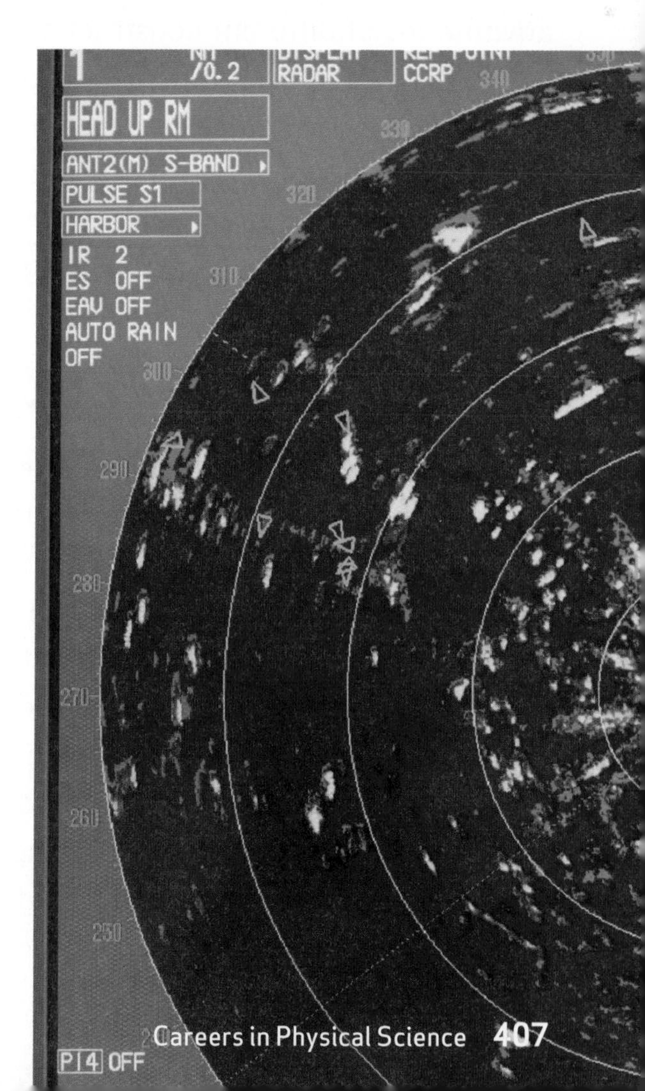

Sonar technicians can detect other vessels, objects on the sea floor, and even ocean animals.

Engineering Careers

Civil Engineer

Do you have an interest in construction, architecture, and design? If so, you might want to consider a career as a civil engineer. Civil engineers design and maintain roads, bridges, water systems, and energy systems. They also design and maintain public facilities such as ports, mass transit systems, and airports. They even design and construct structures in our national parks. In addition to having a good understanding of forces and motion, civil engineers must have knowledge of the advantages of different construction materials and which ones would be best to use for a specific purpose. Civil engineers usually work in an office or at a construction site. Students interested in becoming civil engineers need a solid background in math and the physical sciences. After earning an engineering degree, civil engineers must pass a test to be licensed.

Updating structures such as bridges is an important challenge for civil engineers.

Electrical and Electronics Engineers

Do you have an interest in the science behind electricity, or in the technology that powers electronics such as computers and global positioning systems? If so, you may want to consider electrical or electronics engineering. These engineers plan, design, produce, and test electrical equipment. The equipment could be inside the power plants that generate electricity around the world. They also design communication systems, certain components of automobile engines, and radar and sonar equipment. While earning their degrees, students study math, physics, computer-aided drafting, circuitry, and computer sciences.

Engineers develop, design, test, and manage the manufacture of electrical and electronic components used in solar systems.

Mechanical Engineer

Are you interested in knowing what makes the blades spin on a ceiling fan, or how the parts inside a car work together to put it in motion? How would you like to design and plan the world's tallest or fastest roller coaster? Mechanical engineers use their understanding of energy, materials, forces, and motion to design and manufacture machines and equipment of all types. They design, develop, produce, and test everything from simple batteries to complicated electric generators to robots. They also work in manufacturing and agricultural production and the maintenance of engineered systems. Mechanical engineers must have a strong understanding of physics, technology, and math. They must also be creative and good at solving problems.

Mechanical engineers are involved in the design and testing of wind turbines.

Biomedical Engineer

Biomedical engineers work with doctors, therapists, and researchers to develop systems, equipment, and devices to solve medical problems. They design and develop artificial limbs and other prosthetics. Some work with athletes to analyze injuries to determine how exercise affects healing tissues. Recently, biomedical engineers have developed a fabric that can serve as a base for growing the cartilage that cushions the joints between bones. Tissue engineering is a growing segment of biomedical engineering. Through research and testing, bioengineers hope to manufacture whole organs to replace those that are diseased or injured. Students in biomedical engineering take courses in human body structures and function, biology, communications, and computer science.

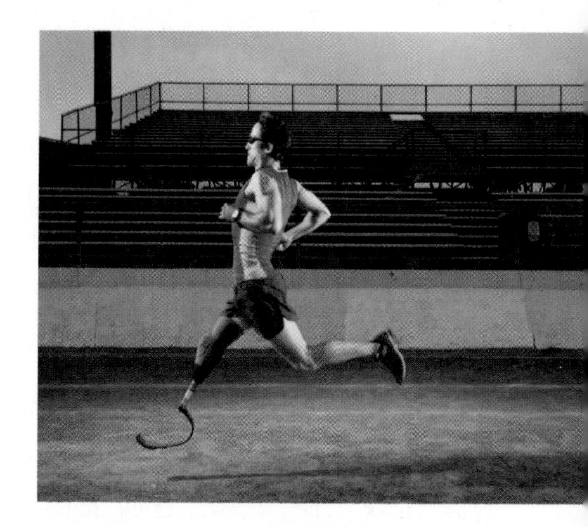

Biomedical engineers use plastics and carbon-fiber composites to make modern replacement limbs.

Manufacturing Engineer

Are you interested in finding ways to complete tasks more efficiently? Manufacturing engineers study production systems, looking for ways to improve efficiency or reduce costs. They also look for ways to reduce the impact of manufacturing on the environment. These engineers have skills that apply to almost any manufacturing situation. They work in the automotive industry, the food processing industry, the oil and gas industry, and the pharmaceutical industry, to name just a few. They often work in teams with other engineers to diagnose and solve problems. Manufacturing engineers major in electrical, mechanical, or production engineering, and often spend time in a cooperative engineering program in which they get on-the-job training.

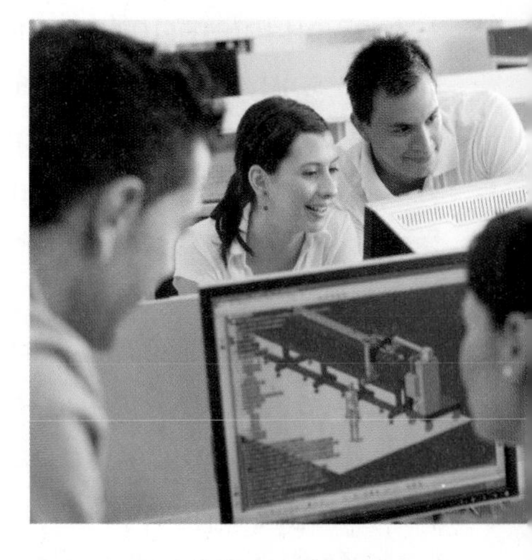

Manufacturing engineers look for new and better ways to produce goods with less waste.

Chemical Engineer

Chemical engineers apply the principles of chemistry, biology, and physics to solve problems involving the production of chemicals, fuel, drugs, food, and other products. For example, a chemical engineer may develop ways to turn recycled materials into shoes or clothing, or work for a pharmaceutical company to develop a coating for pills and capsules. Chemical engineers also work closely with other engineers to solve problems. For example, they work with environmental engineers to help clean up water or land pollution. They work with agricultural engineers to help develop fertilizers and pesticides that are better for the environment. Becoming a chemical engineer requires at least four years of college. In addition to chemistry, students take courses in biology, physics, math, and computer science. Many participate in programs that offer supervised job experience as well as academic training.

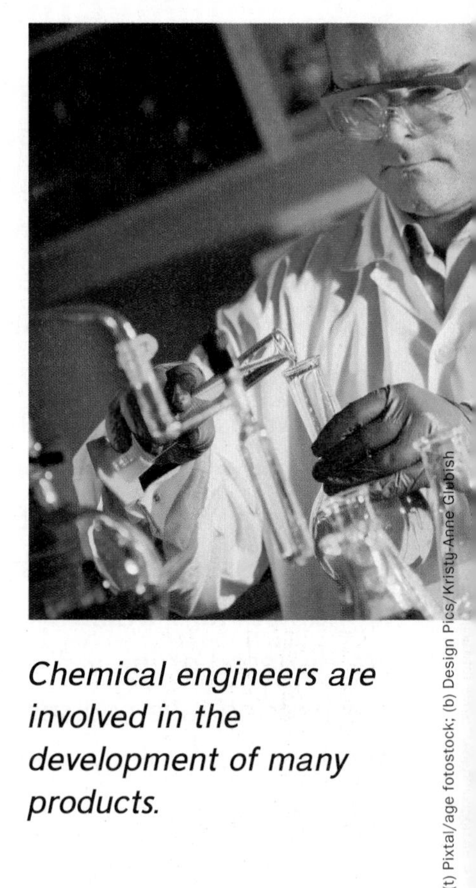

Chemical engineers are involved in the development of many products.

(t) Pixtal/age fotostock; (b) Design Pics/Kristy-Anne Glubish

Automotive Engineer

If you have an interest in ground-based vehicles of all kinds, then you might enjoy a career as an automotive engineer. Automotive engineers design passenger cars, commercial vehicles, and off-highway vehicles. They work on all systems, including the braking, electrical, steering, and fuel systems. The goal of an automotive engineer is to design a vehicle that is comfortable, functional, and fuel-efficient. Most importantly, they want to make vehicles that are safe. Therefore, automotive engineers are also involved in anticipating or detecting problems with a design and taking steps to fix them. In order to become an automotive engineer, you will need a four-year degree in mechanical engineering or a similar field. Automotive engineers have a strong background in math, physics, and chemistry.

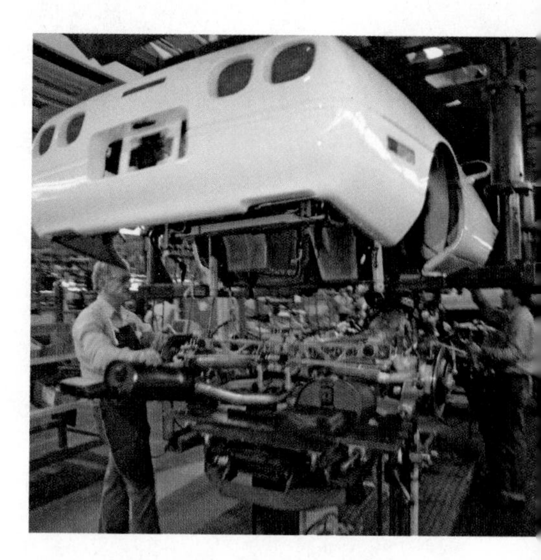

Automotive engineers are involved in planning and designing safe and efficient vehicle production processes.

Agricultural Engineer

Do you want to help develop farming practices that conserve Earth's resources and keep land and water safe? Agricultural engineers are involved in the entire farm-to-table process, from crop production to harvesting, processing, and food storage. They work with both plants and livestock to investigate ways to improve product quality. Some agricultural engineers even help design the equipment that is used for farming, such as irrigation systems and seed planters. Becoming an agricultural engineer requires at least four years of college that includes classroom and laboratory work, as well as fieldwork. Students should also have a background in math, biology, chemistry, and physics.

Agricultural engineers devised irrigation systems that help conserve water.

Overview
Science and Engineering Guide
Science exploration uses many tools to measure data and to interpet information. Learn about them in this section!

Alistair Berg/Digital Vision/Getty Images

Science Guide

Using Maps

A *map* is a drawing that shows an area from above. It represents a large area of Earth's uneven surface on a smaller, flat surface. Maps show the location of surface features. They show natural features like rivers, lakes, valleys, and mountains. They can also show features made by people such as roads, buildings, bridges, and parks.

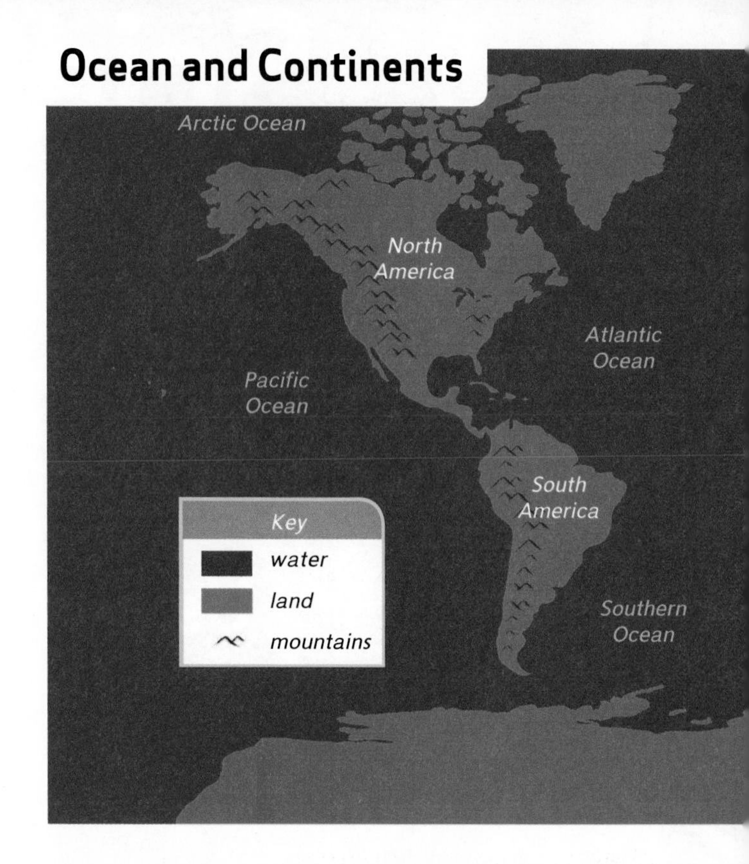

Types of Maps

There are many different kinds of maps.

Physical maps show the features of Earth's land and water. Mountains, deserts, and plains are often shown on a physical map. You might describe the shape of an ocean or identify the name of a continent from a physical map.

Political maps show boundaries. You can tell how places are divided into countries, states, or cities on these maps. They often indicate the location of major cities and large bodies of water.

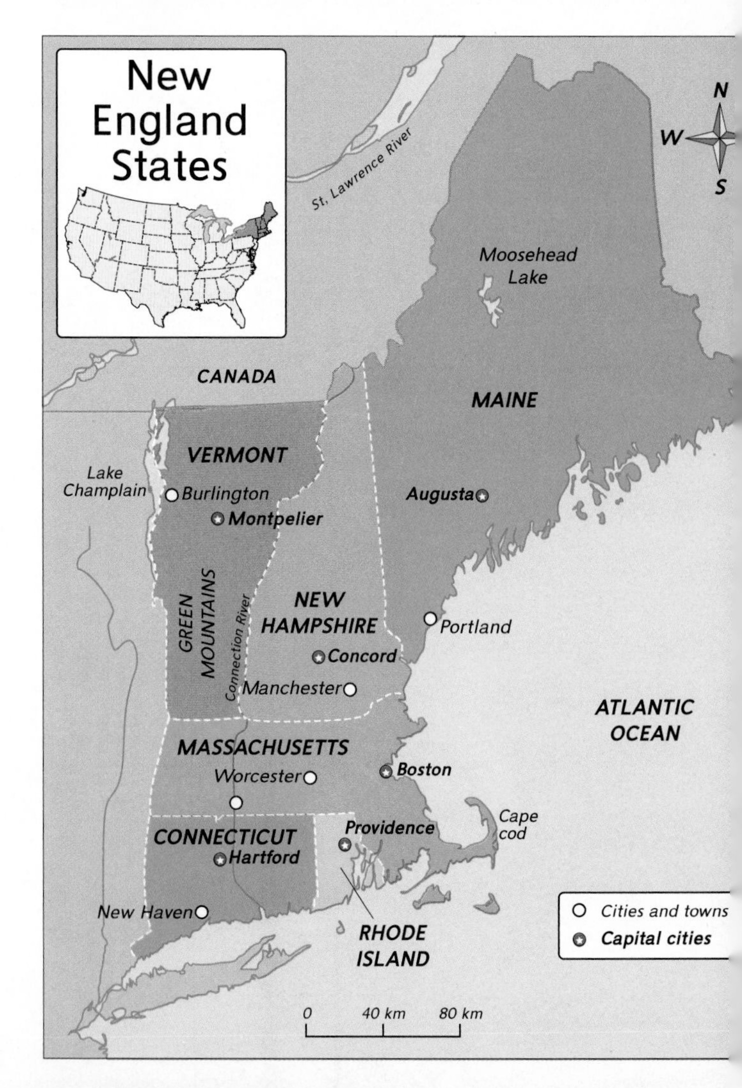

A common type of map is a road map, which shows roads and highways. They are often used to plan routes to travel from one place to another. Some show features of towns and sites such as buildings and parks.

Some maps have special or more specific uses. They might show climate zones of the United States, world religions, or the rides in an amusement park. Some use shading or lines to represent changes in elevation, the height of the land above sea level. A map that uses shading to show elevations is called a relief map. The shading makes the map look as if it has three dimensions: length, width, and height.

A topographical map uses lines to show elevation. Each line represents a different elevation labeled with a number. In addition to elevation, contour lines can tell you how steep or gradual a slope is. Contour lines that are close together represent a rapid change in elevation. Contour lines that are far apart represent a gradual change in elevation.

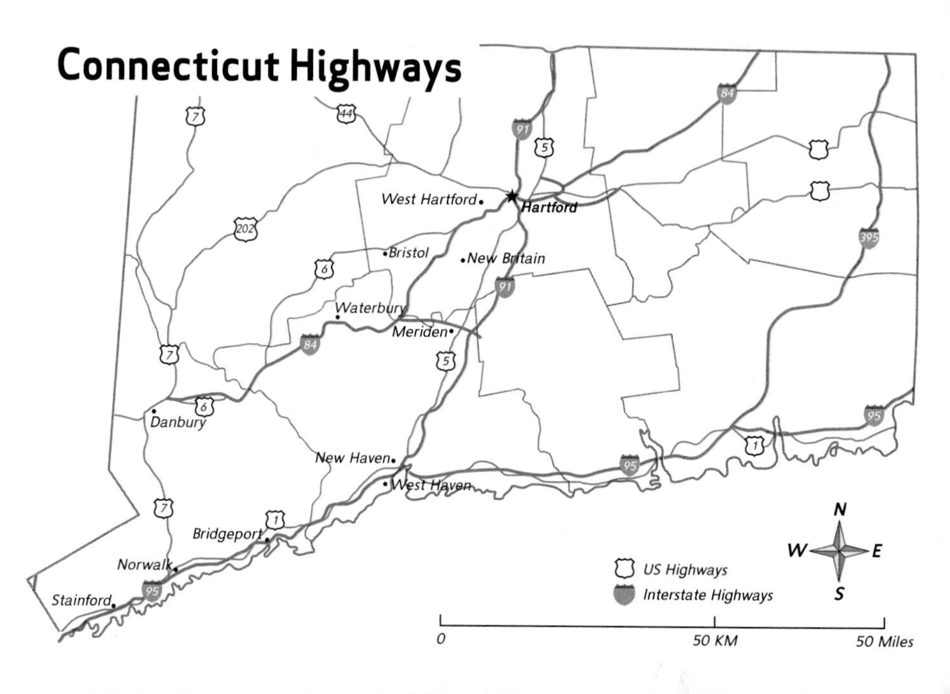

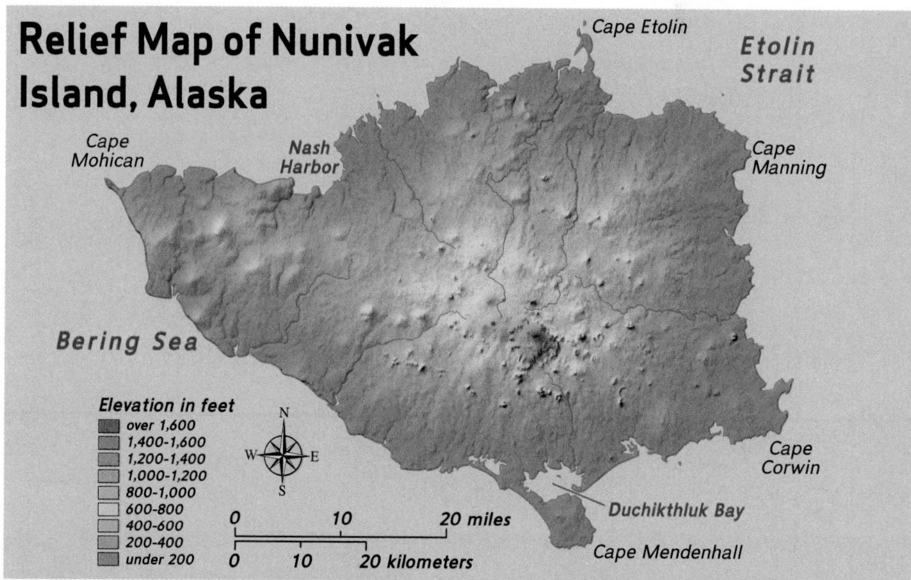

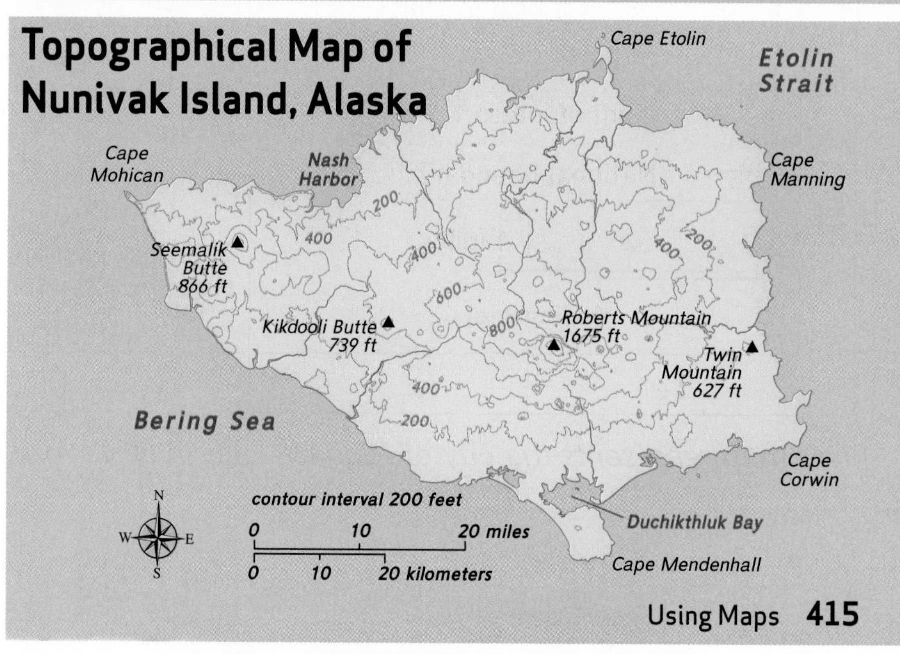

Reading Maps

Maps represent real places on Earth. To read a map, start with the title, which tells you what place on Earth the map represents. From the title you know the map below shows the city of Niagara Falls in New York.

Niagara Falls, New York

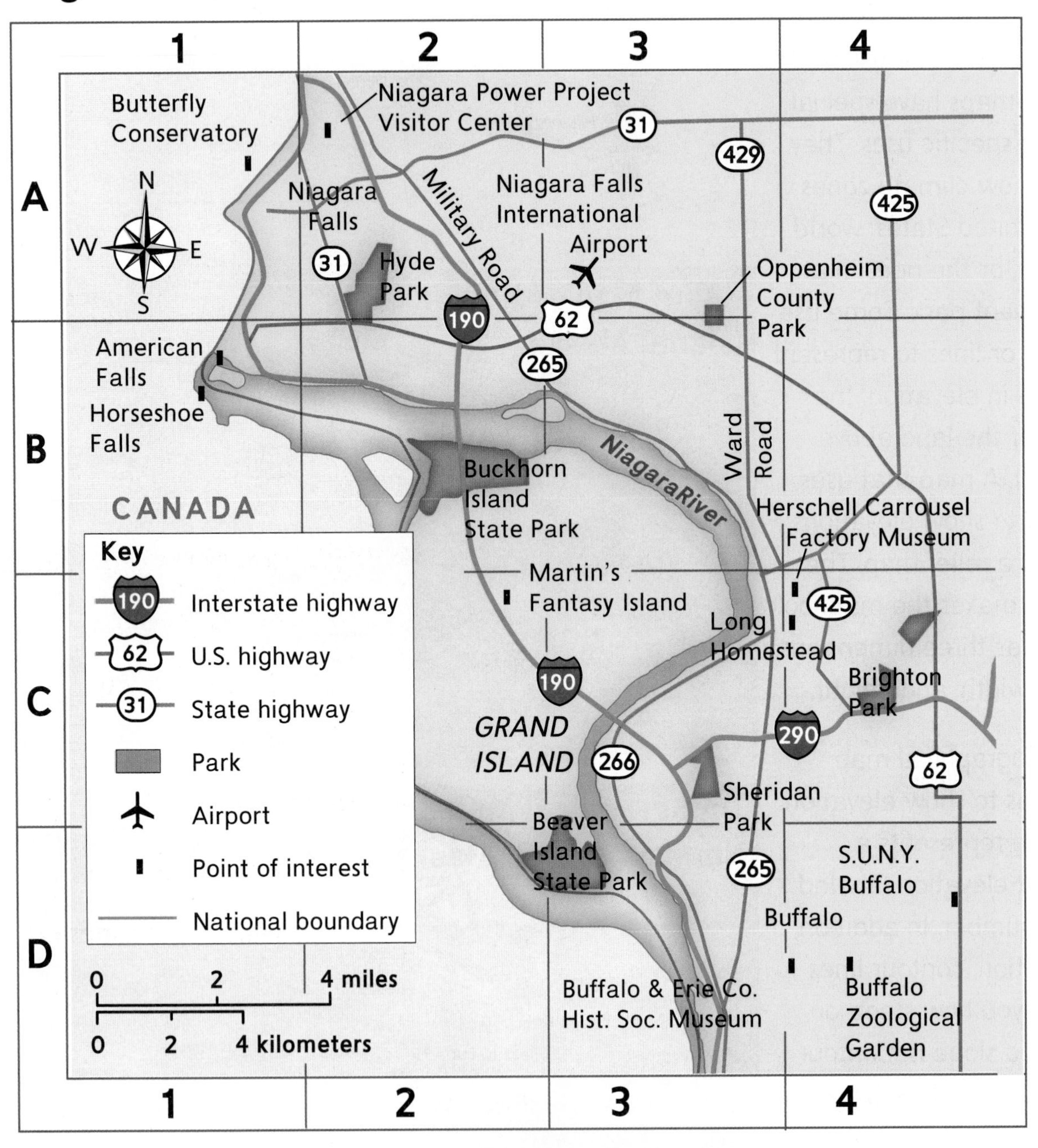

This map represents the city of Niagara Falls in New York.

Map Features

Compass Rose Maps also show the locations of real features. That means they show directions, which include north (N), south (S), east (E), and west (W). Most maps show which way north is. The map shown here indicates all four directions on a compass rose. Notice that the compass rose shows north as up. Using that information, you can tell that American Falls is located north of Horseshoe Falls.

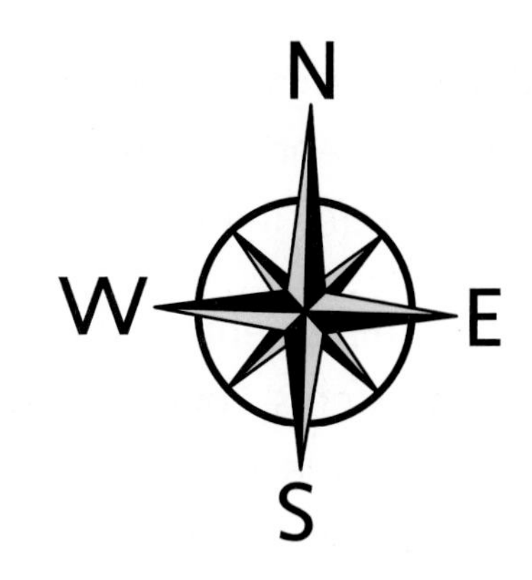

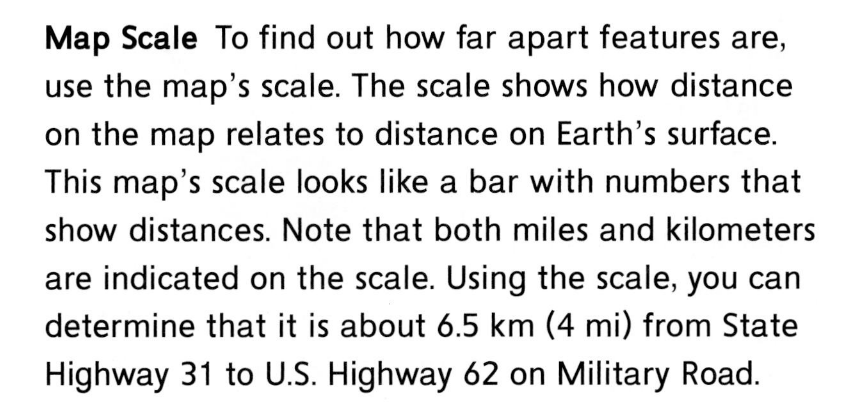

Map Scale To find out how far apart features are, use the map's scale. The scale shows how distance on the map relates to distance on Earth's surface. This map's scale looks like a bar with numbers that show distances. Note that both miles and kilometers are indicated on the scale. Using the scale, you can determine that it is about 6.5 km (4 mi) from State Highway 31 to U.S. Highway 62 on Military Road.

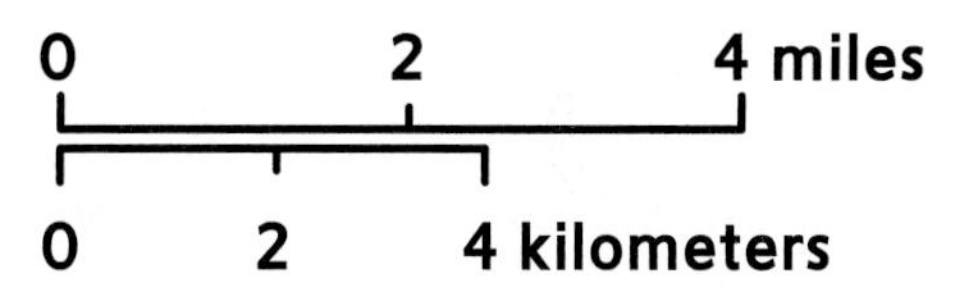

Key The colors and shapes on a map are symbols. A key, or legend, shows what each symbol represents. The key helps you locate features. Many maps use colored lines to show roads. Three different kinds of roads are identified in the key: interstate highways, U.S. highways, and state highways. Green areas indicate parks. An airplane shape represents an airport. Black rectangles indicate points of interest.

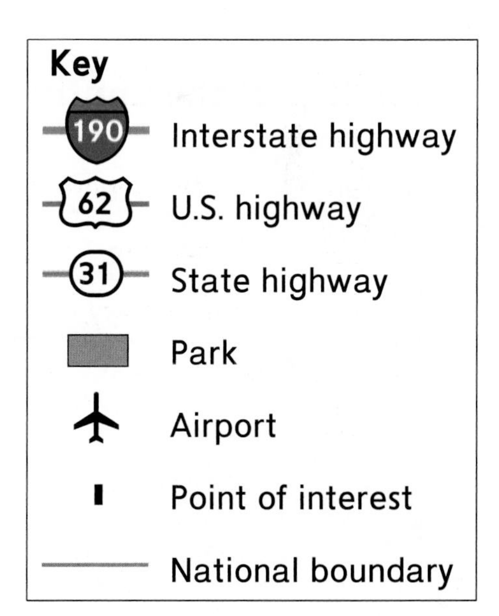

Grid Many maps have numbers and letters along the top and side. The letters and numbers form a grid that can help you find locations. The Buffalo Zoological Garden, for example, is located in section D4 on this map. To find it, place a finger on the letter D along the left side of the map and another finger on the number 4 at the top. Move your fingers straight across and down the map until they meet.

How Scientists Use Maps

Many scientists depend on maps. They use maps to organize and analyze information. They also use maps to share information with others.

Maps help scientists understand organisms and ecosystems. For example, scientists studying lizards in Madagascar make maps to show where the organisms are located. The maps help them compare where they predict lizards are likely to live and where they actually observe them. Their maps help them find patterns in their data and understand what animals need to survive.

Scientists use maps to show where different kinds of landforms, soils, and rocks are found. Analyzing and sharing these maps help scientists better understand Earth's features and how they change.

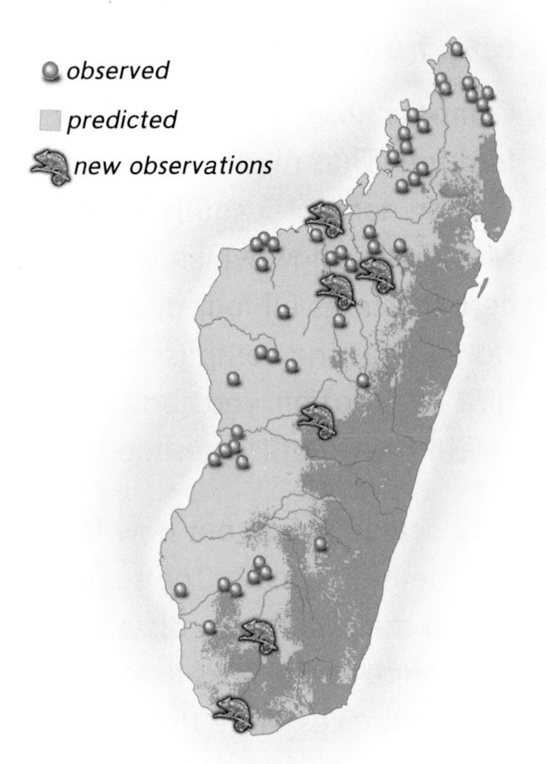

This map of Madagascar helps scientists understand lizards and their ecosystems.

Relief Map of the Continental U.S.

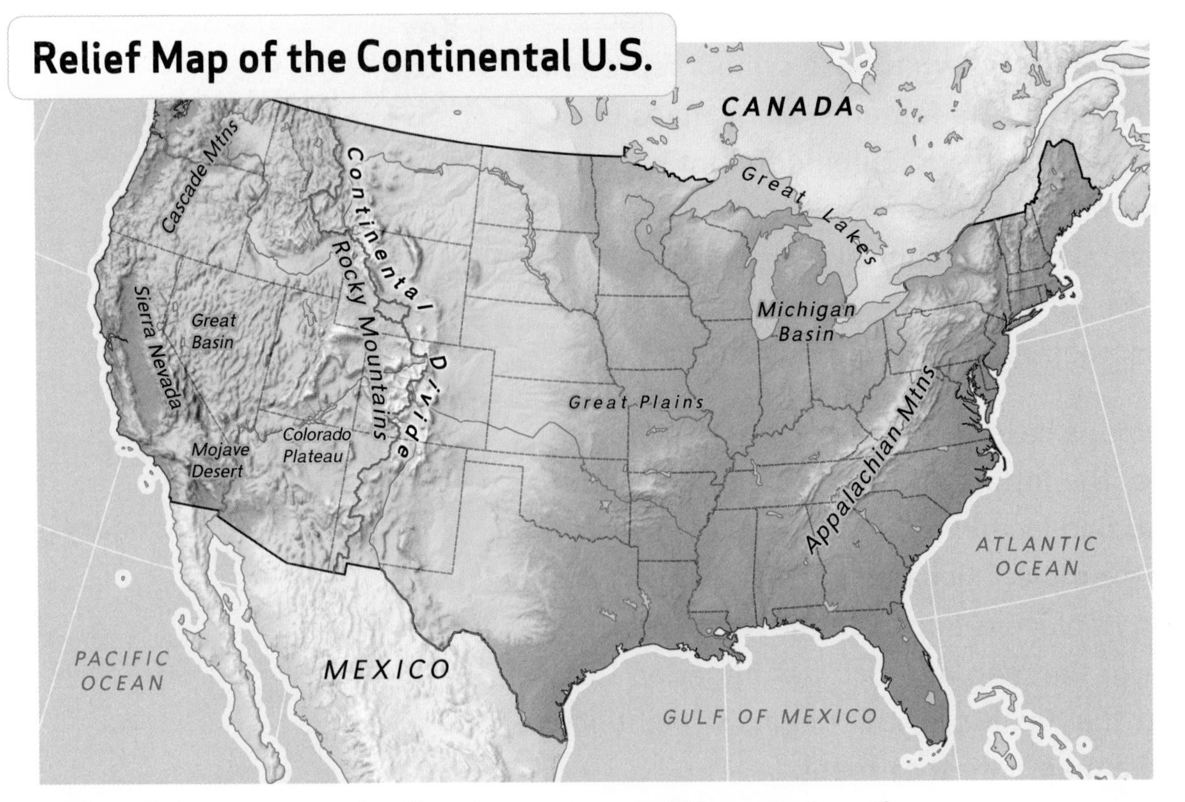

This relief map uses shading to represent different elevations of major landforms. You can use it to find out where most mountains are found in the continental United States.

Meteorologists collect a great deal of data about the weather. They organize the data they collect on weather maps. Weather maps show weather conditions for a certain area. They use symbols to show information about air temperature, precipitation, clouds, and winds. Some show fronts and the direction in which they move. They help scientists analyze their data and forecast the weather. They also help scientists share weather information with others who need it.

Weather maps help farmers decide when to harvest their crops. They help pilots fly airplanes safely and guide tourists on when to travel. They help people decide what to wear and which outdoor activities they can do each day.

Skill Builder

Read a Map

On most weather maps, the symbol for a cold front is a blue line with triangles. The symbol for a warm front is a red line with half circles. The triangles and half circles point in the direction the front is moving.

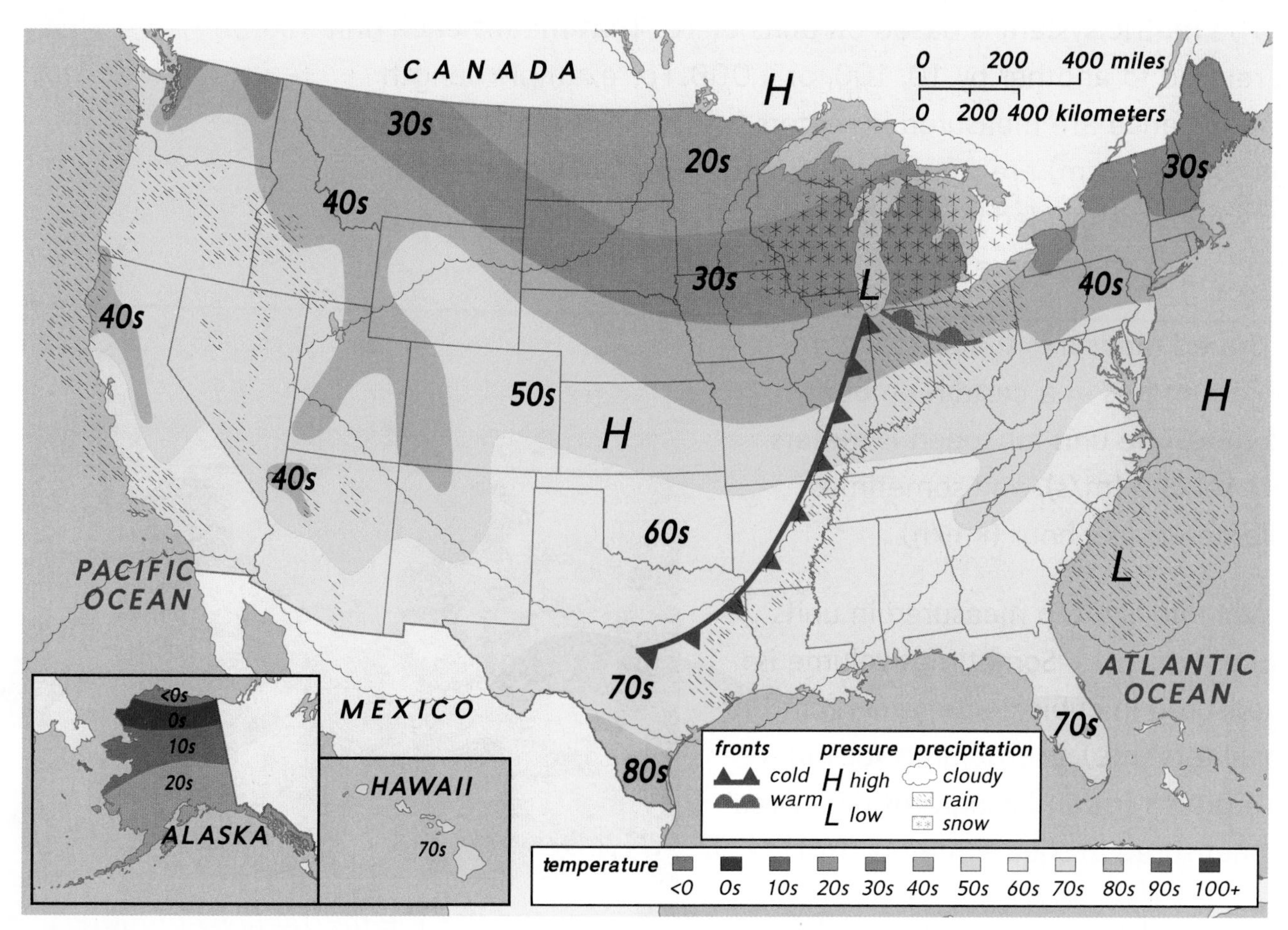

Weather maps help scientists organize weather data and make forecasts.

Using Math

Units of Measurement

Math is an important part of science. One way scientists use math is to measure objects and events. They measure to determine length, volume, area, mass, weight, and temperature.

A measurement includes both a number and a unit. A unit of measurement on which people agree is known as a *standard unit*. Scientists use a common system of standard units called the *metric system*. This system is also called the International System of Units, or SI.

The metric system is based on units of 10. That means each unit is related to another by 10, 100, or 1,000. For example, length and distance are measured in meters (m). Millimeters (mm) and centimeters (cm) are smaller units made from parts of meters. One meter is divided into 100 cm and 1,000 mm. Kilometers (km) are larger units. There are 1,000 m in 1 km.

Speed measures the distance an object moves in a certain amount of time. The unit for speed is meters per second (m/s), and sometimes kilometers per hour (km/h).

Volume is often measured in units called liters (L). Sometimes volume is measured in cubic centimeters (cm^3) or milliliters (mL). Metric units for mass are grams (g) and kilograms (kg). Area is measured in square meters (m^2).

The metric unit for weight and force is the newton (N). Temperature is measured in degrees Celsius (°C).

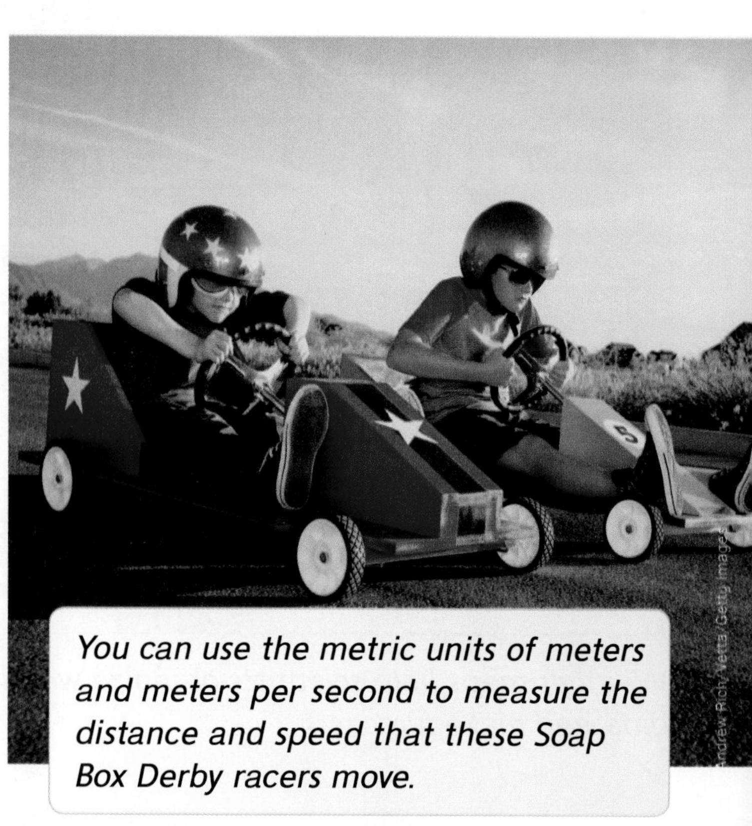

Andrew Rich/Vetta/Getty Images

You can use the metric units of meters and meters per second to measure the distance and speed that these Soap Box Derby racers move.

Another system of standard units used in the United States is the *customary system.* In this system, length and distance are measured in inches (in.), feet (ft), and miles (mi). Fluid ounces (fl oz) are used to measure liquid volume. Temperature is measured in degrees Fahrenheit (°F).

Table of Measurements

International System of Units (SI)	Tools	Customary Units
Temperature Water freezes at 0°C (degrees Celsius) and boils at 100°C.		**Temperature** Water freezes at 32°F (degrees Fahrenheit) and boils at 212°F.
Length and Distance 1,000 meters (m) = 1 kilometer (km) 100 centimeters (cm) = 1 meter (m) 10 millimeters (mm) = 1 centimeter (cm)		**Length and Distance** 5,280 feet (ft) = 1 mile (mi) 3 feet (ft) = 1 yard (yd) 12 inches (in.) = 1 foot (ft)
Volume 1,000 milliliters (mL) = 1 liter (L) 1 cubic centimeter (cm^3) = 1 milliliter (mL)		**Volume** 4 quarts (qt) = 1 gallon (gal) 2 pints (pt) = 1 quart (qt) 2 cups (c) = 1 pint (pt) 8 fluid ounces (oz) = 1 cup (c)
Mass 1,000 grams (g) = 1 kilogram (kg)		**Mass and Weight** 2,000 pounds (lb) = 1 ton (T) 16 ounces (oz) = 1 pound (lb)
Weight 1 kilogram (kg) weighs 9.81 newtons (N).		

Changing Metric Units

Suppose you measure the length of an object as 4 meters, but you need to share your results with the class in centimeters. You can change your measurement from one unit to another. You can do this for different units of the same kind of measurement, such as different units of length. To change metric units, you can multiply or divide by powers of 10.

Use the table below to find out how to change meters to centimeters. Look for a row with "meters" in the first column and "centimeters" in the second column. Then look under "What to Do" in that row. You should multiply by 100, because 1 m equals 100 cm.

4 meters x 100 = 400 centimeters

Did You Know?

Atoms, which are some of the smallest pieces of matter, are too tiny to be measured with millimeters and centimeters. To measure the lengths of atoms, scientists use nanometers. One nanometer is only one billionth of a meter long. One billion nanometers are equal to one meter.

How to Change Metric Units of Length or Distance

Change	To	What to Do	Why
meters (m)	centimeters (cm)	multiply by 100	1 m = 100 cm
centimeters (cm)	meters (m)	divide by 100	
meters (m)	millimeters (mm)	multiply by 1,000	1 m = 1,000 mm
meters (m)	kilometers (km)	divide by 1,000	1,000 m = 1 km
centimeters (cm)	millimeters (mm)	multiply by 10	1 cm = 10 mm
millimeters (mm)	centimeters (cm)	divide by 10	

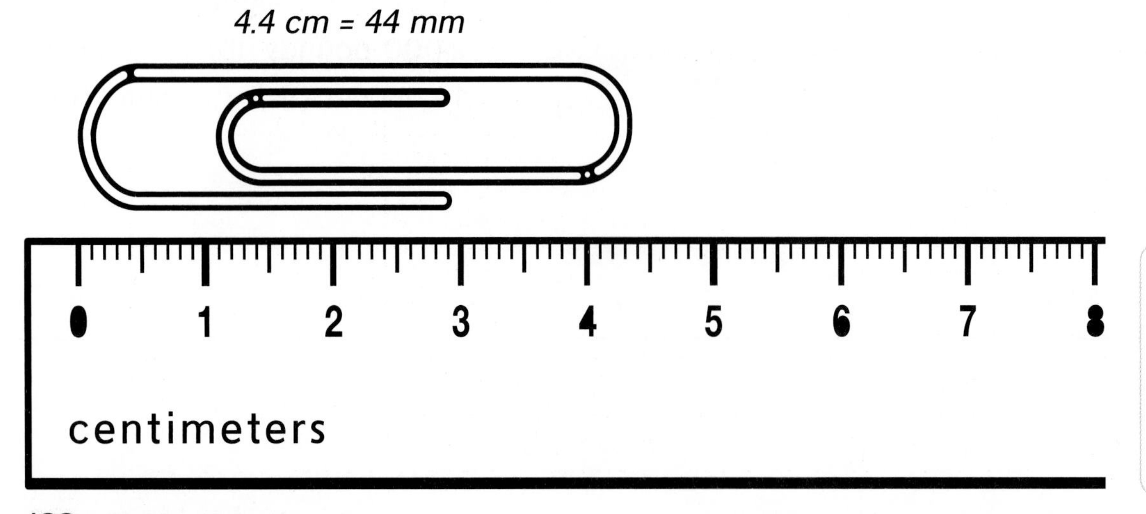

The length of this paperclip is about 4.4 cm.

To change units of other kinds of metric measurements, you also multiply or divide by powers of 10. The table below shows how to change units of volume and mass.

Suppose you find the mass of several rocks to equal 56 grams, but you must report your results in kilograms. Look for a row in the table with "grams" in the first column and "kilograms" in the second column. Then look under "What to Do" in that row. Because 1,000 g equals 1 kg, you should divide the number of grams by 1,000.

$$\frac{56 \text{ grams}}{1{,}000} = 0.0056 \text{ kilograms}$$

Ken Cavanagh/McGraw-Hill Education

How to Change Metric Units of Volume and Mass

Change	To	What to Do	Why
liters (L)	milliliters (mL)	multiply by 1,000	1 L = 1,000 mL
milliliters (mL)	liters (L)	divide by 1,000	
grams (g)	milligrams (mg)	multiply by 1,000	1 g = 1,000 mg
milligrams (mg)	grams (g)	divide by 1,000	
grams (g)	kilograms (kg)	divide by 1,000	1,000 g = 1 kg
kilograms (kg)	grams (g)	multiply by 1,000	

Math Formulas Used in Science

You can use tools such as rulers to measure length and balances to measure mass. Other properties cannot be measured directly. Instead, different quantities must be added, subtracted, multiplied, or divided.

Volume

To find the volume of an oddly shaped object, such as a rock, you follow several steps. First, measure the volume of some water. Then place the object in the water and find the volume again. The difference in the volumes is equal to the rock's volume. Use these measurements and this formula to find an object's volume.

Volume of object and water – *Volume of water* = *Volume of object*

Volume = 860 mL – 550 mL = 310 mL

Suppose you have an object shaped like a rectangle or cube. Then you can use the formula below to find the object's volume. First, measure the length, width, and height of the object. Then multiply these values to calculate the object's volume.

Volume = *length* x *width* x *height*

Volume = 5 cm x 4 cm x 10 cm = 200 cm^3

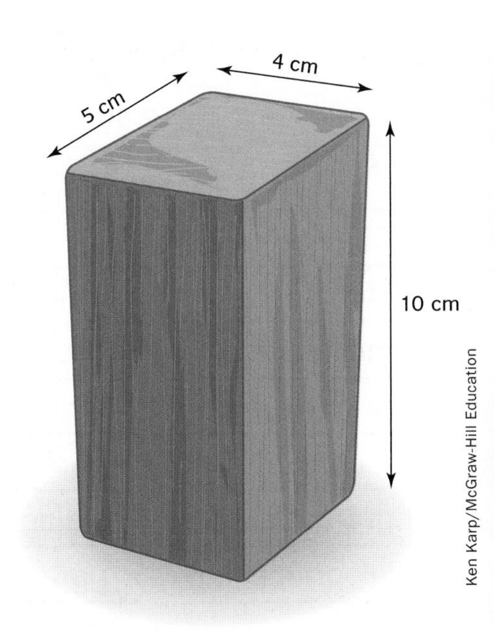

Ken Karp/McGraw-Hill Education

Area

Multiplication is also used to find area. For a rectangle or square, such as a tabletop, measure the length and width. Then use the formula to determine the area.

Area = *length* x *width*

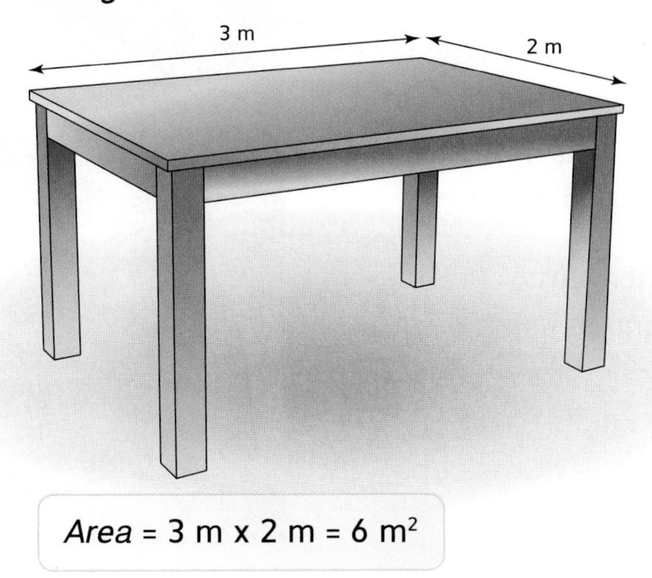

Area = 3 m x 2 m = 6 m^2

Speed

To find the speed of an object, you need to know two things. You need to know the distance the object traveled. You also need to know how much time it took for the object to travel that distance. To find speed, divide the distance by the time, using the formula below.

Speed = *distance/time*

Suppose a student riding a skateboard traveled 100 meters in 50 seconds. You can use the formula to find his speed.

Speed = *distance/time* = 100 m/50 s = 2 m/s

The skateboarder's speed is calculated by dividing the distance traveled by the time spent traveling.

Density

To calculate the density of an object, first measure the object's mass and volume. Then divide the mass by the volume, using the formula below.

Density = *mass/volume*

Using Language

Science Word Parts

When you study science, you encounter many new terms. These terms may seem difficult to learn and remember at first, but there are steps you can take to help you. You can learn and look for common word parts used in science.

Like plants, words have roots and other parts. Knowing about these parts can help you learn and remember science terms. Words have been around for a long time. Many English science words did not start as words in English. They came from older words in other languages, such as Greek and Latin. These older words are called root words.

For example, the root word *thermos* is a Greek word meaning "hot." Knowing this root word may help you remember a word like *thermometer*, a tool that measures how hot something is. The table below lists science root words.

This photograph shows Earth's geosphere, hydrosphere, and atmosphere. All of these words include the Latin root word sphere.

Science Root Words		
Greek or Latin Root	**What It Means**	**Example(s)**
atmos	to blow	atmosphere
bios	life; alive	biosphere, biome
cyclos	wheel; circle, circular	cyclone
geo	Earth	geology, geosphere
hydro	water	hydrosphere, hydroelectric power
metron	measure	meter, kilometer, barometer, thermometer
photo	light	photosynthesis, photon
planeta	to wander	planet, planetary
skopos	seeing	telescope, microscope
solaris	Sun	solar system, solar energy, solar eclipse
sphere	ball or globe	biosphere, atmosphere
therme; thermos	heat; hot	thermal energy, thermometer, thermostat
Vulcan	Roman god of fire	volcano

A prefix is a word part added before a root word. It changes the meaning of the root word. Compare the words *telescope* and *microscope*. Both have the root word *scope* that comes from *skopos*, a Greek word for "seeing." So both the telescope and the microscope are tools used for seeing things. But the prefix *tele-* means "far," so a telescope is a tool for seeing faraway objects. The prefix *micro-* means "very small," so a microscope is a tool for seeing very small objects.

Prefix	What It Means	Example(s)
astro-	of a star or stars	astronaut, astronomy
centi-	one hundredth or 1/100	centimeter
eco-	house/home	ecosystem, ecology
ex-; exo-	out of	excretory system, exoskeleton, extinct
kilo-	1,000	kilometer, kilogram
micro-	very small	microscope, microorganism, microscopic
milli-	one thousandth or 1/1,000	millimeter, milliliter
non-	not	nonrenewable
re-	again	recycle, resource, renewable, reflect, repel
tele-	far	telescope, television

A suffix is a word part added after a root word. It also changes the meaning of the root word. Consider the root word *bio*, which means "life." Add the suffix *-ology,* which means "to study", and you get *biology*, the study of living things. But if you use the suffix *-logist* instead, then you have *biologist*, a person who studies living things.

Science Suffixes		
Suffix	**What It Means**	**Example(s)**
-able	that can or will	renewable, nonrenewable, permeable
-ation	action or process	reaction, evaporation, conservation
-logist; -ist	a person who does an action	biologist, geologist, chemist
-ology	study of	biology, geology, ecology

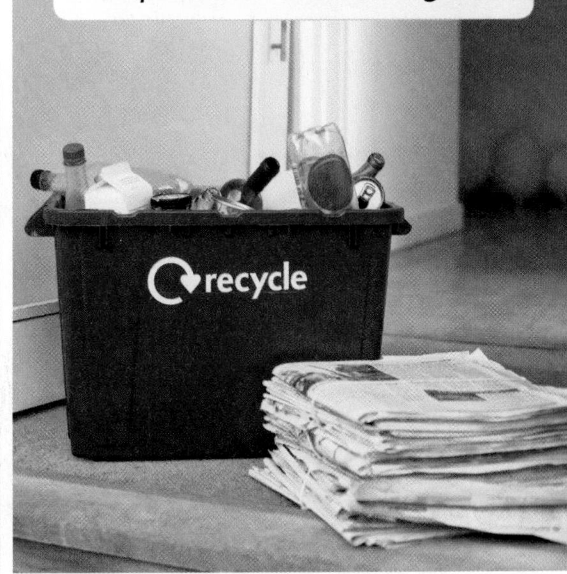

The recycle logo clearly shows what recycling is all about. The word recycle comes from the root word cycle, *which means "circle." The prefix* re- *means "again."*

Research Skills

When scientists ask questions, they conduct research to find answers. To research means to study and gather information about a topic. Scientists use what others have already discovered about a subject to form the basis of their own original investigations. A modern way to conduct research is to search for information on the Internet.

Tips for Using the Internet

The Internet connects your computer with computers around the world, so you can use it to collect a wide range of information. But not all information on the Internet is correct, and not all Web sites are safe. Use these tips to help you when you conduct Internet research.

When using the Internet, only visit sites that are safe and reliable. A reliable Web site is trustworthy. The information on reliable Web sites has been checked for accuracy. It does not contain bias, or favoritism, toward a particular idea. Reliable information is based on facts, not on opinion or what people believe or feel. Science facts are backed up by observations and data. Your teacher can help you find safe and reliable sites to use.

You can use the Internet to gather information about many different science topics.

Look for Web sites with information written by real scientists. They include sites that end in *.gov*, run by the U.S. and state governments. You might also try sites that end in *.edu*, run by colleges and universities. Sites that end in *.org* that are run by museums and professional science groups are also good choices.

New science data are collected every day. To make sure you use the latest information available, look for the date on the Web site. A reliable Web site will tell you when it was last updated. Information that hasn't been updated in several years may no longer be correct.

Check information on more than one reliable Web site. If you find the same information on two or more government, museum, or university sites, it is more likely to be accurate.

Avoid Web sites that try to sell products or ask for personal information. These sites may not be trustworthy or safe. Look for sites with the main purpose of conducting science research and sharing verified scientific information.

Did You Know?

The Internet is not the same thing as the World Wide Web. The Internet is a massive infrastructure used for communication and information that bridges millions of computers throughout the globe. The World Wide Web is a vast system of interlinked hypertext documents accessed on the Internet.

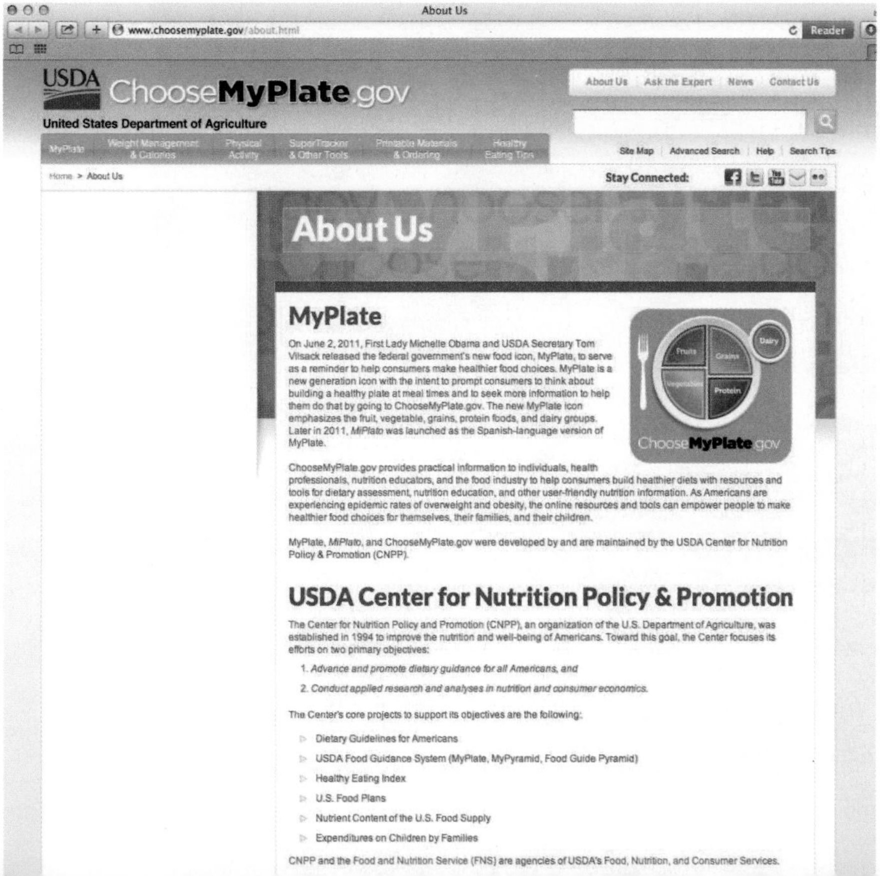

USDA's Center for Nutrition Policy and Promotion

Science information on government Web sites, such as those of the USDA, NASA, or the U.S. Geological Survey, is information written by real scientists. That means the information is more likely to be reliable.

Finding the Right Materials

The Internet is only one of many different sources where you can find science information. You can also do research with materials in your school media center or library.

Research materials may be on paper, like hardcover books or newspapers. They might be digital, such as e-books, online newspapers, compact discs (CDs), or digital video discs (DVDs). No matter what kinds of materials you find, look at their publication dates. Always use the newest materials in your media center or library. Recent information is more likely to include new discoveries and to be the most accurate.

Suppose you are researching the igneous rock granite. To start looking, use your topic or key words to search in your library catalog. Start with a broad topic like "rocks" instead of a specific one like "igneous rocks" or "granite." You will find more possible materials that way.

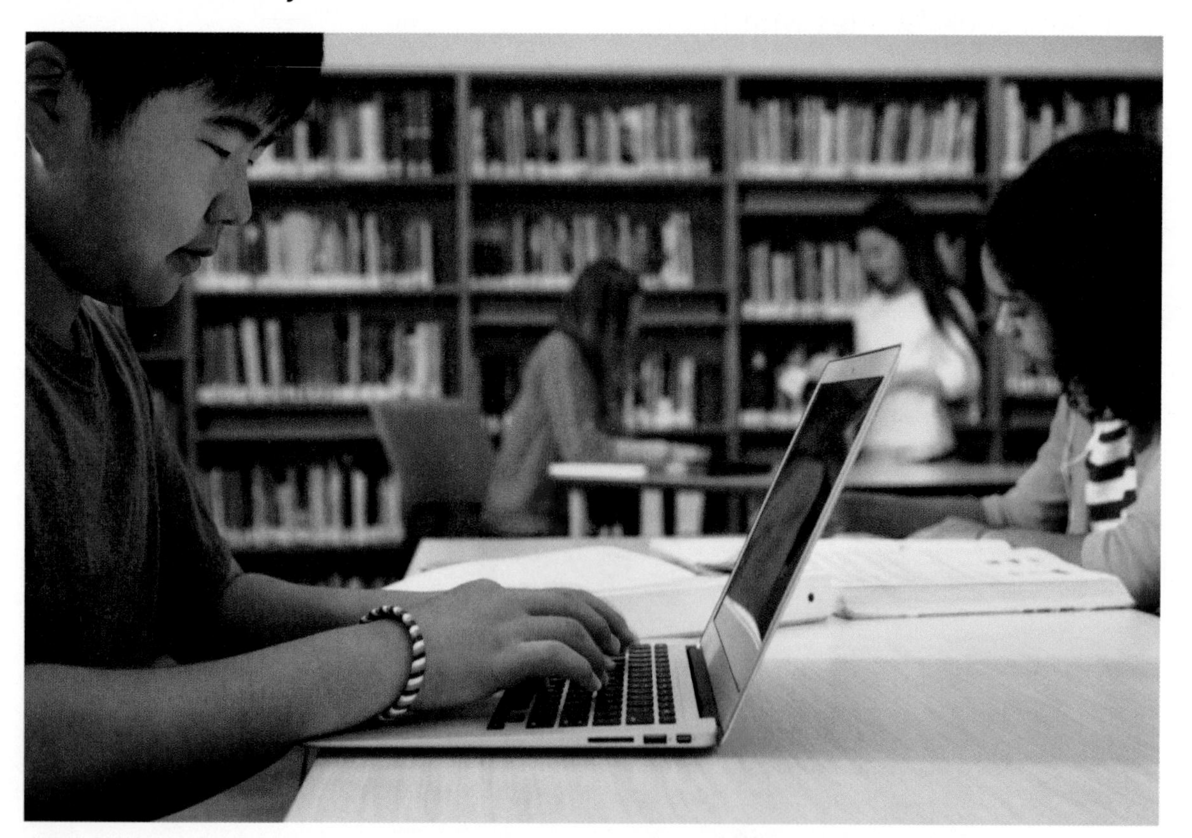

Your school media center or library has a variety of materials you can use to conduct science research.

From there you can narrow down your search. Look for nonfiction books and articles written by scientists for readers your age. You should find details about the writer on a book's back cover or at the beginning or end of an article.

You might find science articles in newspapers or in science magazines called periodicals or journals. Books and magazines have a table of contents in the front pages and an index in the back pages. These sections can help you find your topic quickly. They can also help you rule out materials that do not cover your topic or are too technical.

If you cannot find the information you need, ask a teacher or librarian for help. They can suggest other research materials to use. You might even try asking a scientist or other expert. Ask your teacher to help you find and contact a scientist either in your community or even far away.

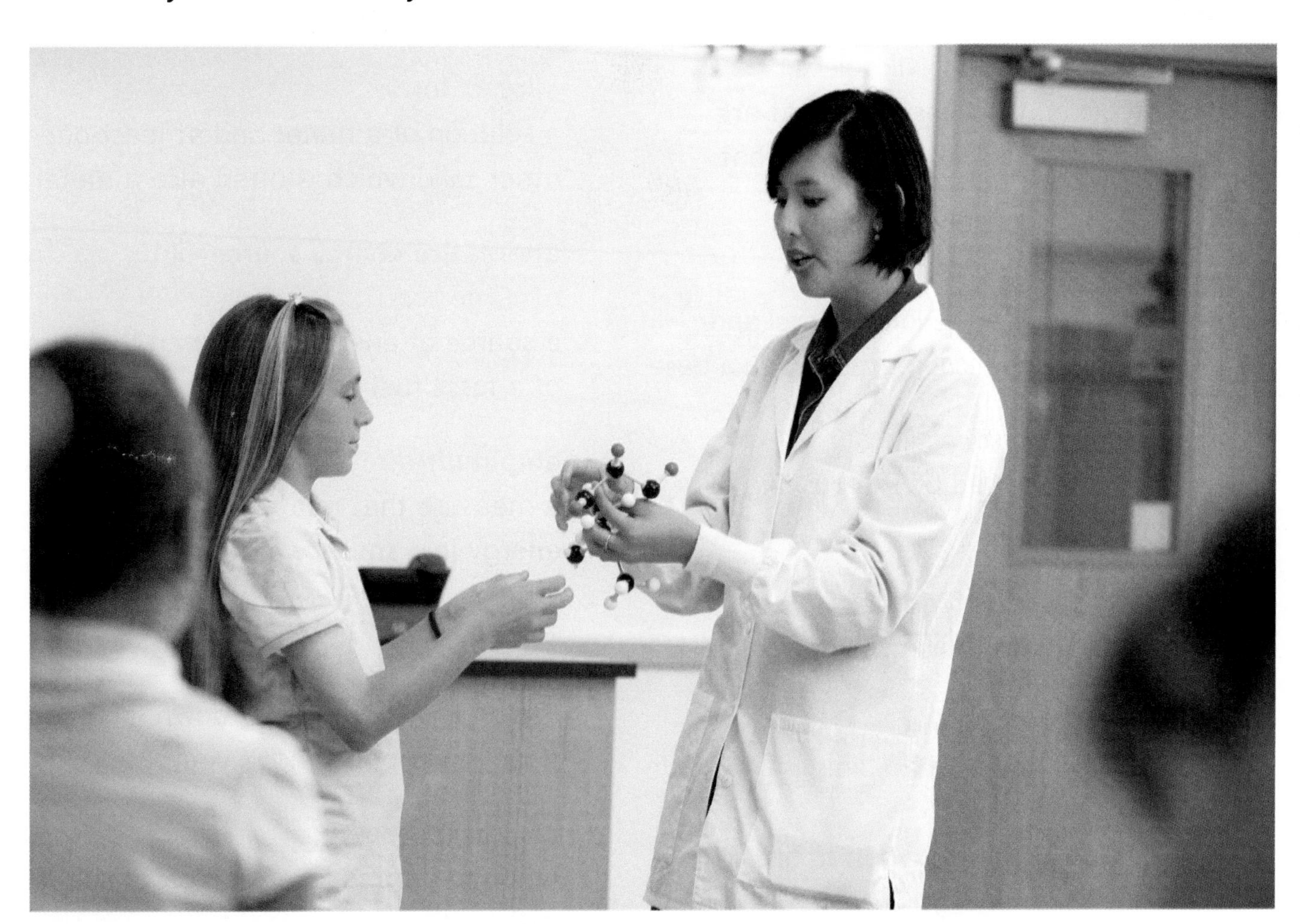

Steve Debenport/iStock/Getty Images

A scientist in your community may be able to answer your research questions.

Glossary

Pronunciation Key

The following symbols are used throughout this glossary.

a	at	e	end	o	soft	u	up	hw	white	ə	about
ā	ape	ē	me	ō	go	ū	use	ng	song		taken
ä	farther	i	tip	ôr	fork	ü	rule	th	thin		pencil
âr	care		ice	oi	oil	ù	pull	th	this		lemon
ô	law	îr	fear	ou	out	ûr	turn	yü	rule		circus
								zh	measure		

´ = primary accent; shows which syllable takes the main stress, such as **at** in **atmosphere** (at´mə·sfîr´)
´ = secondary accent; shows which syllables take lighter stresses, such as **sphere** in **atmosphere**

abiotic factors ā´bī·o´tik fak´tər
the effects on the ecosystem that are a result of the nonliving parts of that ecosystem

acid a´səd
a substance that has a low pH level and forms salts in chemical reactions with a base

acid rain a´səd rān
harmful rain caused by the burning of fossil fuels

adaptation a·dap´tā´shən
a trait that helps a living thing survive in its environment

aerobic respiration ei·ro´bihk res´pə·rā´shən
the process of using oxygen to break down food into energy

alkalinity al´kə ·li´nə·tē
the strength of a base

alloy a´loi
a solution of a metal and at least one other solid which is often also a metal

alternative energy source ôl·tûr´nə ·tiv e´nûr·jē sôrs
a source of energy other than the burning of a fossil fuel

amplitude am´pli·tüd´
a measure that relates to the amount of energy in a sound wave

anaerobic respiration ae·nə·ro´bihk res´pə·rā´shən
the process of breaking down food into energy without using oxygen

aquifer a´kwə·fû r
an underground layer of rock or soil filled with water

arthropod är´thrə·pod´
an invertebrate with jointed legs, a body that is divided into sections, and an exoskeleton

asteroid as´tə·roid´
one of many small, rocky objects between Mars and Jupiter

atmosphere at´mə·sfîr´
the layers of gases that surround Earth

atom a´təm
the smallest unit of an element that has the properties of that element

avalanche a´və·lanch´
a large, sudden movement of ice and snow down a hill or mountain

balanced forces ba´lənst fôrs´ə z
forces that cancel each other out when acting together on a single object

base bās
a substance that has a high pH level and forms salts in chemical reactions with an acid

biomass bī´ō·mas´
plant and animal wastes that can be processed to make fuel

biome bī´ōm´
one of Earth's large ecosystems, each with its own kind of climate, soil, plants, and animals

biomimicry bī´ō·mi´mi·krē
the act of using nature as a model for human inventions

biosphere bī´ōə·sfîr´
the part of Earth in which living things exist and interact

biotic factors bī·o´tik fak´tûr
a living thing in an ecosystem, such as a plant, an animal, or a bacterium

black hole blak hōl
an object whose gravity is so strong that light cannot escape it

boiling point boi´ling point
the temperature at which a substance changes state from a liquid into a gas

buoyancy boi´ən·sē
the upward push of a liquid or a gas on an object placed in it

cell sel
the smallest unit of living matter

cellular respiration sel´ū·lər res´pə·rā´shən
the process of releasing energy from food molecules, such as glucose, which takes place in the mitochondria of a cell

chemical change kə´mi·kəl chānj
a change of matter that occurs when atoms link together in a new way, creating a new substance different from the original substances

chemical reaction ke´mi·kəl rē·ak´shən
event that occurs as two or more substances are combined to form a new substance

chlorophyll klôr´ə·fil´
a green chemical in plant cells that allows plants to use the Sun's energy to make food

chloroplast klo´rə·plaest´
part of a plant cell that contains chlorophyll; where photosynthesis takes place in the cell

circuit sûr´kət
a path through which electric current can flow

climate klī´mət
the average weather pattern of a region over time

colloid kō´loid
a type of mixture in which the particles of one material are scattered through another and block the passage of light without settling out

comet ko´mət
a mixture of ice, frozen gases, rock, and dust left over from the formation of the solar system

commensalism kə·men´sə·li·zəm
a relationship between two kinds of organisms that benefits one without harming the other

compound kom·pound´
a substance that is formed by the chemical combination of two or more elements

condensation kon´den·sā´shən
the changing of a gas into a liquid as heat is removed

conduction kən·duk´shə n
the transfer of thermal energy between two objects that are touching

conductivity kən·duk´tiv·ə·tē
the property of a material to be able to transmit heat and electricity

conductor kən·duk´tər
a material through which heat or electricity flows easily

conservation kon´sú r·vā´shən
saving, protecting, or using natural resources wisely

conservation of energy kon´sûr·vā´shən uv e´nər·jē
a principle in physics that states that energy can neither be created nor destroyed and that the total energy of a system by itself remains constant

conservation of mass kon´sûr·vā´shən uv mas
a physical law that states that matter is not created or destroyed during a chemical reaction

constellation kon´stə·lā´shən
any of the patterns that are formed by a group of stars in the night sky

constraint kən´strānt
something that limits or restricts someone or something

contact force / input force kän´takt´ fôrs / inpût fôrs)
force that results from the interaction of two objects in contact with each other

continental drift kon´tə·nen´təl drift
the slow movement of the continents over many years

convection kən·vek´shən
the transfer of thermal energy by flowing gases or liquid, such as the rising of warm air from a heater

core kôr
the central part of Earth

corrosion kə·rō´zhən
the gradual wearing away of a metal as it combines with nonmetals in its environment

criteria krī´tir´ē·a
a standard on which a judgment or decision may be based

crust krust
Earth's solid, rocky surface

density den´sə·tē
the amount of matter in a certain volume of a substance; found by dividing the mass of an object by its volume

deposition de´pə·zi´shən
the dropping off of eroded soil and bits of rock

desalination di·sal´ih·neit´shən
the process of removing salt from salt water in order to be used by living things

design process di·zīn´ prä´ses
a series of steps that engineers follow to come up with a solution to a problem

distillation dis´tə´lā´shən
the process of separating the parts of a mixture by evaporation and condensation

drought drout
when there is no rain in an area for a long period of time

ductility duk´ti´lə·tē
the ability to be pulled into thin wires without breaking

earthquake ûrth´kwāk´
a sudden shaking of the rock that makes up Earth's crust

echo e´kō
a repetition of a specific sound produced by reflection of sound waves from a surface

echolocation e´kō·lō·kā´shən
the process of finding an object by using reflected sound

electric current i·lek´trik kûr´ənt
a flow of electricity through a conductor

electromagnet i·lek´trō·mag´nət
a magnet formed when electric current flows through wire wrapped in coils around an iron bar

electromagnetic spectrum i·lek´trō·mag·net´ik spek´trəm
a range of all light waves of varying wavelengths, including the visible spectrum

element e´lə·mənt
a pure substance that cannot be broken down into any simpler substances through chemical reactions

endoskeleton en´dō·ske´lə·tən
an internal supporting structure

energy e´nər·jē
the ability to perform work or change an object

energy pyramid e´nər·jē pîr´ə·mid´
a diagram that shows the amount of energy available at each level of an ecosystem

energy transfer e´nûr·jē tran(t)s·fər´
the movement of energy from one object to another or the change of energy from one form to another

erosion i·rō´zhən
the process of carrying away soil or pieces of rock

estuary es´chə·wer´ē
the boundary where a freshwater ecosystem meets a saltwater ecosystem

evaporation i·va´pə·rā´shən
the slow changing of a liquid into a gas below the boiling point

exoskeleton ek´sō·ske´lə·tən
a hard covering that protects the bodies of some invertebrates

fault line fôlt
a pattern where a break or crack in the rocks of Earth's crust where movements can take place

flood flud
a great flow of water over land that is usually dry

floodplain flud´plān
land near a river that is likely to be under water during a flood

food chain füd chān
the path that energy and nutrients follow in an ecosystem

food web füd web
the overlapping food chains in an ecosystem

force fôrs
any push or a pull by one object on another

fossil fo´səl
any evidence of an organism that lived in the past

fossil fuel fo´səl fūl
a source of energy made from the remains of ancient, once-living things

freezing point frēz´ing point
the temperature at which a substance changes state from a liquid to a solid

frequency frē´kwən·sē
the number of wavelengths that pass a reference point in a given amount of time

front frənt
the boundary between two air masses; examples include cold fronts, warm fronts, and stationary fronts

G

generator je´nə·rā´tər
a device that produces alternating current by spinning an electric coil between the poles of a magnet; changes motion into electrical energy

geosphere jē´o´sfîr´
the layers of rock, dirt, and soil on Earth, including the mantle, cores, and crust

geothermal jē´ō·thûr´məl
using the heat of Earth's interior

gravity gra´və·tē
the force of attraction between any two objects due to their mass

grounding groun´ding
connecting an object to Earth with a conducting wire to prevent the buildup of static electricity

groundwater ground´wô´tər
precipitation that seeps into the ground and is stored in tiny holes, or pores, in soil and rocks

heat / thermal energy hēt / thûr´məl e´nûr·jē)
the movement of thermal energy from a warmer object to a cooler object

hurricane hû´rə·kān´
a very large, swirling storm with strong winds and heavy rains

hydroelectricity hī´drō·i·lek´tri´si·tē
relating to production of electricity by waterpower

hydrosphere hī´drə ·sfîr´
Earth's water, whether found on land or in oceans, including the freshwater in ice, lakes, rivers, and underground

image i´mij
a "picture" of the light source that light rays make in bouncing off a polished, shiny surface

inertia i·nûr´shə
the tendency of a moving object to keep moving in a straight line or of any object to resist a change in motion

insolation in´sō·lā´shən
the amount of the Sun´s energy that reaches Earth

insulator in´sə·lā´tər
a material that slows or stops the flow of energy, such as thermal energy, electricity, and sound

invasive species ihn·vā´sihv spi·siz
a specific classification of a plant or animal that is not native to an ecosystem

invertebrate in·vûr´tə·brət
an animal without a backbone

kinetic energy kə·ne´tik e´nər·jē
the energy an object has because it is moving

kingdom king´dəm
the largest group into which an organism can be classified

landform land´fôrm´
a physical feature on Earth's surface

landslide land´slīd´
the rapid movement of rocks and soil down a hill

light-year līt´yîr´
the distance light travels in a year

limiting factors li´mə·ting fak´tər
anything that controls the growth or survival of a population

longitudinal wave lon´jə·tüd´nəl wāv
a wave that moves matter left and right as it travels through a medium

lunar eclipse lü´nûr i·klips´
a situation that occurs when Earth, the Sun, the Moon are in a straight line and Earth's shadow falls across the Moon

magnetism mag´nə·ti´zəm
the ability of an object to push or pull on another object that has the magnetic property

malleability ma´lē·ə·bi´lə·tē
the ability to be bent, flattened, hammered, or pressed into new shapes without breaking

mantle man´təl
the layer of Earth beneath the crust

mass mas
a measure of the amount of matter in an object

matter ma´tər
anything that has mass and takes up space

melting point mel´ting point
the particular temperature at which a substance changes state from a solid into a liquid

meteor mē´tē·ər
a chunk of rock from space that travels through Earth's atmosphere

mitochondria mī·tuh·kon´dree·uh
oval parts of a cell that supply energy to the cell

mixture miks´chûr
a physical combination of two or more substances that are blended together without forming new substances

molecule mo´li·kūl´
a particle that contains more than one atom joined together

motion mō´shən
a change in an object's position over time

mutualism mū´chə·wə·li´zm
a relationship between two kinds of organisms that benefits both

natural resource na´chə·rəl rē´sôrs´ə
something that is found in nature and is valuable to humans

nebula ne´byə·lə
a huge cloud of gas and dust in space that is the first stage of star formation

nervous system nûr´vəs sis´təm
the set of organs that uses information from the senses to control all body systems

neutralization nü´trə·lə·zā´shən
the chemical change of an acid and a base into water and a salt

niche nich
the role of an organism in an ecosystem

nitrogen cycle nī´trə·jən sī´kəl
the continuous trapping of nitrogen gas into compounds in the soil and its return to the air

nonrenewable resource non´rē·nü´ə·bəl rē´sôrs´
a resource that cannot be replaced within a short period of time or at all

orbit ôr´bət
the path one object travels around another object

organ ôr´gən
a group of tissues working together to do a certain job

output force / reaction force out´pùt fôrs / rē·ak´shən fôrs
the push or pull of a second object back on the object that started the push or pull

oxygen-carbon dioxide cycle äk´si·jən-kär´bən dī´äk·sīd sī´kəl
the continuous exchange of carbon dioxide and oxygen among living things

ozone ō´zōn´
a form of oxygen gas that makes a layer in the atmosphere that screens out much of the Sun's ultraviolet rays

parasitism per´ə·sə·ti´zəm
a relationship in which one organism lives in or on another organism and benefits from that relationship while the host organism is harmed by it

peripheral nerve pə·rif´ər·əl nərv
a nerve that is not part of the central nervous system and receives sensory information from cells in the body

phase fāz
the appearance of the shape of the Moon at a particular time

phloem flō´em´
the tissue through which food from the leaves moves throughout the rest of a plant

photon fō´ton
a tiny bundle of energy through which light travels

photosynthesis fō´tō·sin´thə·səs
the food-making process in green plants that uses sunlight

phototropism fō´tō´trō´pi´zəm
a movement or growing in a particular direction that is made by a living thing in response to light

physical change fi´zi·kəl chānj
a change of matter in size, shape, or state that does not change the type of matter

pitch pich
how high or low a sound is as determined by its frequency

plant tropism plaent trō´pi´zəm
the response of a plant toward or away from a stimulus

plate tectonics plāt tek·to´niks
a scientific theory that Earth's crust is made of moving plates

potential energy pə·ten´shə l e´nə r·jē
the energy that is stored inside an object

precipitation prisi´pətā´shən
water that falls from clouds to the ground in the form of rain, sleet, hail, or snow

product pro´dukt
a substance at the end of a chemical reaction

prototype prō·tō´tīp
an original or first model of something from which other forms are copied or developed

radiation rā´dē·ā´shən
the transfer of energy through space

reactant rē·ak´tənt
an original substance at the beginning of a chemical reaction

reflection ri·flek´shən
the bouncing of light waves off a surface

reflex rē´fleks
an action or movement of the body that happens automatically as a reaction to something

refraction ri·frak´shən
the bending of light as it passes from one transparent material into another

renewable resource rē·nü´ə·bəl rē´sôrs´
a resource that can be replanted or replaced naturally in a short period of time

respiration res´pə·rā´shən
the using and releasing of energy in a cell

respiratory system res´prə·tôr´ē sis´təm
the organ system that brings oxygen to body cells and removes waste gas

response ri´späns
a reaction to a stimulus

revolution re´və·lü´shən
one complete trip of one object around another object

rotation rō·tā´shən
a complete spin on an axis

runoff run´ôf´
precipitation that flows across the land's surface or falls into rivers and streams

satellite sa´tə·līt
a natural or artificial object in space that circles around another object

sediment se´də·mənt
the particles of soil or rock that may be eroded and deposited

sedimentary rock se´də·men´tə·rē rok
a rock that forms when small bits of matter are pressed together in layers

seismic wave sīz´mik wāv
a vibration caused by an earthquake

seismograph sīzmə·graf´
an instrument that detects and records earthquakes; shows seismic waves as curvy lines along a graph

solar cell sō´lər sel
a device that uses light or heat from the sun to produce electricity

solar eclipse sō´lûr i·klips´
a blocking of the Sun's light that happens when Earth passes through the Moon's shadow

solubility sol´yə·bi´lə·tē
the maximum amount of a substance that can be dissolved by another substance

solution sə lü´shən
a mixture of substances that are blended so completely that the mixture looks the same everywhere

sound wave sound wāv
a wave that transfers energy through matter and spreads outward in all directions from a vibration

species spē´shēz
a group of similar organisms in a genus that can reproduce more of their own kind

speed spēd
the distance an object moves in an amount of time

star stär
an object in space that produces its own energy, including heat and light

stimulus stim´yə·ləs
something in the environment that causes a living thing to respond

stomata stō´mə·tə
pores in the bottom of leaves that open and close to let in air or give off water vapor

succession sək·se´shən
the process of one ecosystem changing into a new and different ecosystem

supernova sü´pər·nō´və
a star that has produced more energy than gravity can hold together and has exploded

symbiosis sim´bē·ō´səs
a relationship between two kinds of organisms over time

tectonic plates tek·to´nik plāts
a scientific theory that Earth's crust is made of moving plates

tide tīd
the regular rise and fall of the water level along a shoreline

tissue ti´shü
a group of similar cells that work together at the same job

topographical map to´pə·gra´fi·kəl map
a map that shows the elevation of an area of Earth's surface using contour lines

transpiration trans´pə ·rā´shə n
the loss of water through a plant's leaves

transverse wave trans·vûrs´ wāv
a wave that moves matter up and down as it travels through a medium

tsunami sù·nä´mē
a huge wave caused by an earthquake under the ocean

unbalanced forces un·ba´lənst fôrs´əz
forces that do not cancel each other out when acting together on a single object

vegetation ve·jə´tā·shən
plants that cover a particular area

vertebrate vûr´tə·brət´
an animal with a backbone

volcano vol·kā´nō
a mountain that builds up around an opening in Earth's crust

volume vol´yəm
how loud or soft a sound is based on its amplitude

volume vol´yəm
a measure of how much space an object takes up

water cycle wô´tər sī´kəl
the continuous movement of water between Earth's surface and the air, changing from liquid into gas into liquid

water vapor wô´tər veipər
water in the form of a gas that is present during evaporation

watershed wô´tûr·shed´
the region that contributes water to a river or a river system

wavelength wāv´length´
the distance from the top of one wave to the top of the next

weathering weth´ring
the breaking down of rocks into smaller pieces

weight wāt
a measure of how gravity pulls on an object

white dwarf hwīt dwôrf
a star that can no longer turn helium into carbon; it cools and shrinks, becoming very small and dense

xylem zī´ləm
the tissue through which water and minerals move up in a plant

Index

A

C

D

J

K

L

M

N

O

P

S

T

U

V

W